APPLICATIVE ASPECTS OF POLY(ORGANOPHOSPHAZENES)

APPLICATIVE ASPECTS OF POLY(ORGANOPHOSPHAZENES)

MARIO GLERIA AND ROGER DE JAEGER
EDITORS

Nova Science Publishers, Inc.
New York

Coordinating Editor: Tatiana Shohov
Senior Editors: Susan Boriotti and Donna Dennis
Office Manager: Annette Hellinger
Graphics: Matt Dallow
Editorial Production: Alexandra Columbus, Maya Columbus, Alexis Klestov, Vladimir Klestov, Matthew Kozlowski and Lorna Loperfido
Circulation: Ave Maria Gonzalez, Vera Popovic, Luis Aviles, Sean Corkery, Raymond Davis, Melissa Diaz, Meagan Flaherty, Magdalena Nuñez, Marlene Nuñez, Jeannie Pappas and Frankie Punger
Marketing: Cathy DeGregory

Library of Congress Cataloging-in-Publication Data
Available upon request

ISBN 1-59454-025-X

400 Oser Ave, Suite 1600
Hauppauge, New York 11788-3619
Tele. 631-231-7269 Fax 631-231-8175
e-mail: Novascience@earthlink.net
Web Site: http://www.novapublishers.com

Printed in the United States of America

To Carla, Luca and Fabio

To Monique, and all those I hold dear

for their love and patience

CONTENTS

PREFACE

The main aim of the book "Phosphazenes. A Worldwide Insight" was to provide the scientific community with the most complete picture about the research on cyclo- and poly-phosphazenes presently carried out all over the world. We invited therefore the most eminent scientists active in this field to contribute to this enterprise, writing chapters on their own activity. Their enthusiastic answer made this book possible.

However, as the field of phosphazene materials is very wide, it was decided to start a parallel edition of this book in three volumes entitled respectively:

Volume I: Synthesis and Characterization of Poly(organophosphazenes);
Volume II: Applicative Aspects of Poly(organophosphazenes);
Volume III: Applicative Aspects of Cyclophosphazenes.

The present book is the second volume of this series.

The interest on poly(organophosphazenes) dates back to 1965, the official date of the synthesis of linear polydichlorophosphazene, and expanded rapidly to the synthesis and characterization of poly(organophosphazenes), and their potential and practical use in different fields.

Nowadays, phosphazene substrates occupy an important position in the panorama of inorganic-organic materials, somewhere at the borderline among the inorganic, organometallic and organic chemistry, and among polymeric, ceramic and metallic materials. Furthermore, theoretical aspects on their structures are still of importance in the scientific debate and their intriguing characteristics deeply investigated.

This book is prevalently dedicated to the applicative aspects of poly(organophosphazenes), covering their use in biology (Chapters 1 and 2), photochemistry (Chapter 3), and high energy radiation chemistry (Chapter 4). Their utilization as hybrid materials (Chapter 5), flame and fire retardants (Chapter 6), blend components (Chapter 7), ionic conductors (Chapter 8), membranes (Chapters 9, 10 and 11), and catalysts (Chapter 12) is also addressed, together with a series of mixed additional applications (Chapter 13).

We sincerely hope that this book will be useful both to scientists that have been active in the phosphazene field for long time and to newcomers in this domain that need to know more about the potential and practical exploitation of phosphazene polymers. We tried, in fact, to collect a great deal of information on this argument just in one place, and to present them in an harmonic and logical way.

We feel deeply grateful to the authors of this book. Their hard work and fine contributions provided an excellent compilation of the different aspects of polyphosphazene utilization, also reflected by the great number of papers and patents which have recently appeared in this domain.

M. G. and R. D. J.

In: Applicative Aspects of Poly(organophosphazenes)
Editors: M. Gleria and R. De Jaeger, pp. 1-32
ISBN 1-59454-025-X

Chapter 1

BIODEGRADABLE POLYPHOSPHAZENES FOR BIOMEDICAL APPLICATIONS

M. Heyde and E. Schacht
Polymer Materials Research, Ghent University
Krijgslaan 281, S4bis
9000 Ghent, Belgium
e-mail: etienne.schacht@rug.ac.be

1. INTRODUCTION

Poly[(organo)phosphazenes] are polymers with an inorganic backbone, consisting of alternating nitrogen and phosphorous atoms linked by alternating single and double bonds. Each phosphorous bears two side groups (R), R can be an alkoxy, aryloxy, amino, alkyl, heterocyclic ring, inorganic or organometallic unit. The synthesis of poly[(organo)phosphazenes] was first extensively studied by Allcock and co-workers. Most polyphosphazene derivatives are prepared by nucleophilic replacement of chlorine from a precursor polymer, poly[(dichloro)phosphazene] (fig 1). In this way a wide range of polymers can be synthesised having different properties[1]

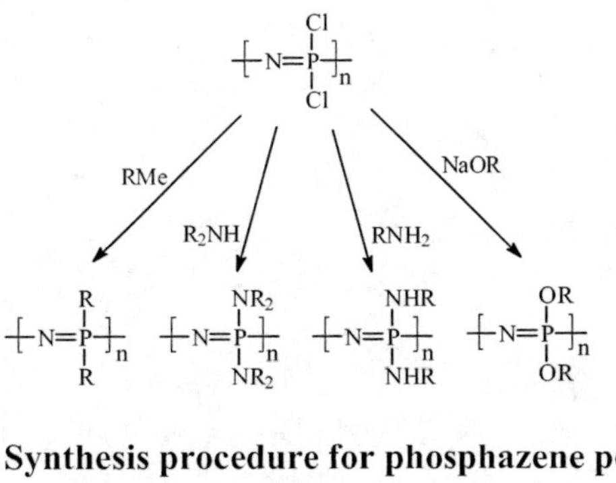

Figure 1: Synthesis procedure for phosphazene polymers

The most important methods for the preparation of poly[(dichloro)phosphazene] are discussed below.

2. SYNTHESIS OF POLY[(DICHLORO)PHOSPHAZENE]

2.1. Synthesis by Ring-Opening Polymerisation

Hexachlorocyclotriphosphazene a starting product for the preparation of poly[(dichloro)phosphazene] was first prepared in 1834 by Wohler and Rose[2,3], by reaction of phosphorous pentachloride with ammonia. In 1897, Stokes discovered that the reaction actually resulted in a mixture of compounds of which the main product was a cyclic trimer, shown in fig 2. Stokes also described the formation of an inorganic rubber when the cyclic trimer was heated[4]. This material received little interest during the next 60 years, probably because of its insolubility and its tendency to hydrolyse to phosphate, ammonia and hydrochloric acid when exposed to moisture.

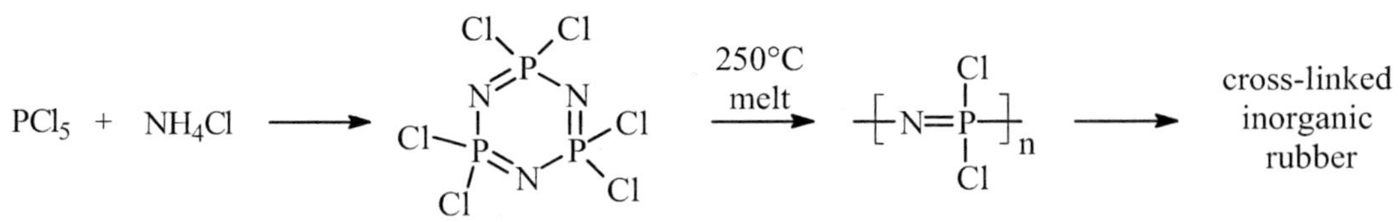

Figure 2: Formation of an inorganic rubber

In 1965, Allcock succeeded in creating a soluble and useful poly[(dichloro)phosphazene] by careful control of the polymerisation conditions[5,6,7]. Further heating of this polymer caused cross-linking and yielded in an insoluble inorganic rubber as described by Stokes. At temperatures above 350°C, the polymer depolymerised to cyclic oligomers. The uncross-linked product was a highly reactive polymeric precursor for a wide range of derivatives.

This polymerisation reaction has been extensively studied and several mechanisms have been proposed[8,9,10,11] It was originally reported as being radical[12] or anionic[13] Currently, it is accepted that the process occurs via a cationic mechanism[14,15] initiated by the heterolytic cleavage of a P-Cl bond caused by heating the hexachlorocyclotriphosphazene at 250°C (fig 3).

The thermal polymerisation seems to be an irreproducible reaction[16,17] and is affected by traces of impurities. Some can act as accelerators and others as inhibitors'[18]. In order to avoid the cross-linking reactions, low trimer conversion degrees (50-60%) are recommended. A typical uncatalysed bulk polymerisation at 250°C yields very high molecular weight polymers (M_w 2000 kDa) with a broad M_w distribution (d = 5-15).

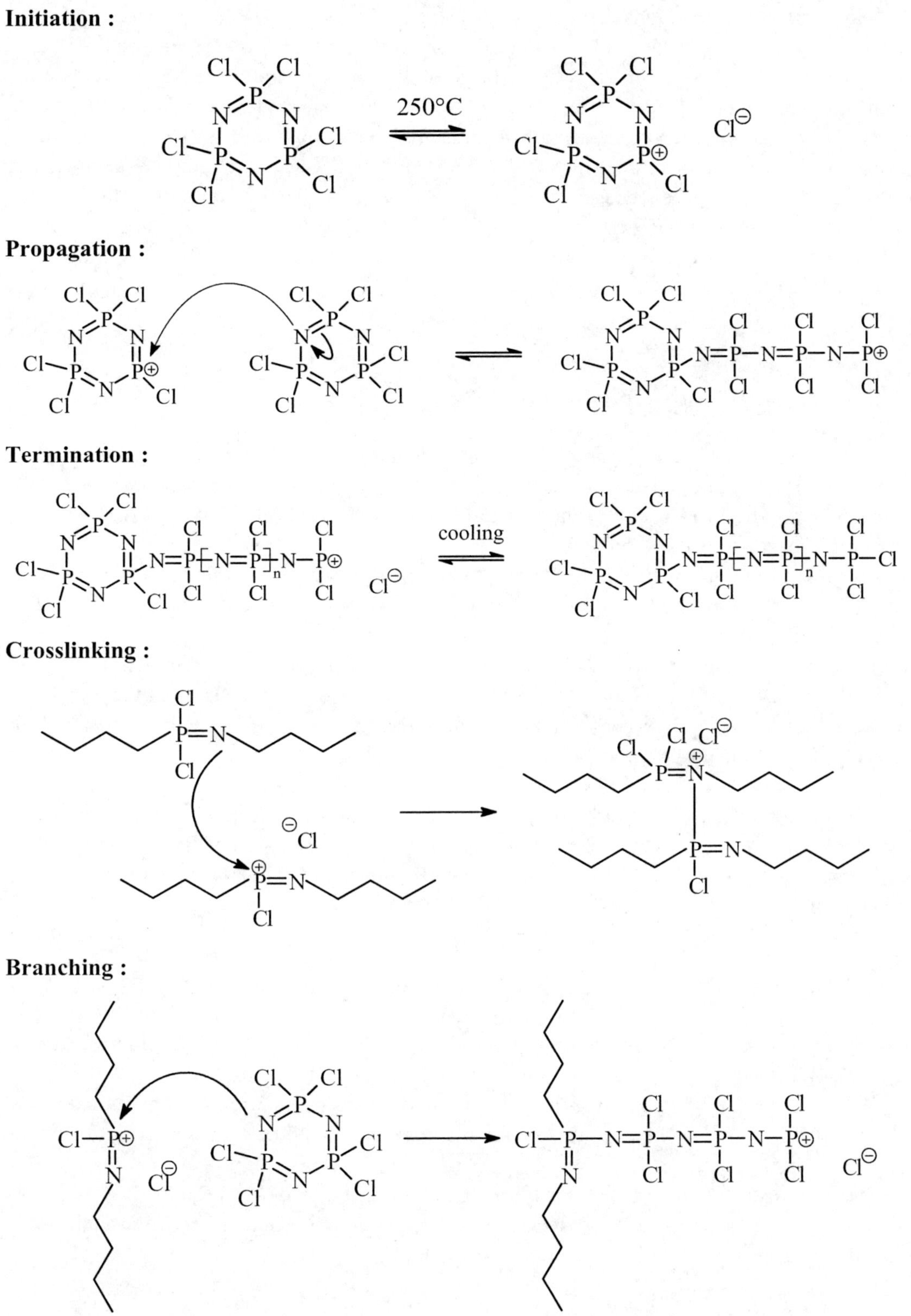

Figure 3: Ring-opening polymerisation

Solution polymerisations have been investigated as well. The advantage of using a solvent is a lower viscosity and a better control of the polymer chain structure. The main problem is the choice of a good solvent, because at high temperatures many solvents react with the chlorophosphazenes. This was observed for toluene and xylene[19,12]. With carbon tetrachloride as solvent, polymers with a molecular weight as high as 130 kDa could be obtained. However an insoluble fraction was formed as well. Successful polymerisations have been carried out using benzene, in presence of accelerators[20], or in carbon disulfide at high pressure.[21] Mujumdar *et al.* described the use of 1,2,4-trichlorobenzene with sulfamic acid as catalyst and $CaSO_4.2H_2O$ as promotor[22]. The polymerisation proceeded faster in solution than in the molten state, but the resulting molecular weight of the polymers were significantly lower.

2.2. Synthesis by Polycondensation

A new method for the preparation of poly[(dichloro)phosphazene] was described by De Jaeger and D'Halluin. Bulk or solution condensation of the phosphoranimine, N-(dichloro)phosphoryl-P-(trichloro)mono-phosphazene with the elimination of phosphoryl trichloride leaded to the desired polymer[23,24,25]

In bulk, polymers with a low degree of polymerisation were formed. When the same reaction was accomplished in solution, the molecular weights were much higher (up to 1000 kDa). Trichlorodiphenyl was used as solvent. It is chemically inert with respect to phosphochlorinated compounds and can be separated from phosphoryl trichloride.

The mechanism for the bulk polymerisation has been studied in more detail and revealed an intermediate mechanism between a stepwise and a random chain growth. A nucleophilic attack of the nitrogen of the monomer occured on the phosphorous atom of the electron-poor trichlorophosphazene (fig 4). Reaction between two monomers proceeded slowly compared to reactions between the monomer and the oligomer or between oligomers. Rings were formed at the beginning of the reaction and their concentration remained constant. This formation could be explained by intramolecular reactions. Cross-linking might occur at high conversion but could be reduced by the use of a solvent.

2.3. Synthesis by Polymerisation of Phosphoranimines

Reaction of a substituted fluorophosphorane with lithium bis(trimethylsilyl)amide leaded to the formation of silylaminophosphorane (fig 5, (I)).

The latter compound was not isolated. It was converted by elimination of fluorotrimethylsilane to the phosphoranimine, which then could be transformed into trimeric and tetrameric cyclophosphazene.[26]

New routes for preparing N-silyl-phophoranimines have been studied[27]. The first polydialkylphosphazene was obtained with a phosphoranimine having a trifluoroethoxy as leaving group (fig 6).

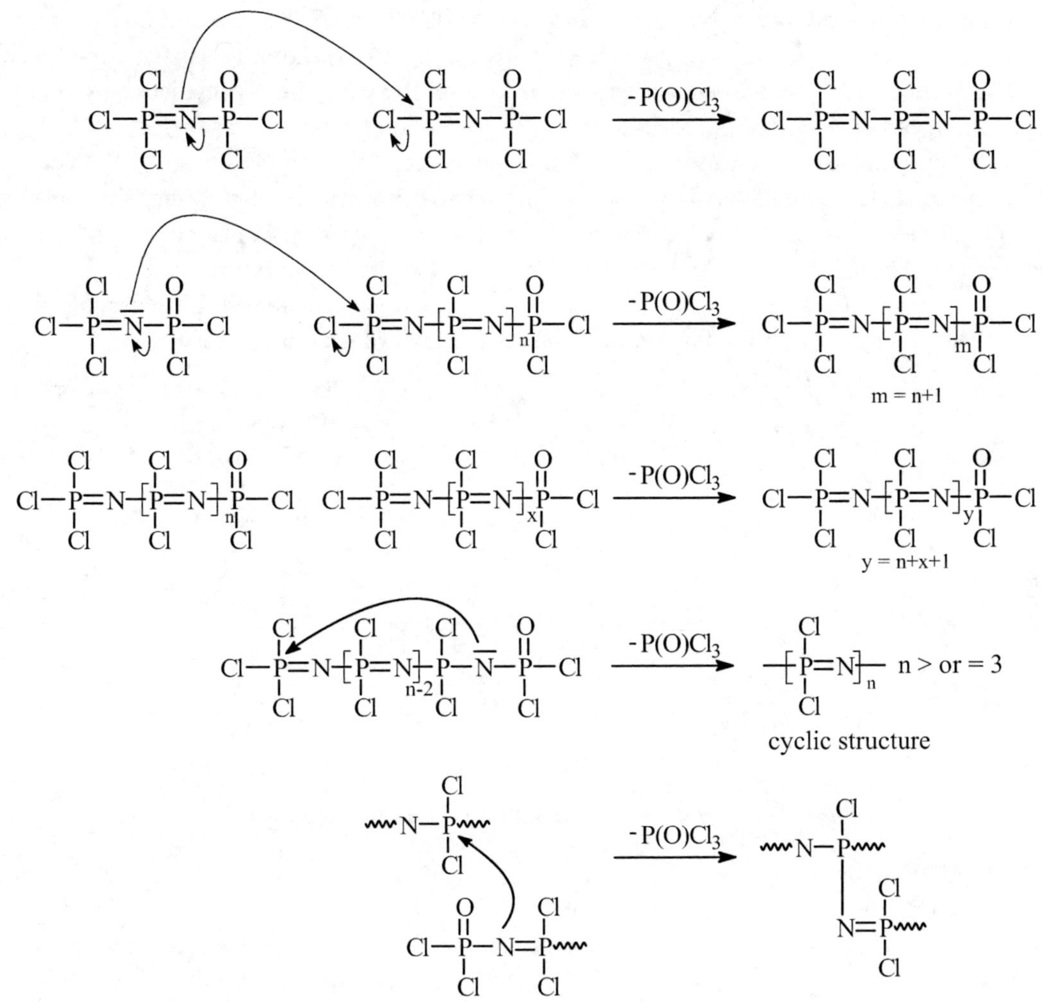

Figure 4: Synthesis of poly[(dichloro)phosphazene] by polycondensation

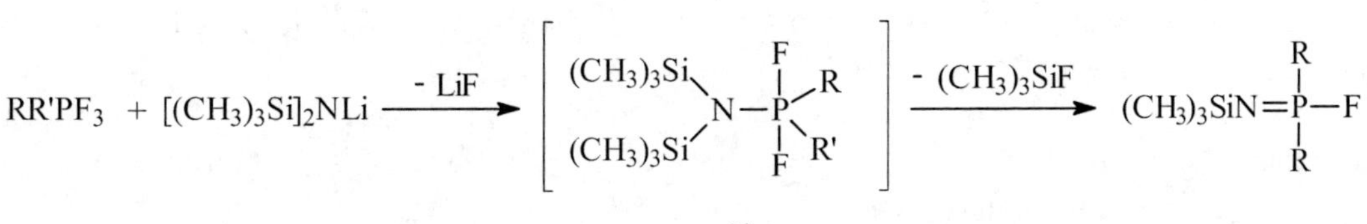

(I)

Figure 5: Synthesis of a silylaminophosphorane

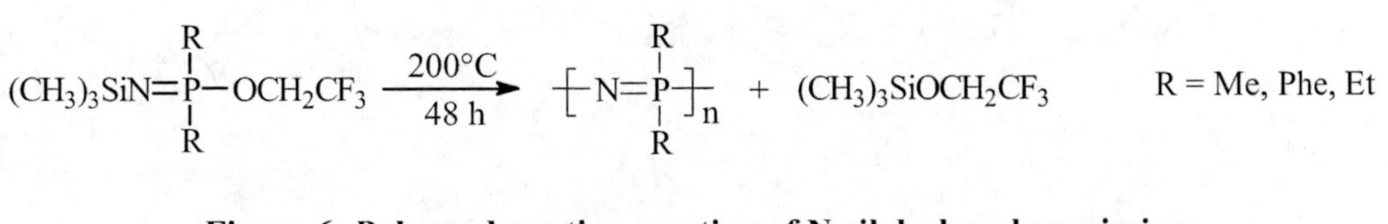

Figure 6: Polycondensation reaction of N-silyl-phosphoranimines

Using this method, it was possible to incorporate the desired substituents directly into the starting silyl compound (alkyl- or aryl-side groups). In the same way a variety of copolymers have been prepared[27]. Molecular weights of 50 kDa to 250 kDa and a polydispersity of 1.5-3.0 could be obtained with this procedure.

This polymerisation procedure was re-investigated by Matyjaszewski[28,29,30]. It was demonstrated that the condensation of tris(trifluoroethoxy)-N-trimethylsilyl-phosphoranimine could be catalysed by a fluoride initiator (fig 7). The polymerisation occured via an anionic mechanism. The reaction could be carried out at lower temperatures in bulk (100°C, 90 minutes) or in solution (diglyme or toluene, 125°C, 4 hours). This catalysed reaction has been extended to other phosphoranimines and to the synthesis of block copolymers[31].

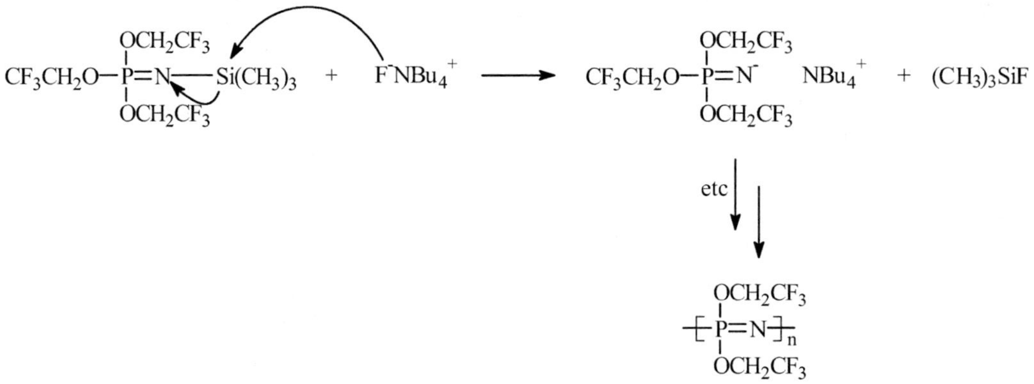

Figure 7: Anionic polymerisation with a fluoride initiator

Recently, Honeyman *et al.* proposed a new method for the preparation of poly[(dichloro)phosphazene] starting from N-trimethylsilyl-trichloro-phosphoranimine and the addition of trace amounts of phosphorous pentachloride[32,33]. It was important to start from a pure phosphoranimine. Reaction of phosphorous pentachloride and lithium bis(trimethylsilyl)amide however leaded to the production of N-trimethylsilyl-trichloro-phosphoranimine and chloro bis(trimethylsilyl)amide[34,35]. These reaction products could not be separated from each other by distillation. The latter by-product acted as an inhibitor so a purification step was needed. The overall yield was generally low. Allcock *et al.* recently reported the synthesis of a pure phosphoranimine by reaction of phosphorous pentachloride and tris(trimethylsilyl)amine. By changing the reaction conditions, hexachlorocyclotriphosphazene could be prepared[36] (fig 8). Hence this method is a new way to synthesise the trimer monomer.

A living cationic polymerisation of N-trimethylsilyl-trichloro-phosphoranimine could be carried out at room temperature in bulk or in solution. Various solvents were investigated. The polymerisation proceeded with no side reactions in methylene chloride and cyclohexane[37] High molecular weights with a polydispersity of 1.3 could be obtained. This process could also be used for the preparation of block[38] or star-copolymers[39,40].

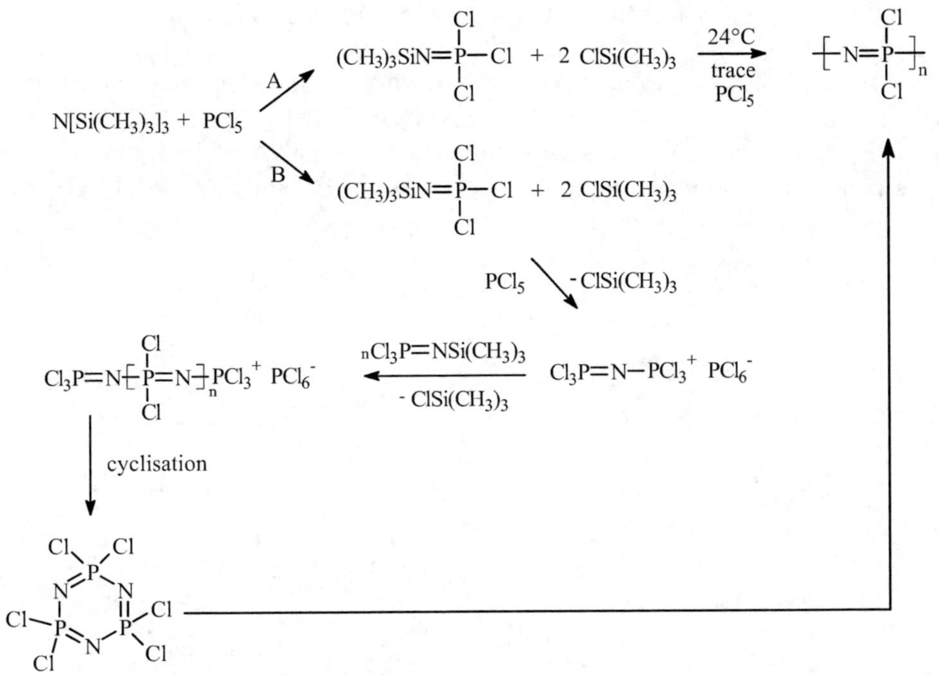

A: fast addition of tris(trimethylsilyl)amine to PCl_5 in CH_2Cl_2
B: slow addition of PCl_5 to tris(trimethylsilyl)amine, reflux 8 hours

Figure 8: Synthesis of N-trimethylsilyl-trichloro-phosphoranimine and hexachlorocyclotriphosphazene

In addition to the main procedures for the preparation of poly[(dichloro)phosphazene] described above, additional methods have been described in literature. Hornbaker and Li described the direct formation of poly[(dichloro)phosphazene] starting from PCl_3, Cl_2 and NH_4Cl without isolation of an intermediate or precursor compound[41].

3. Biodegradable Polyphosphazenes

It has been demonstrated that some polyphosphazenes can hydrolyse in the presence of water leading to a polymer with hydroxyl side groups directly linked to the phosphorous atom. This product is sensitive to two consecutive reactions (fig 9). The first one is a rearrangement of the proton from oxygen to the nitrogen in the skeleton. The second is an attack of water on the skeleton. Examples of side groups that induce hydrolytic degradation include amino acid esters, imidazolyl, glucosyl, glyceryl, glycolate, and lactate.

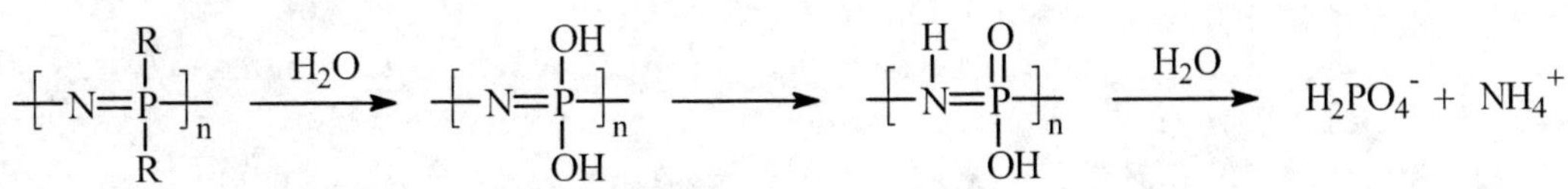

R = amino acid ester, imidazolyl, glucosyl, glyceryl, glycolate, lactate, ...

Figure 9: Degradation of polyphosphazenes

3.1. Polyphosphazenes having Amino Acid Ester Side Groups

The first class of biodegradable polyphosphazenes are the ones substituted with amino acid esters. Allcock *et al.* used ethyl and/or methylesters of glycine, leucine, alanine and phenylalanine[42]. These polymers were made by reaction of poly[(dichloro)phosphazene] with the corresponding amino acid ester. A general scheme is shown in fig 10. Only glycine ethylester and alanine methylester gave total replacement of chlorine, the others were prepared by adding methylamine as co-substituent.

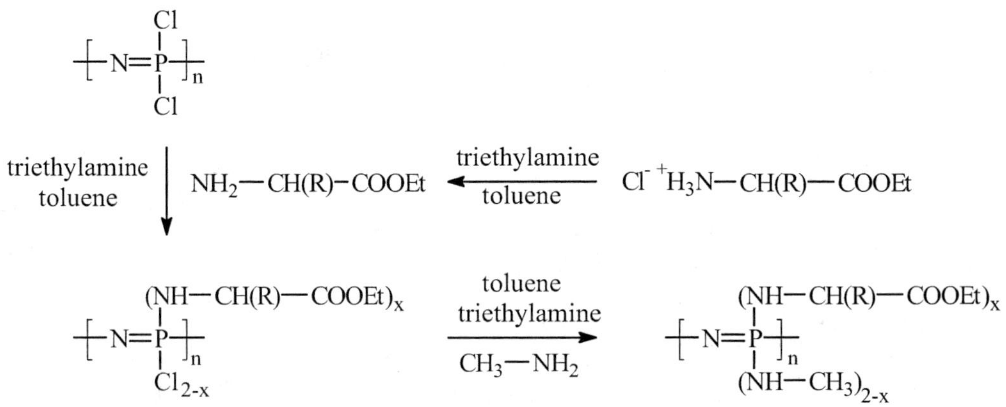

Figure 10: Synthesis of poly[(amino acid ester)-co-(methylamine)phosphazenes]

The amino acid containing polymers are sensitive to hydrolysis, leading ultimately to harmless products: amino acid, phosphate, ammonia and alcohols, which can be metabolised or excreted. *In vitro* studies showed that polyphosphazenes with different amino acid esters degraded over a period of several months, the degradation rate strongly depended upon the nature of the amino acids present along the polymer backbone[43,44,45].

Comparing three different poly[(amino acid ester)phosphazenes] (glycine ethylester, alanine ethylester, alanine benzylester) Allcock showed that poly[bis(glycine ethylester)phosphazene] underwent the fastest loss of weight.[45]. This side group provided the least steric hindrance against attack of water on the skeleton. The alanine ethylester had a slower degradation profile in comparison with glycine. The presence of a bulky, hydrophobic benzyl group instead of an ethylester on alanine reduced the degradation even more.

Phenylalanine has also been used as amino acid ester[46,47,48,49]. Due to the presence of this hydrophobic ester group the degradation was even slower than for alanine ethyl ester.

The rate of degradation could also be controlled by introducing different amino acid esters in different ratios on the same polyphosphazene or by adding in addition to the amino acid ester another component. Goedemoed[46] described the synthesis of copolymers containing glycine ethylester and glutamic acid diethylester (fig 11).

Polymers containing glycine ethylester and the hydrophobic *p*-methylphenoxy (50:50) were prepared by Ibim (fig 12). The degradation proceeded relatively slowly in neutral and basic solutions (± 20% after 35 days). Higher degradation levels were obtained in acidic media (63% after 35 days).[50]

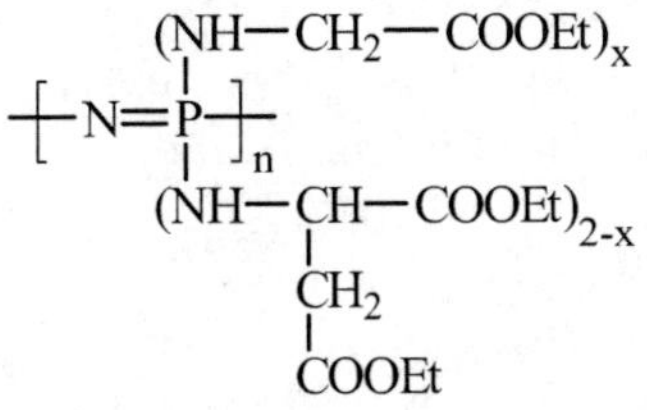

Figure 11: Poly[(glutamic acid diethylester)-co-(glycine ethylester)phosphazene]

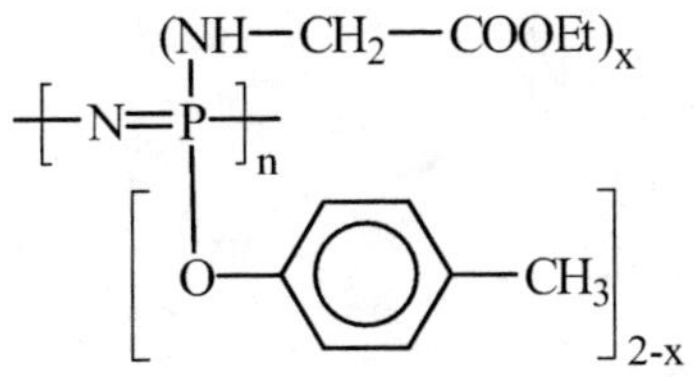

Figure 12: Poly[(p-methylphenoxy)-co-(glycine ethylester)phosphazene]

Copolymers containing glycine ethylester and α-amino-ω-methoxy-polyethylene oxide (M_w 5 kDa) were prepared by Vandorpe[51] (fig 13). It was demonstrated that the presence of the hydrophilic side groups increased the degradation rate.

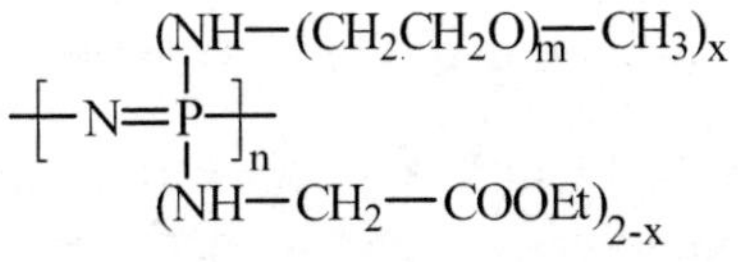

Figure 13: Poly[(α-amino-ω-methoxy-PEO)-co-(glycine ethylester)phosphazene]

Polyphosphazenes having methoxy-poly(ethylene glycol) (MPEG) side groups are thermosensitive materials[52] (fig 14). The polymers had a lower critical solution temperature (LCST) in the range of 25-99°C, depending on the composition of the substituents, the chain length of MPEG, the type of amino acid and the nature of the ester group. The LCST was high when a more hydrophilic composition was used. It was also affected by the presence of salts[53] and solvents[54]. Since these polymers were also biodegradable, they could be good candidates for drug delivery materials. The degradation rate could be controlled by changing the amino acid, by using different ester groups or by changing the molar ratio of the substituents[52].

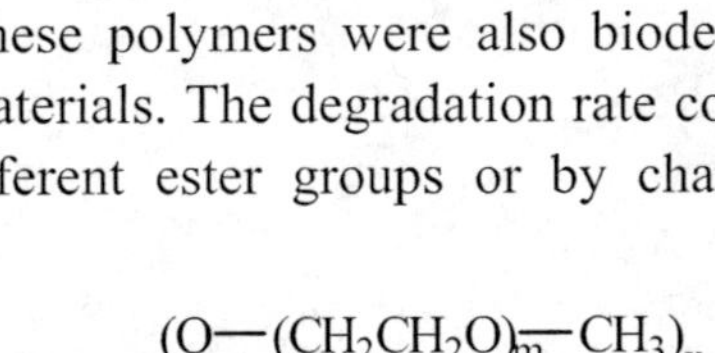

Figure 14: Poly[(methoxy-PEO)-co-(amino acid ester)phosphazene]

Several mechanisms have been proposed for the degradation process[55,56]. Allcock postulated that the ester was first hydrolysed generating a free carboxylic acid which further catalysed degradation of the backbone (fig 15).

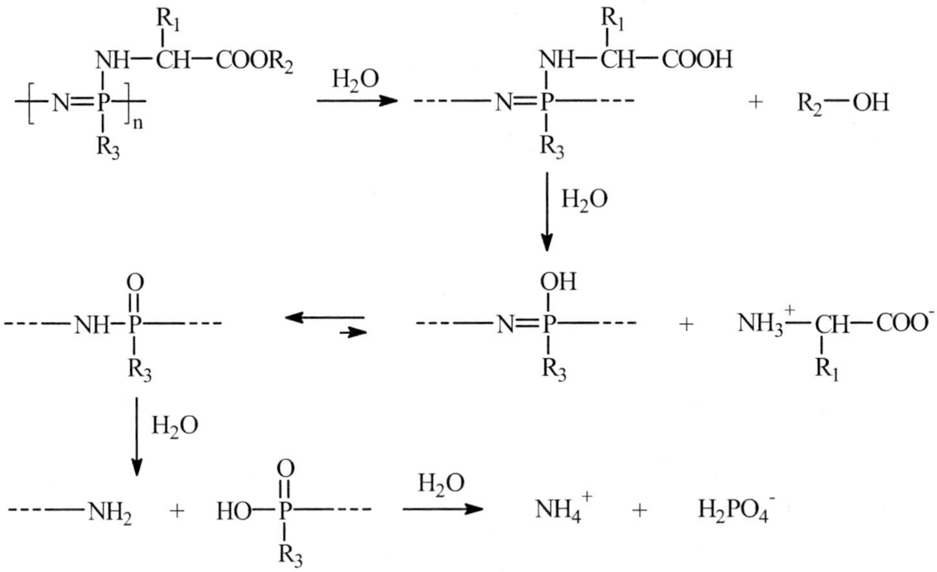

Figure 15: Degradation mechanism proposed by Allcock

Detailed analysis of the degradation products by Crommen showed the presence of amino acid esters and no sign of ethanol[55]. This indicated that the first step was the release of the amino acid ester, which subsequently hydrolysed to form the amino acid and the alcohol (fig 16). However, initial hydrolysis of the ester bond could not be excluded, but was considered to be less probable.

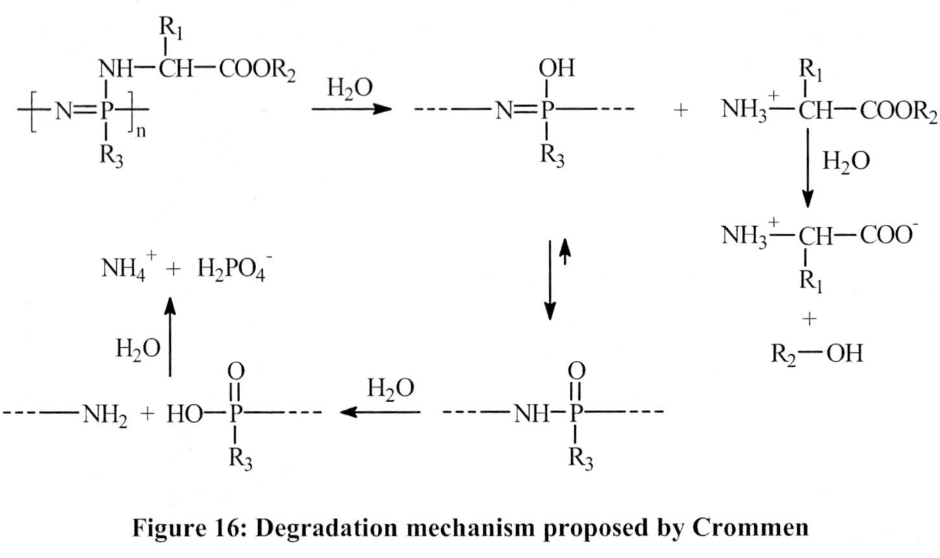

Figure 16: Degradation mechanism proposed by Crommen

The degradation of the polyphosphazenes was proven to be catalysed by acids[56]. This led Crommen to the idea of incorporating a controlled amount of hydrolysis sensitive amino acid ester side groups on the phosphazenes. Hence the so-called depsipeptide esters were introduced as co-substituents[57]. Depsipeptides are composed of an amino acid and a glycolic or lactic acid ester. The ester linkage between the amino acid and the lactide (or glycolic) component is sensitive towards hydrolysis, leading to the formation of a carboxylic acid, promoting degradation (fig 17). Introduction of a small amount of this depsipeptide resulted in a significant increase of the degradation. The rate could be varied by changing the content of the depsipeptide or by changing the chemical composition, glycolide has a better leaving capacity than lactide[57].

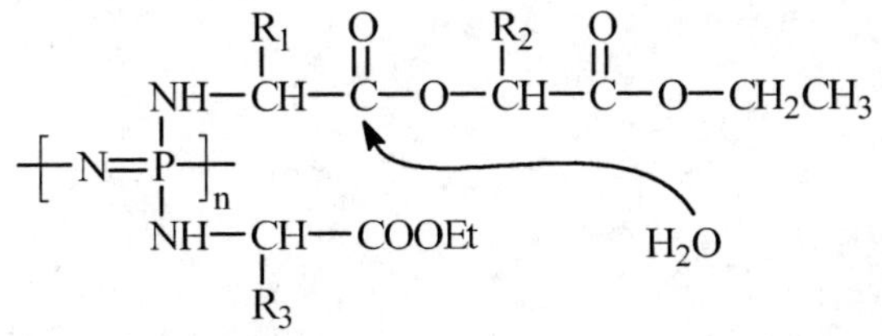

Figure 17: Poly[(depsipeptide ester)-co-(amino acid ester)phosphazene]

Polymers containing controlled amounts of depsipeptides and polyethylene oxide side groups have been prepared by Vandorpe[58] The rate of degradation could be controlled by variation of the amount of poly(ethylene oxide) substituted, or variation of the molecular weight of the poly(ethylene oxide) side-groups.

3.2. Imidazolyl-Substituted Polyphosphazenes

The imidazolyl group is very sensitive to hydrolysis. Polyphosphazenes containing only imidazole side groups decomposed in moist air to imidazole, phosphate and ammonia[59]. When a hydrophobic co-substituent was used, the degradation could be slowed down (fig 18). Another advantage of introducing a hydrophobic substituent was the possibility to achieve surface erosion. The degradation rate could be controlled by the ratio of both side groups.[60,61]

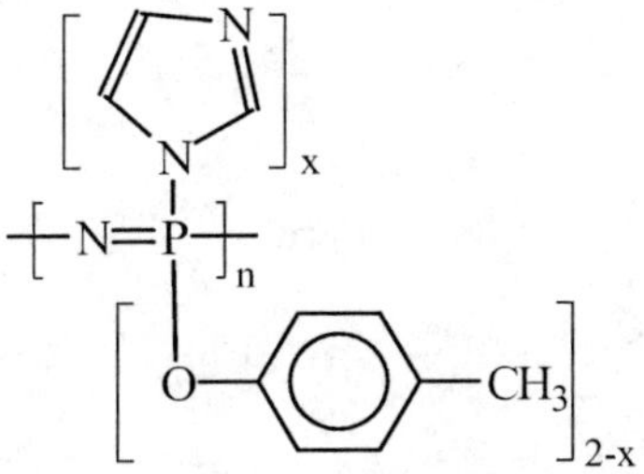

Figure 18: Poly[(imidazole)-co-(p-methylphenoxy)phosphazene]

Polyphosphazenes containing imidazolyl and amino acid ester side groups have been synthesised by Caliceti.[62] (fig 19). Degradation studies of poly[(alanine ethylester)-co-(imidazole)phosphazenes] (90:10 and 70:30) and poly[(phenylalanine ethylester)-co-

(imidazole)phosphazenes] (90:10 and 70:30) in a borate buffer pH = 7.4 at 37°C showed that the degradation proceeded slower for the phenylalanine substituted polymers. Enhancing the hydrophobic character decreased the degradation rate. On the other hand increasing the amounts of imidazole increased the degradation rate. A polyphosphazene containing phenylalanine ethylester, imidazole and a small amount of residual chlorine has been used as matrix for the release of naproxen[63,64]. Polymers containing alanine ethylester and imidazole (70:30) have been used to produce nerve growth guiding tubes[65]. Microcapsules made from polyphosphazenes bearing phenylalanine and imidazole (80:20) were used for the treatment of periodontal diseases.

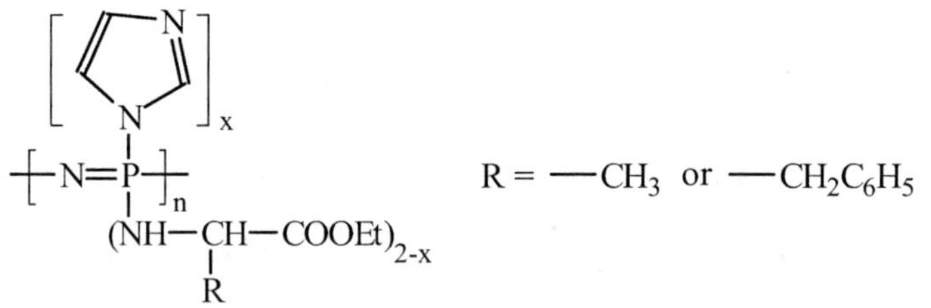

Figure 19: Poly[(imidazole)-co-(amino acid ester)phosphazenes]

3.3. Glyceryl and Glucosyl Containing Polymers

Allcock *et al.* prepared phosphazenes having glyceryl side groups[66]. Nucleophilic replacement of the chlorine atoms with glycerol would lead to cross-linking hence, two of the three hydroxyl groups were protected (fig 20). The deprotected polymer was soluble in water and hydrolysed slowly releasing glycerol, phosphate and ammonia. The presence of two remaining hydroxyl groups made it possible to attach bioactive components or to produce hydrogels by cross-linking.

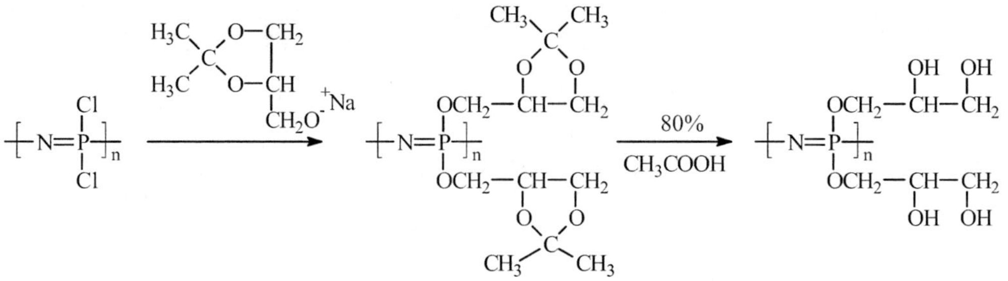

Figure 20: Synthesis of poly[di(glyceryl)phosphazene]

In a comparable way α-D-glucose could be attached to a polyphosphazene[67]. The first step was the protection of four hydroxyl groups of the glucose with aceton. Coupling of the diacetone glucose with poly[(dichloro)phosphazene] leaded to a soluble polymer. The diacetone glucose group is rather bulky, so complete substitution of all the chlorines was rather difficult. Instead of using forced conditions it was better to use a second, less hindered side group, e.g. methylamine (fig 21), trifluoroethoxy or (methoxyethoxy)ethoxy side groups (fig 22)[68]. Poly[(α-D-glucose)-co-(methylamino)phosphazenes] has a half-life of 165-175 hours at human body temperature and pH = 7.4. The hydrolysis products were phosphate, glucose, ammonia and methylamine. This last product might be unacceptable in biomedical applications, but instead co-substituents such as amino acid esters could be used.

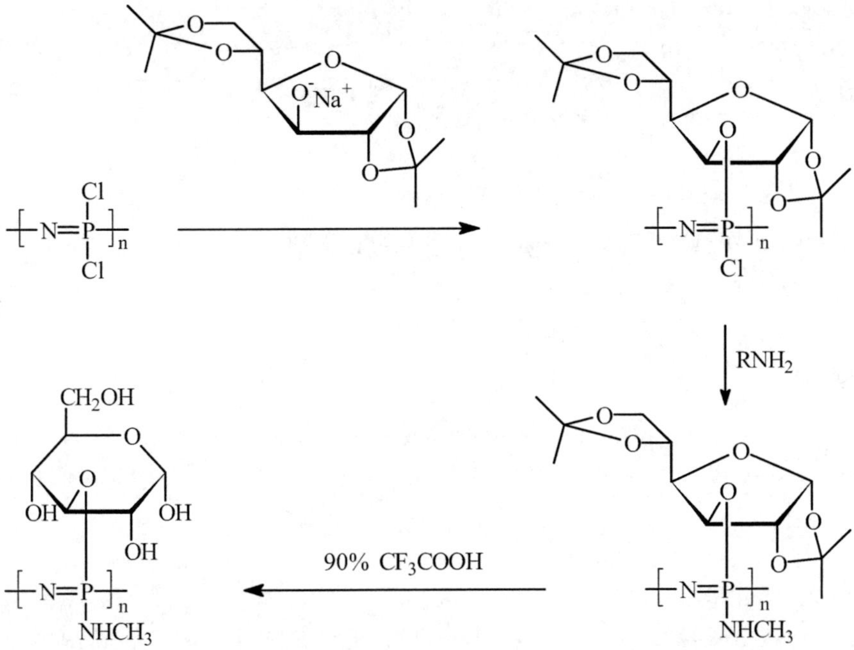

Figure 21: Synthesis of poly[(α-D-glucose)-co-(methylamino)phosphazene]

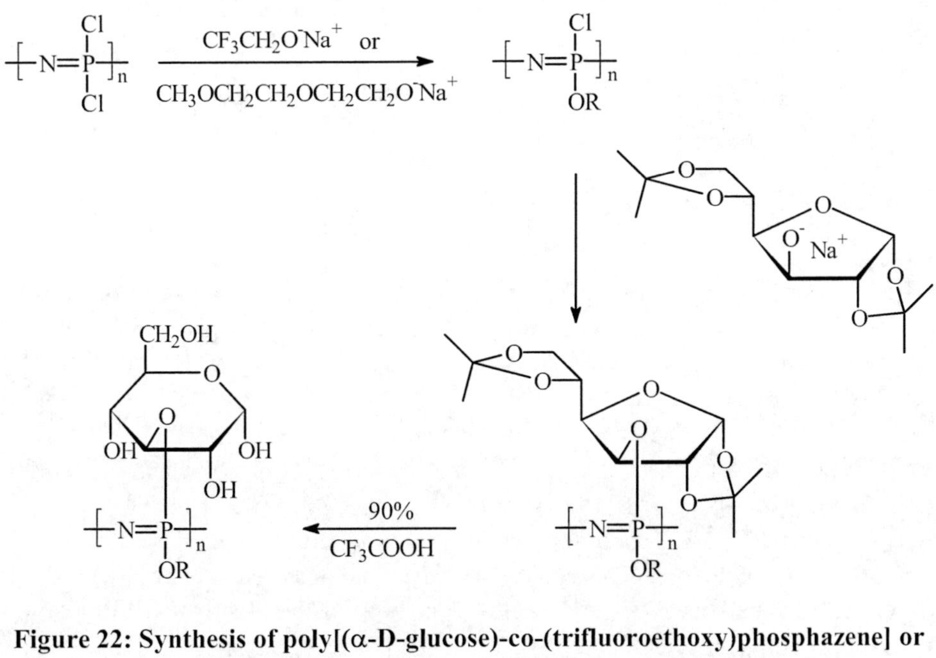

Figure 22: Synthesis of poly[(α-D-glucose)-co-(trifluoroethoxy)phosphazene] or poly[(α-D-glucose)-co-(methoxyethoxyethoxy)phosphazene]

3.4. Glycolate and Lactate Substituted Polyphosphazenes

The synthesis of polyphosphazenes bearing esters (ethyl or benzyl) of glycolic or lactic acid has been described by Allcock *et al.*[69] (fig 23). It was shown that the benzyl esters degraded slower than the ethyl esters. As expected, the polymers bearing the lactide side group degraded slower than the ones with the glycolic side group. The presence of a methyl group on the α-carbon made the polymers less accessible to hydrolytic degradation. The degradation products were benzyl alcohol or ethanol, glycolic or lactic acid, phosphate and ammonia and are considered non-toxic when present in small quantities.

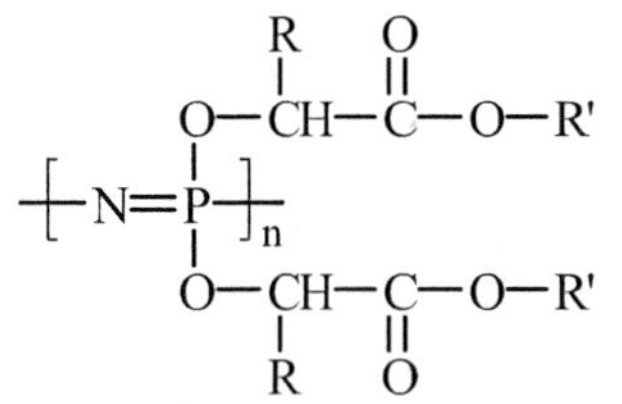

glycolide : R = H and R' = $—CH_2CH_3$ or $—CH_2C_6H_5$

lactide : R = CH_3 and R' = $—CH_2CH_3$ or $—CH_2C_6H_5$

Figure 23: Poly[bis(glycolide ethylester or benzylester)phosphazene] and poly[(bis(lactide ethylester or benzylester)phosphazene]

Blends of polyphosphazenes and poly(lactide-co-glycolide) have been made[70] (fig 24). This is a way to preserve the desired properties of the two polymers: the well known biocompatibility and the numerous applications of the polyesters and the hydrophobicity of a poly[(glycine ethylester)-co-(*p*-methylphenoxy)phosphazene] (PPHOS) which undergoes a surface erosion.

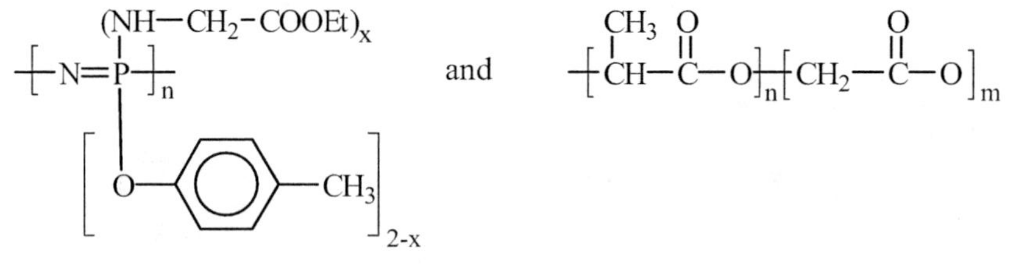

Figure 24: Blends of poly[(glycine ethylester)-co-(p-methylphenoxy)phosphazene] and poly(lactide-co-glycolide)

PPHOS has been blended in a 50:50 ratio with a 50:50 poly(lactide-co-glycolide). The two polymers showed no phase separation. The degradation rate could be controlled by changing the amount of glycine ethylester. A higher content resulted in a more rapid degradation, due to the hydrolytically sensitive group.

Blends of poly[bis(glycine ethylester)phosphazene] (PGP) and poly(D,L-lactide) (PLA), poly(lactide-co-glycolide) (80:20) (PLGA) and poly(sebacic anhydride-co-trimellitylimidoglycine)-block-poly(ethylene glycol) (30:50:20) (PSTP) have been prepared[71] (fig 25). Blends of PLA and PGP were completely incompatible, while PLGA and PSTP were partially compatible with PGP. Degradation experiments showed that the degradation rate could be varied by changing the PLGA or PSTP content of the blends. PGP/PLGA took 120 days to disappear completely while PGP/PSTP needed only 20 days.

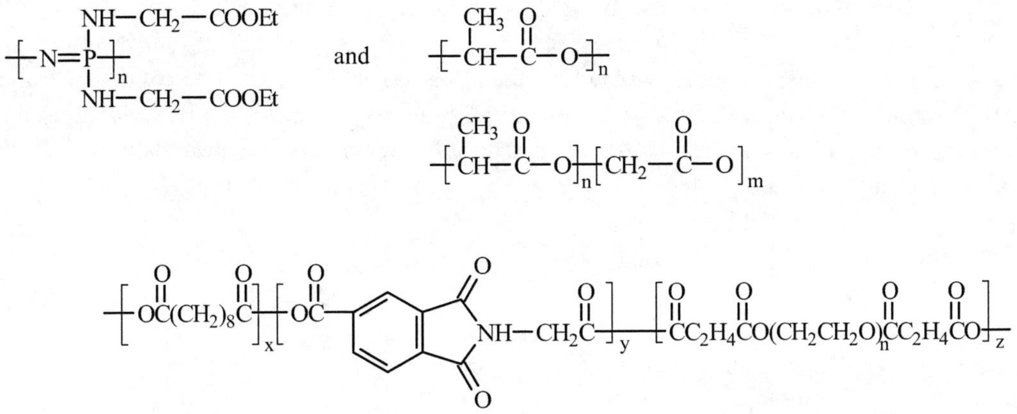

Figure 25: Blends of poly[bis(glycine ethylester)phosphazene] and poly(D,L-lactide), poly(lactide-co-glycolide) and poly(sebacic anhydride-co-trimellityl-imidoglycine)-block-poly(ethylene glycol)

3.5. Other Biodegradable Polyphosphazenes

In addition to soluble polymers, hydrogels based on polyphosphazenes have been prepared. Such materials can swell substantually in water and degrade in a more or less controlled manner. Poly[bis(carboxylatophenoxy)phosphazene] (PCPP) could be gelled by treating it with divalent cations in an aqueous media (fig 26). This hydrogel was slowly degradable. Introduction of the hydrolytically sensitive glycinato side group in addition to the carboxylato groups increased the degradation rate (PCGPP). The degradation rate could also be varied by coating the PCPP or PCGPP microspheres with poly-L-lysine[72]

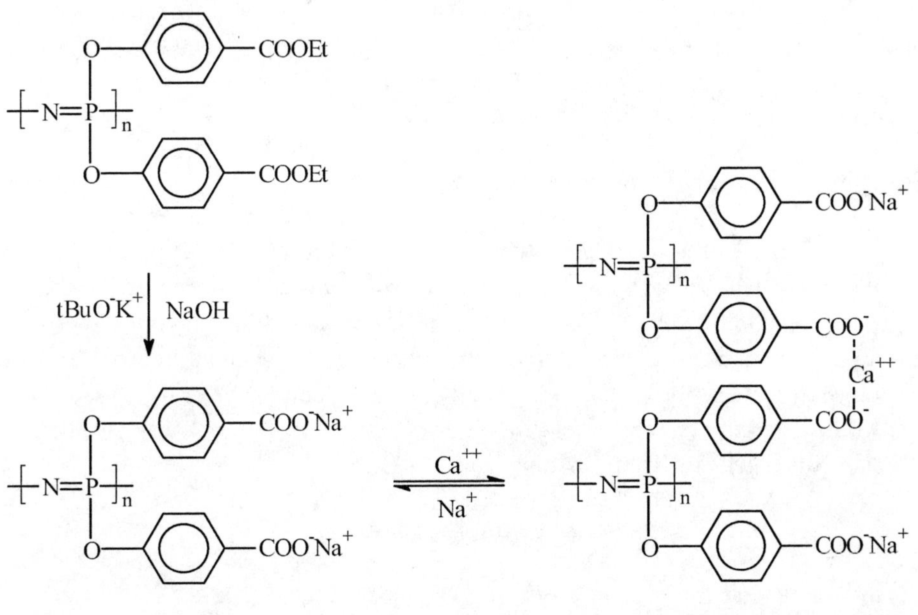

Figure 26: Formation of a poly[bis(carboxylatophenoxy)phosphazene] hydrogel

Degradable hydrogels have also been synthesised using a degradable cross-linker based on phosphazene derivatives. The trimeric phosphazene cross-linker had allylamine side groups, to provide polymerisable double bonds and imidazole groups to control the hydrolytic stability of the cross-linker (fig 27). Free radical polymerised poly(acrylic acid) gels have been synthesised with this cross-linker. Degradation occured over a period of one to two months in an aqueous media[73].

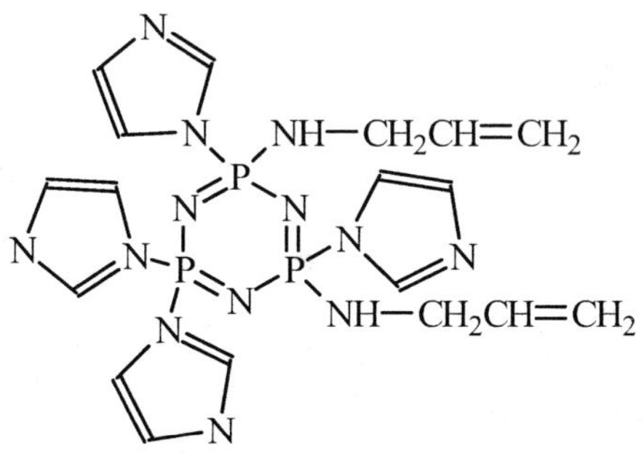

Figure 27: A trimeric cross-linker

4. Polyphosphazenes Used in Drug Delivery Systems

Biodegradable polymers are of great interest for biomedical applications. Many biodegradable polymers have been investigated for the development of release systems of pharmaceuticals or other bioactive agents. The release of guest molecules is achieved by the hydrolytic degradation of the polymer. It is very important that the degradation product are harmless, easily metabolised and excreted by the body. Phosphazenes satisfy these requirements. There are two types of drug delivery systems, the drug can be covalently attached to the polymer backbone or the polymer may be used as a matrix in which the drug is physically dispersed.

4.1. Covalent Polymer-Drug Combinations

A large variety of drugs and bioactive molecules have been covalently attached to the polyphosphazene chain. Steroids, such as desoxoestrone (fig 28), estrone, 17β-estradiol, 17α-ethylylestradiol, estradiol 3-methyl ether and 1,4-dihydroestradiol 3-methyl ether, have been linked to a polyphosphazene chain via the sodium salt of the steroidal hydroxy function. The residual chlorines were replaced by methylamine, glycine ethylester or n-butylamine[74]. These types had of low toxicity, were soluble in water or organic solvents and were biodegradable when glycine ethylester was used as side group.

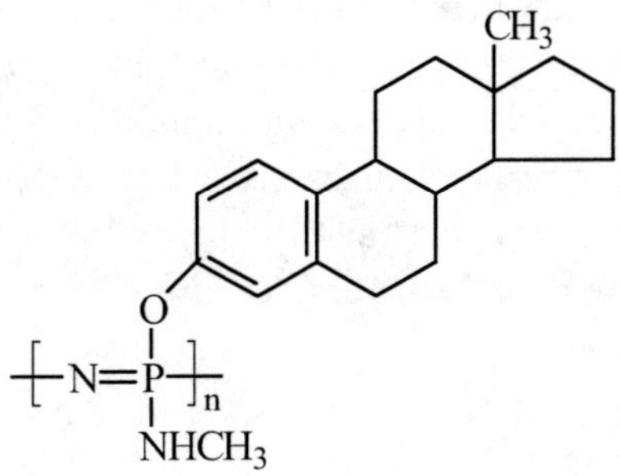

Figure 28: Poly[(desoxoestrone)-co-(methylamine)phosphazene]

Local anaesthetics produce loss of sensation and motor activity. Undesired side effects are caused by their high concentration in the blood. So there is a need for local anaesthetics which can be administered in a prolonged slow release way. Benzocaine, chloroprocaine, butyl *p*-aminobenzoate and 2-amino-4-picoline (fig 29) have been attached to polyphosphazenes by Allcock *et al.*

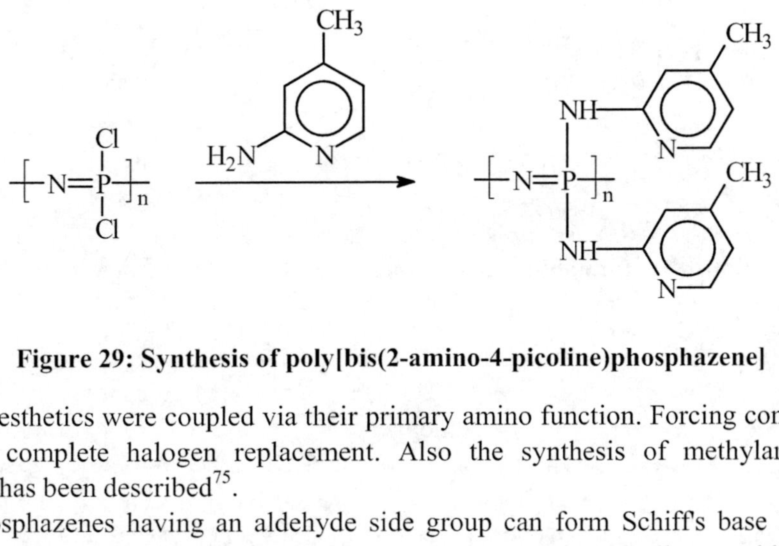

Figure 29: Synthesis of poly[bis(2-amino-4-picoline)phosphazene]

The anaesthetics were coupled via their primary amino function. Forcing conditions were needed for complete halogen replacement. Also the synthesis of methylamine as co-substituents has been described[75].

Polyphosphazenes having an aldehyde side group can form Schiff's base linkage with biologically active amines (fig 30). These linkages are hydrolytically unstable, hence the bioactive agent can be released when exposed to water. Examples of drug molecules that have been linked via Schiff's base coupling are sulfadiazine (antibiotic agent), dopamine (neurotransmitter), amino-4-picoline (analgesic agent) and hydralazine (antihypertensive agent)[76].

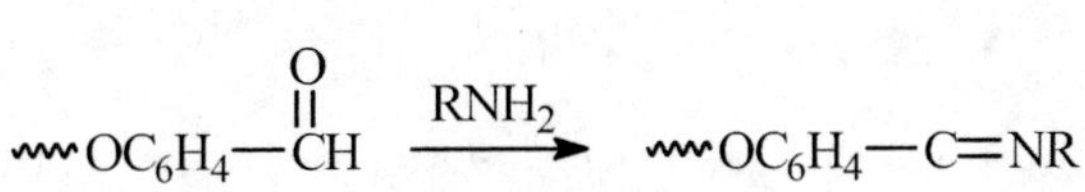

Figure 30: Schiffs' base linkage

Glucose-6-phosphate and trypsin have been immobilised on poly[bis(phenoxy)-phosphazene]. Porous alumina particles were coated with poly[bis(phenoxy)phosphazene], the surface was then nitrated and reduced to the amino derivative. An activated support was created by reaction of the amino groups with cyanogen bromide, nitrous acid or glutaric dialdehyde, followed by treatment with the enzymes (fig 31). The activity of the free glucose-6-phosphate was lost after 11 days, the immobilised enzyme preserved a high degree of activity for at least 89 days. For trypsin, most of the activity was lost after 3 days, but the immobilised one preserved more than 75% after 40 days[77].

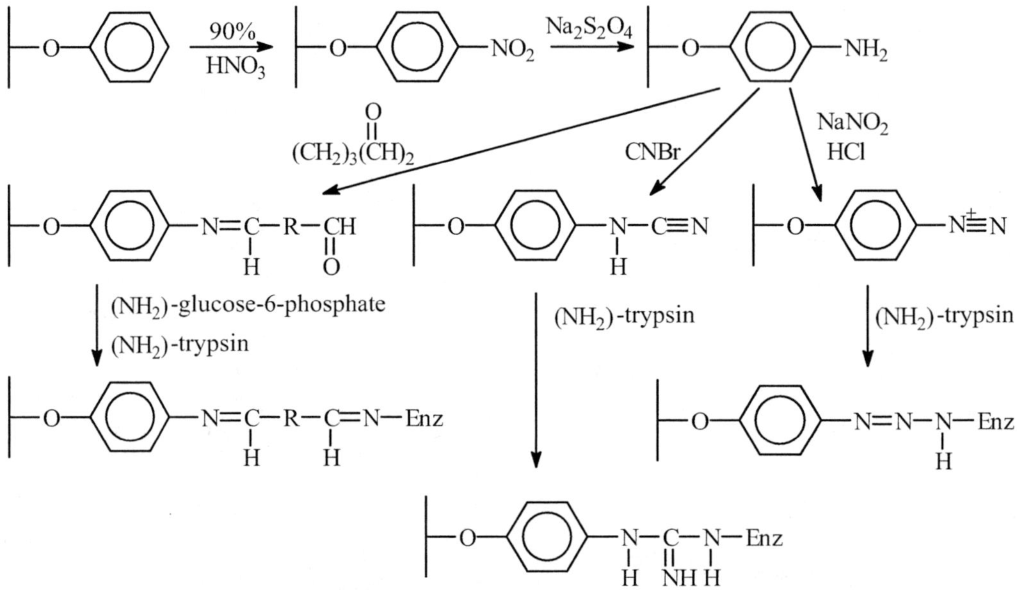

Figure 31: Immobilisation of glucose-6-phosphate or trypsin to poly[(bis(phenoxy)phosphazene]

Bioactive agents having a free carboxyl group have been coupled to polyphosphazenes containing amino side group[78] (fig 32). The latter was prepared by reaction of poly[(dichloro)phosphazene] with 4-cyanophenol. The cyano group was subsequently reduced to an aminomethyl group. Coupling occured using carbodiimide agents. Nicotinic acid (vitamin), N-acetyl-DL-penicillamine (antiviral agent) and p-(dipropylsulfamoyl)benzoic acid (diuretic agent) have been coupled via this approach. The use of amino acid esters or imidazolyl as co-substituent could make these polymers biodegradable. Introduction of methylamino side groups could make them water soluble.

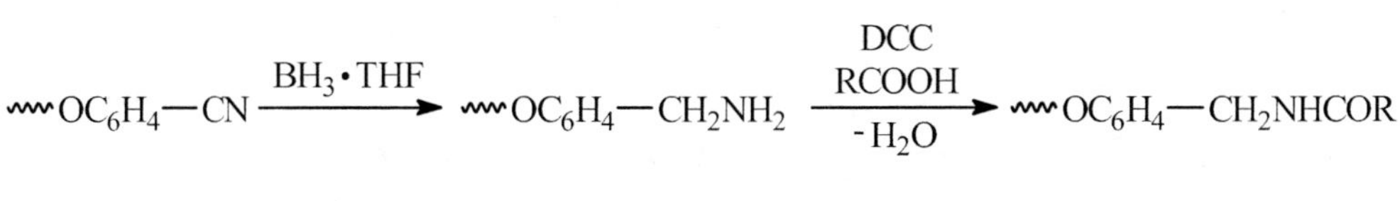

Figure 32: Amide linked polymer-drug conjugates

Catecholamines, such as dopamine, DL-epinephrine and DL-norepinephrine have been coupled to amino-aryloxy side groups by diazo-coupling techniques (fig 33). The polymer-bound dopamine could inhibit the prolactin release from rat pituitary cells in culture for 48 hours[79].

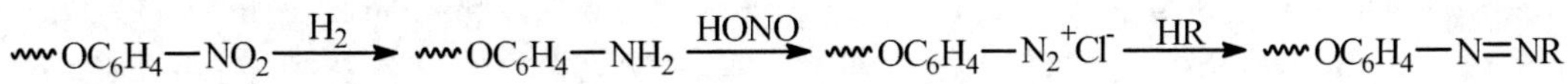

Figure 33: Diazo-coupling

The anticoagulant heparin has been ionically coupled to a phosphazene. Thermally initiated free radical bromination of a *p*-methylphenoxyphosphazene, followed by treatment with a tertiary amine, yielded a quaternised polymer. This water soluble salt, formed a stable polyelectrolyte complex when treated with an aqueous solution of sodium heparin (fig 34). The surface of the polymer film showed anti-thrombogenic activity[80].

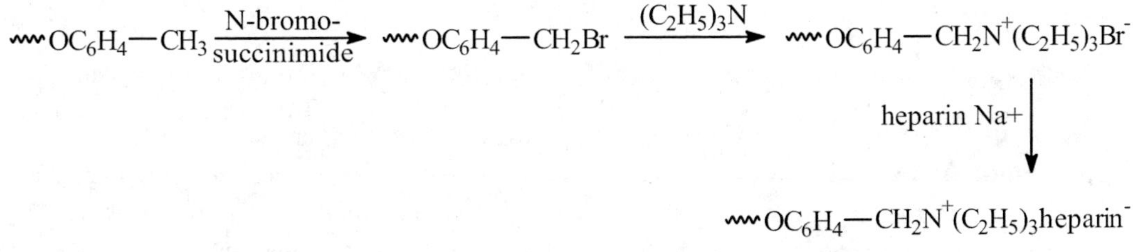

Figure 34: Ionic coupling of heparin to a phosphazene

Grolleman *et al.* covalently attached the anti-inflammatory agent naproxen via a spacer: L-lysine ethylester. Because of the bulkiness of the drug, only incomplete substitution could be achieved[81]. Hence the residual chlorines were replaced by glycine ethylester (fig 35). *In vitro* studies showed that at 37°C the release was slow and nearly constant[82]. The material was subcutaneously implanted in rats. Except for a slight encapsulation, no foreign body reactions were observed. These materials seemed acceptable as implant materials. The rate of naproxen release from the implanted device was however extremely low[83].

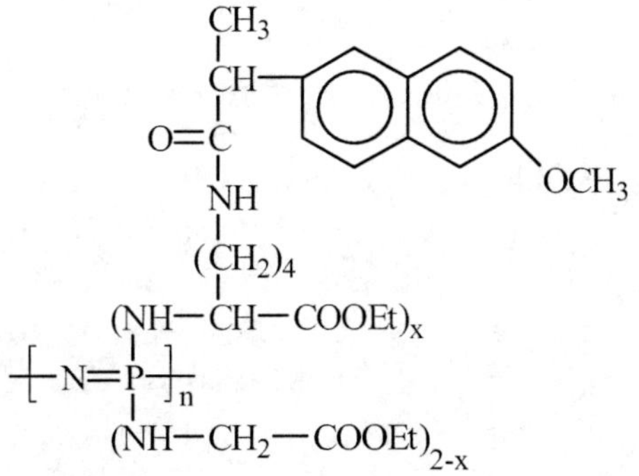

Figure 35: Poly[(glycine ethylester)-co-(L-lysine-naproxen)phosphazene]

The platinum complex, $(NH_3)_2PtCl_2$ is a well known antitumor agent. This drug is soluble in water and easily excreted through the kidneys and can therefore cause severe damage.

Linking this drug to a nonexcretable, but water-soluble polymer, can improve the effectiveness of the chemotherapy and decrease the toxicity. Poly[bis(methylamino-phosphazene] formed a coordination complex with $PtCl_2$ through the nitrogen in the backbone (fig 36)[84]. The polymer was a base that could replace the ammonia ligands in $(NH_3)_2PtCl_2$. Biological data demonstrated that a part of the activity could be preserved in the polymer-drug conjugate.

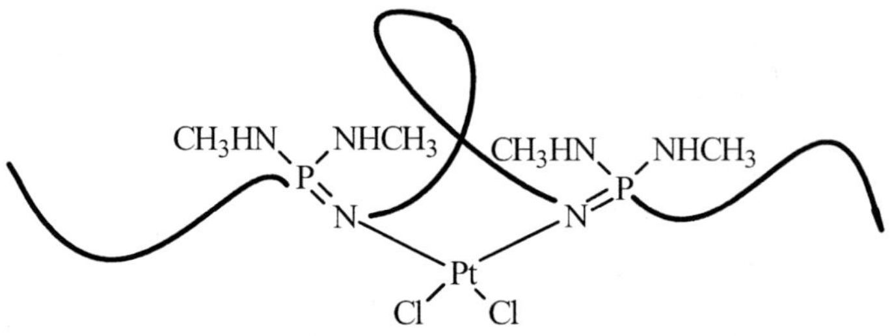

Figure 36: Coordination complex of $PtCl_2$ with poly[bis(methylamine)phosphazene]

Sohn *et al.*[85] reported the introduction of the antitumor agent (diamine)platinum to polyphosphazenes by using dicarboxylate spacer, glutamate or aspartate. Solubilising groups were introduced into the polymer backbone (fig 37). The use of different spacer groups did not give rise to differences in activity, but solubilising groups and amine ligands had a big influence on the antitumor activity. The solubilising side groups influenced the degradation rate. The best activity was observed with (DACH)Pt (trans(±)-1,2-diaminocylcohexane) and (DAMCB)Pt (1,1-diaminomethylcyclobutane).

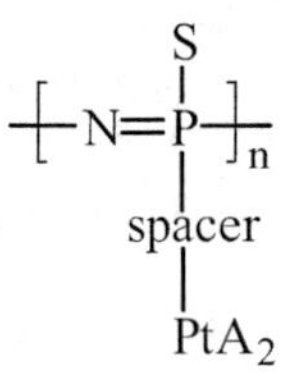

S = OH, OCH_3, $NHCH_3$ or $N(CH_3)_2$

spacer = aspartate or glutamate

A = NH_3, CH_3NH_2

A_2 = trans(±)-1,2-diaminocyclohexane (DACH)
1,1-diaminomethylcyclobutane (DAMCB)
tetrahydro-4H-pyran-4,4-dimethaneamine (THPDMA)

Figure 37: Polyphosphazene with different antitumor agents

Song *et al.*[86] introduced in addition to (DACH)Pt also a liver targeting β-galactosyl group to a 5-carboxyl-1-pentoxy group attached to polyphosphazene by carboxyl activation with a mixed anhydride method (fig 38). The platinum agent was released at a moderate rate when incubated in acidic or basic buffered solutions. Better antitumor activity was observed *in vivo* when (DACH)Pt was used instead of cisplatin or carboplatin.

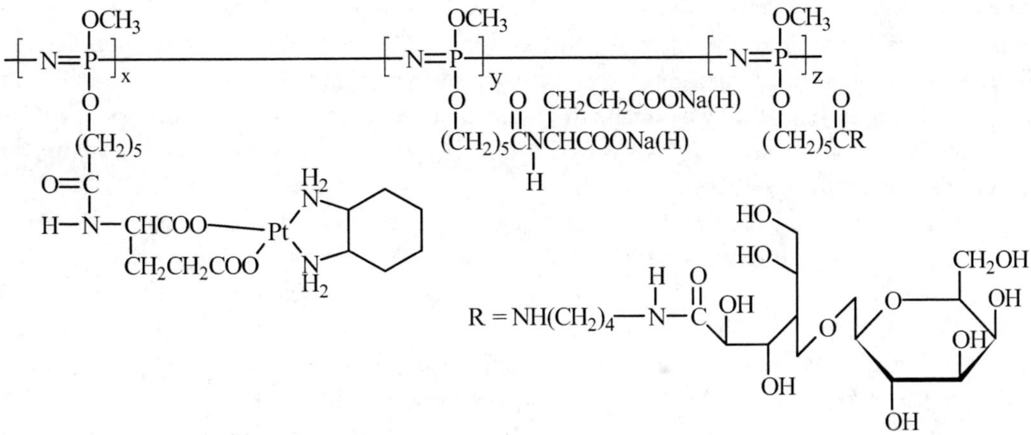

Figure 38: Polyphosphazene/(DACH)Pt conjugate having β-galactosyl moiety

Finally (DACH)Pt and doxorubicin have been linked to polyphosphazenes via a L-glutamic acid spacer. A polymeric conjugate having 12.3 mol % doxorubicin and 47.6 mol % (DACH)Pt showed the highest antitumor activity compared to polymer conjugates having only cisplatin, carboplatin or doxorubicin[87].

4.2. Physical Polymer-Drug Combinations

Controlled release devices can also be obtained by physical incorporation of a bioactive compound into an implantable or injectable matrix system. Incorporation of drugs in a matrix having a certain shape and size, can offer diverse advantages over covalently bound combinations. The fabrication is relatively easy and there is no need for chemical modifications of the incorporated drug.

Laurencin *et al.* have used polyphosphazene derivatives with imidazole and methylphenoxy side groups as drug delivery device for the controlled release of progesterone and bovine serum albumin (fig 18). The release of 5 and 10 % loaded progesterone from a 20% imidazole substituted polyphosphazene took place over 42 days *in vitro*. *In vivo* the release showed a lag period of ± 3 days, followed by a period of constant release for 37 days. A burst release of 25% BSA was observed from a 2.5% loaded BSA polyphosphazene. This effect could be reduced to 5% by dipping the matrices in a 20% solution of a 20% imidazole substituted polyphosphazene. At 42 days 45% of the protein was released.

Matrix devices containing the anticancer agent melphalan have been prepared by Goedemoed *et al.*[88,89]. Three different polyphosphazenes were used as matrix materials for this cytotoxic drug, a poly[bis(glycine ethylester)phosphazene], a poly[(glutamic acid diethylester)-co-(glycine ethylester)phosphazene] (50:50) (fig 11) and a poly[(phenylalanine ethylester)-co-(glycine ethylester)phosphazene] (50:50). The results showed that effective concentrations of melphalan could be maintained for longer periods[88,89]. Different methods of preparations were investigated using poly[bis(glycine ethylester)phosphazene] and poly[(glutamic acid diethylester)-co-(glycine ethylester)phosphazene]. The release profile was influenced by these different methods. An *in vivo* tumor model was used to investigate various drug matrix preparations. It was indicated that the chances of survival were better when a slow and gradual release was used instead of injections[88]. A solvent evaporation technique was used to make microspheres of poly[(phenylalanine ethylester)-co-(glycine ethylester)-phosphazene] containing melphalan methylester and melphalan. A gradual and well sustained release profile was obtained for melphalan methylester but a very poor release for melphalan. *In vivo* experiments showed promising therapeutic results for the melphalan methylester loaded microspheres[89].

Subdermal implants containing trypanocidal drugs have been prepared and evaluated *in vivo*. Films of poly[bis(glycine ethylester)phosphazene] containing 25% ethidium were grounded into a powder. The powder was melt extruded into strands of 1.2-2 mm diameter, rods of 1 cm were cut off. These were subcutaneously implanted by injection in the thorax of rabbits. The rabbits were infected with T. congolense trypanosomes (sleeping disease). *In vivo* results showed that the ethidium is therapeutically effective against this infection. The drug was efficient over a significant period of time. The infected rabbits treated with a blank polymer died 21 days after infection[90].

Polyphosphazene hydrogels have also been used in controlled delivery systems. Ionic hydrogels are particularly important because their equilibrium degree of swelling is affected by changes in pH or ionic strength. Poly[bis(carboxylatophenoxy)phosphazene] (PCPP) was an ionic hydrogel when treated in an aqueous media with salts of divalent cations (fig 26). Proteins such as albumin and β-galactosidase or 30-nm polystyrene beads could be trapped into PCPP gel microspheres and could be released in a controlled way. Increasing the concentration of the polymer or the calcium ions, gave rise to a decreased diffusion of the proteins and the polystyrene beads. However when the calcium ion concentration exceeded 12 % w/w a reversed phenomena (higher release) was observed. Coating the microspheres with a polyelectrolyte complex, formed by PCPP and poly-L-lysine, extended the release of β-galactosidase, but not that of albumin. This indicated the selectivity of PCPP-PLL membranes for macromolecules with different molecular weights. By adding a second layer of PCPP on the PCPP-PLL microsphere a significant decrease in membrane permeability was observed[91,92]. PCPP could also be used as a vaccine delivery vehicle. Antigens could be mixed with the soluble PCPP and then injected directly into an animal for parenteral immunisation. The combination of the PCPP/antigen solution could also be cross-linked with divalent cations. The microspheres could then be used for parenteral or mucosal immunisation.

pH-Sensitive hydrogels have been designed by Allcock *et al.*[93]. These polyphosphazenes contained different ratios of sodium oxybenzoate (1) and methoxyethoxyethoxy (2) side groups. The polymers were cross-linked through the methoxyethoxyethoxy side group by ^{60}Co gamma irradiation (fig 39). The release of the dye Biebrich Scarlet was studied. At pH 7.4 the dye was completely released in 4 to 12 hours due to the high swellability of the gels. However at pH 2 only a small fraction was released after 48 hours. The rate could also be varied by changing the ratio of the side groups. At pH 2 a significant higher release was observed from a phosphazene hydrogel containing 30 % (1) and 70% (2) than from a 76% (1) and 24% (2) one. The larger number of carboxylate groups will be protonated at pH 2, leading to a more pronounced interaction of the hydrogel with the azo, hydroxy and sulphonate groups of the dye.

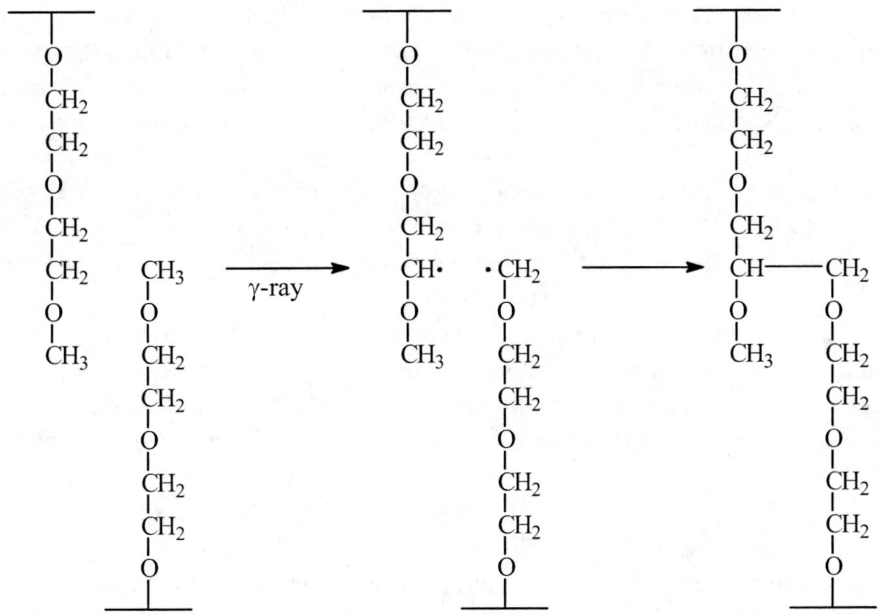

Figure 39: Cross-linking of the methoxyethoxyethoxy side groups

Poly[(glycine ethylester)-co-(*p*-methylphenoxy)phosphazene] (50:50) has been synthesised by Ibim *et al.* to incorporate ^{14}C-labeled inulin (polysaccharide)[70]. The release was characterised by a burst effect followed by a period of near constant release kinetics. The release was highest in acidic media. Increasing the drug loading resulted in an increased hydrophilicity of the matrix, which in turn resulted in an increased degradation. Hence the release could be varied by changing the pH and the drug loading.

Allcock *et al.* have synthesised different polyphosphazenes bearing amino acid esters: glycine ethylester, alanine ethylester and alanine benzylester. Solutions of these polymers were mixed with ethacrynic acid (diuretic) or Biebrich Scarlet (azo dye). The initial release occured through diffusion of the drug through the polymer matrix. At a certain time an increase in release was observed caused by the degradation of the polymer. For the different polymers this time point was different as the different side groups have different degradation rates.

Schacht *et al.* have described the synthesis of biodegradable polyphosphazenes substituted with amino acid esters (glycine ethylester and a combination of glycine- and phenylalanine ethylester) which contained the antitumor agent mitomycin C (MMC). The release rates of MMC could be varied by changing the side groups or by changing the ratio of the side groups. When the hydrophobicity of the matrix was increased, the water uptake was decreased and so the drug diffusion decreased. Polymer disintegration occured after the tablets had released all MMC. Tablets were implanted in the peritoneal cavity of mice bearing i.p. inoculated C26 murine colorectal cells. Matrices having 1.6 mg MMC showed promising results, especially with a poly[(glycine ethylester)-co-(phenylalanine ethylester)phosphazene] (65:35)[94].

Nanoparticles have been made from poly[(glycine ethylester)-co-(phenylalanine ethylester)phosphazene] using a polyphosphazene with polyethylene oxide side groups as suspending agent[51]. These particles proved to be of interest for the site-specific drug delivery.

Intravenous administration in rats resulted in a prolonged blood circulation and a reduced liver uptake. Studies in rabbits showed the same results and a significant amount of the nanoparticles were targeted to the bone marrow[48]. These results indicated that polyphosphazene nanoparticles having a hydrophilic coat could be used as long circulating drug delivery systems.

Tantalum balloon expandable coronary stents have been coated with polyphosphazenes, substituted with amino acid esters (a combination of glycine- and phenylalanine ethylester), and loaded with 5% methylprednisolon. This corticosteroid has anti-inflammatory and immunosuppresive effects. *In vitro* results showed that the methylprednisolon coated stents released 96% of the drug within 24 hours. Stents were implanted in the right coronary artery of pigs and evaluated after 6 weeks. The lumen diameter of the coated stent with methylprednisolon was significantly larger than a coated stent with no drug[95].

Polyphosphazenes with amino acid esters (alanine ethylester) and a combination of amino acid esters and imidazole (alanine ethylester/imidazole (90:10 and 70:30), and phenylalanine/imidazole (90:10 and 70:30)) (fig 19) have been used by Caliceti *et al.* Matrices were prepared in the form of films by solvent evaporation and spheres by solvent extraction. Naproxen (anti-inflammatory drug), acetyl tryptophanamide (peptide analogue) and narciclasine (antitumor activity) were evaluated as model drugs. The release rate depended on the polymer composition. The conclusion of this study was that films were more suitable for drug release.[63]

Caliceti *et al.* have also prepared a polyphosphazene containing, in addition to phenylalanine ethylester and imidazole, also a small amount of chlorines on the polymer for the controlled release of naproxen. Films were obtained by solvent evaporation techniques. The release depended upon the thickness of the matrix and the concentration of the entrapped drug. The naproxen loaded films were suitable for acute or chronic inflammation, depending upon the preparation conditions. Subcutaneous implantations of thin matrices in rats were able to reduce carrageenan oedema. The use of thick matrices showed a reduction in chronic inflammation caused by adjuvant arthritis. These positive results encouraged the authors to study a controlled release system that was more suitable for use in humans; e.g. microspheres that could be injected by syringe. When the previously described polyphosphazene-naproxen loaded matrices were utilised, minor surgery was needed. To overcome this problem microspheres could be synthesised. The previously described phosphazenes could not be used to prepare microspheres, instead poly[(phenylalanine ethylester)-co-(imidazole)-phosphazenes] were used. The microspheres were prepared by three methods: spray-drying, solvent evaporation and solvent evaporation-extraction. When a spray-drying technique was used, smaller microspheres could be obtained but the drug release was very rapid. The other two techniques gave rise to larger and less uniform microspheres, but the release was slow and continuously over a longer time period both *in vitro* and *in vivo*[96].

Microspheres made from poly[(phenylalanine ethylester)-co-(imidazole)phosphazenes] have also been used for the delivery of insulin. Three different techniques were applied: suspension-solvent evaporation, double emulsion-solvent evaporation and suspension/double emulsion-solvent evaporation. The preparation method affected the mean particle size and the particle distribution profile. *In vivo* result showed that the glucose level could be controlled for over one month when microspheres were made by the double emulsion-solvent evaporation[97].

Osteoporosis is a bone disease affecting a big part of the ageing population. It can be treated by calcitonin, but the low stability, the poor absorption and the short half-life in plasma are some drawbacks of this polypeptide. To ensure a prolonged calcitonin delivery, a polyphosphazene (phenylalanine ethylester and imidazole) matrix was evaluated by Caliceti *et al.* A therapeutic effect was confirmed by the improved densitometric values observed in the femoral bone[98]. In another study these matrices were implanted in osteopenic female rats and after 30 days a second series was implanted. Excellent *in vivo* results were obtained and a therapeutic effect of the released calcitonin in the osteopenic conditions was observed[99].

Amphiphilic diblock copolyphosphazenes have been designed by Chang et al.[100] via the controlled cationic block copolymerisation of phosphoranimines. The polymer consisted of a hydrophilic part, poly[bis(methoxyethoxyethoxy)phosphazene] (MEEP) and a hydrophobic one, poly[(phenoxy)-co-(methoxyethoxyethoxy)phosphazene] (fig 40).

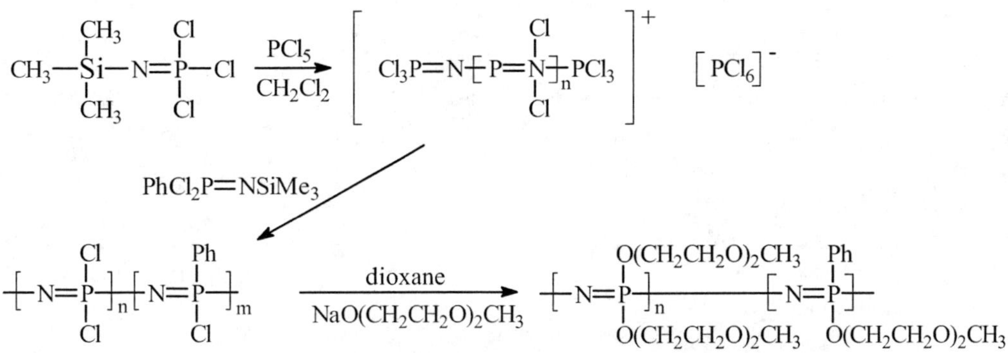

Figure 40: Synthesis of amphiphilic diblock copolyphosphazenes

This block copolymer formed micelles with a spherical shape, useful for drug delivery applications. The outer shell of the micelles was formed by the hydrophilic MEEP. MEEP has a lower critical solution temperature (LCST)[101], consequently the micellar solutions were expected to show a temperature dependent phase transition. An increase of turbidity was observed above 57°C. Polymers with LCST properties could be used in thermosensitive drug release devices.

A fast controlled release is required for pulsatile drug delivery systems. So unlike the most of the known biodegradable polyphosphazenes, a novel fast degradable polymer bearing glycine ethylester and benzylester of amino acethydroxamic acid (40 and 80 %) (PEBP) was synthesised by Qiu *et al.*[102] (fig 41).

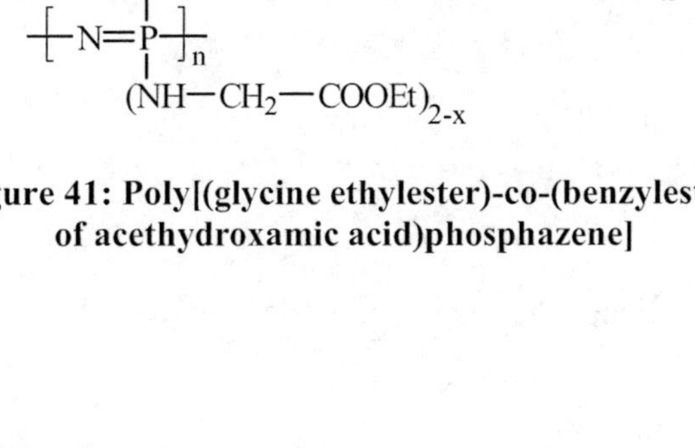

Figure 41: Poly[(glycine ethylester)-co-(benzylester of acethydroxamic acid)phosphazene]

Degradation experiments indicated that the degradation of this polymer was pH sensitive. Further investigations showed a fast hydrolysis of the side groups followed by a slow cleavage of the backbone. The initial hydrolysis of hydroxamic acid ester was base-catalysed. This explained why the polymer was more stable at low pH (fig 42).

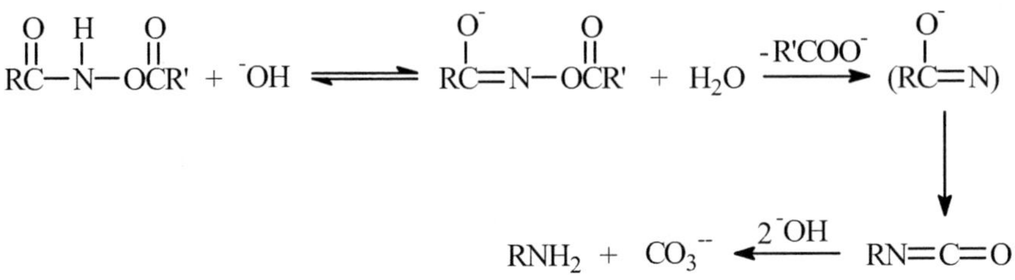

Figure 42: Base-catalysed hydrolysis of hydroxamic acid

These polymers could be useful as drug delivery materials, preliminary experiments were carried out with FITC-dextran and myoglobin as model drugs (10%). FITC-dextran was released at all pH's, the rate was controlled by both diffusion and erosion depending on the pH of the release media. Myoglobin was released at pH = 8.0 and 7.4, but at pH = 5.0 and 6.0 there was no release observed[102].

n another study by Qiu, a core shelled cylindrical polymer was made out of a biodegradable hydrophobic coating and a laminated core of polyanhydrides and polyphosphazenes (fig 43).

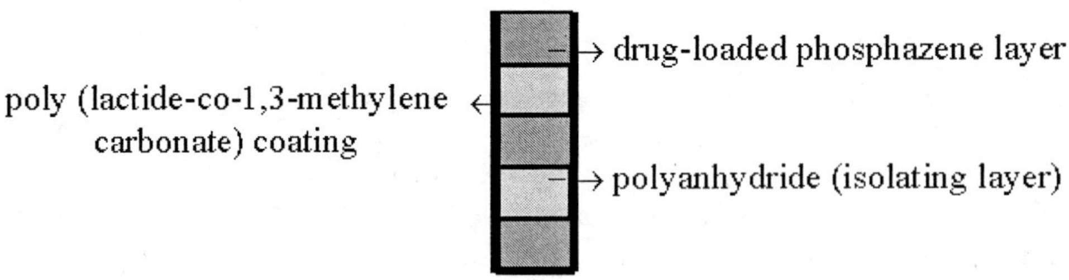

Figure 43: Core shelled cylindrical polymer

Because of the slow degradation profile, (10% weight loss within 3 months) poly(lactide-co-1,3-trimethylene carbonate) was used as coating. Polyanhydrides were used to induce a lag time of drug release. As drug loaded polyphosphazene, PEBP (fig 41) was used. An open end was needed to start the degradation of the core, while the cylindrical device kept its shape. The pulsatile drug release was obtained by the alternating layers. The lag time and the duration of the release could be varied by changing the type or quantity of the polyanhydrides and by changing the quantity of the polyphosphazene[103].

5. Use of Polyphosphazenes in Tissue Engineering

Loss of human tissue or organ is a devastating problem for a patient and their family. Treatment options include reconstructive or transplant surgery. Biomedical materials or tissue implants can be used, they have saved and improved a great number of lives, but it is still not the perfect solution. Transplant surgery is limited by the shortage of donor tissues and organs. Tissue engineering however is a potential alternative or complementary solution. The basic idea behind this field is a combination of a scaffold material with cells from a specific tissue source. Bio-engineered organs may be grown in a laboratory and can be transplanted afterwards into people. Scaffolds provide a mechanical stability of the construct and serve as a template for the three dimensional organisation of the tissue. Once the cells start to produce their own extracellular matrix, the scaffold is no longer needed, so the scaffold needs to be degradable[104,105].

It would be ideal if cells can be obtained from the patient. However this is not always possible. When allogenic or xenogenic cells are used, an immuno-isolative barrier is needed to protect the transplanted foreign cells from the host's immune attack. Poly[bis(carboxylatophenoxy)phosphazene] (PCPP) (fig 26) or poly[(glycinato)-co-(carboxylatophenoxy)phosphazene] (PCGPP) had the ability to gel when treated in an aqueous media with divalent cations. These polymers were compatible for several different cell types. The microspheres could be coated with poly-L-lysine, this allowed inward diffusion of nutrients for the entrapped cells, and made them impermeable to antibodies[106].

Biodegradable polyphosphazenes were studied as potential supports for cells in tissue regeneration. Poly[(imidazole)-co-(methylphenoxy)phosphazene] (fig 18) and poly[(glycine ethylester)-co-(methylphenoxy)phosphazene] have been examined as support for the osteoblast growth. Both polymers seemed to support the growth of MC3T3-E1 cells (osteogenic cell line), but results with the glycine substituted polymers were better. When the imidazolyl content increased, a reduction of the cell attachment and growth was observed. However when the content of glycine was increased above 10% an increased cell attachment and growth was seen. These *in vitro* results suggested that these polyphosphazenes were suitable biomaterials for the construction of a cell-polymer matrix for tissue regeneration.

Polyphosphazenes were also tested in nerve regeneration experiments. The use of autologous nerve grafts is the most frequently used surgical technique for nerve reconstruction. However this procedure has some disadvantages. An alternative is the use of tubular prostheses to bridge a nerve gap. Langone used poly[(alanine ethylester)-co-(imidazole)phosphazenes] (PEIP)(70:30) to create nerve guides by deposition of the dissolved polymer on a glass capillary tube, followed by evaporation of the solvent. *In vitro* results showed that the PEIP guide was more biocompatible than silicon tubes. A tissue cable was found through the PEIP nerve after 45 days, and it had the morphological features of a healthy regenerated nerve[109].

A longer lasting polymer, poly[bis(alanine ethylester)phosphazene] has also been evaluated as a guide for nerve regeneration. Analysis after 30, 90 and 180 days showed gradual degradation without any signs of toxicity. The number of myelinated axons in the lumen of the tube was almost the same as the myelinated axons found in the autologous graft. The reaction of the surrounding tissue was rather mild[107,108].

Growth factors are sometimes required to promote tissue regeneration. Direct injection of growth factors in solution is not very effective, because of the rapid diffusion out of the site within one day. A more promising way to make use of growth factors at the specific site is the controlled release of these factors over a long period of time. When neurite promoting factors could be trapped in the above mentioned polymer, an 'activated' polyphosphazene conduit could be made.

It is also possible to incorporate anti-inflammatory or antibacterial drugs into matrices which can be released in a controlled way. Membranes of polyphosphazenes bearing alanine ethylester and microcapsules bearing phenylalanine ethylester and imidazole have been made. These polyphosphazenes had a degradation rate that corresponded to the healing of the bone defect. Trimethoprim, succinylsulphathiozole (antibacterial drugs) and naproxen (anti-inflammatory drug) could be entrapped in the membranes. These membranes did not cause any inflammatory responses and regenerated bone was observed in close contact of the remaining membrane. The entrapped drugs could be released at a rate that ensured therapeutic concentrations in the surrounding tissue. The microspheres were entrapped with the same drugs except trimethoprim, because of the unsatisfactory release profile. The microspheres could be mixed with hydroxyapatite, a compound that enhances bone formation[109].

In another study, bone defects were performed in the tibia of rabbits, the defects were covered with the previously described polphosphazene membranes and with a poly(tetrafluoroethylene) membrane. Satisfactory results were obtained with the polyphosphazenes, the PTFE membranes gave less sufficient results. Also surgical defects were made in the mandible of animals. Polyphosphazene membranes were placed over the breaches, some of which were filled with microspheres, alone or mixed with hydroxyapatite. Healing occurred in all the cases. The entrapment of drugs did not interfere with the bone healing[110].

6. General Conclusions

Starting from poly[(dichloro)phosphazene], it is possible to synthesise a wide range of polyphosphazenes with different side groups. Depending upon the nature of the side group, stable or hydrolytically degradable polymers can be obtained. Biodegradable poly[(organo)phosphazenes] can be used in drug delivery systems or as scaffold for tissue engineering. The mechanical properties and the rate of degradation can be varied by a proper choice of the chemical composition of the polymer backbone.

7. Acknowledgment

The authors are grateful to the Flemish Institute for Science and Technology (IWT) for providing a research grand to Heyde M.

References

1. Mark, J.E., Allcock, H.R., West, R., *Inorganic Polymers*, Prentice Hall, Englewood Cliffs, New Jersey, Chapter 3, (1992)
2. Liebig, J.,Wohler, J., *Ann. Chem.*, **11**, 139, (1834)
3. Rose, H., *Ann. Chem.*, **11**, 131, (1834)
4. Stokes, H.N., *Am. Chem.*, **19**, 782, (1897)
5. Allcock, H.R., Kugel, R.L., *J. Am. Chem. Soc.*, **87**, 4216, (1965)
6. Allcock, H.R., Kugel, R.L., *Inorg. Chem.*, **5(10)**, 1709, (1966)

7. Allcock, H.R., Kugel, R.L., *Inorg. Chem.*, **5(10)**, 1716, (1966)
8. Allcock, H.R., *Chem. Rev.*, **72**, 315, (1972)
9. Emsley, J., Udy, P.B., *Polymer*, **13**, 593, (1972)
10. Maccallum, J.R., Wernick, A., *J. Polym. Sci., Part A-1*, **5**, 3016, (1967)
11. Singler, R.E., Schneider, N.S., Hagnauer, G.L., Polym. Engin. Sci., 5, 321, (1975)
12. Patat, F., Kollinski, F., *Makromol Chem.*, **6**, 292, (1951)
13. Gimblett, F.G.R., *J. Polym. Sci.*, **60**, S29, (1962)
14. Allcock, H.R., *Polymer*, **21**, 673, (1980)
15. Gleria, M., Audisio, G., Daolio, S., Tralde, P., Vecchi, E., *Macromolecules*, **17**, 1230, (1984)
16. Allcock, H.R., Gardner, J.E., Smeltz, K.M. *Macromolecules* **8**, 36, (1975)
17. Hagnauer, G.L. *Journal Of Macromolecular Science-Chemistry* **A16**, 385, (1981)
18. Ganapathiappan, S., Dhathathreyan, K.S., Krishnamurthy, S.S., *Macromolecules*, **20**, 1501, (1987)
19. Patat, F., *Angew.Chem*, **65**, 173, (1953)
20. Konecny, J.O., Douglas, C.M., Gray, M.Y. *Journal of polymer science* **42**, 408, (1981)
21. Scopelianos, A.G., Allcock, H.R. *Macromolecules*, **20**, 432, (1987)
22. Mujumdar, A.N., Young, S.G., Merker, R.L., Magill, J.H., *Macromolecules*, **23**, 14, (1990)
23. Helioui, M., De Jaeger, R., Puskaric, E., Heubel, J., *Makromol. Chem.*, **183**, 1137, (1982)
24. D'halluin, G., De Jaeger, R., Potin, Ph., *Bull. Soc. Chim. Belge*, **98(9-10)**, 653, (1989)
25. D'halluin, G., De Jaeger, R. Chambrette, J.P., Potin, Ph., *Macromolecules*, **25**, 1254, (1992)
26. Wisian-Neilson, P., Neilson, R.H., Cowley, A.H., *Inorg. Chem.*, **16**, 1460 (1977)
27. Neilson, R.H., Wisian-Neilson, P., *Journal of Macromolecular Science-Chemistry*, **A16**, 425, (1981)
28. Matyjaszewski, K., Cypryk, M., Dauth, J., Montague, R., White, M., *Makromolekulaire Chemie-Macromolecular Symposia*, **54(5)**, 13, (1992)
29. Matyjaszewski, K., Lindenberg, M.S., Moore, M.K., White, M.L., *Journal of Polymer Science, Part A, Polymer Chemistry*, **32**, 465, (1994)
30. Matyjaszewski, K., Moore, M.K., White, M.L., *Macromolecules*, **26**, 6741, (1993)
31. White M.L., Montague, R.A., Matyjaszewski, K., Moore, M.K., Pakula, K., *Polymer*, **36**, 3493, (1995)
32. Honeyman, C.H., Manners, I., Morrissey, C.T., Allcock, H.R. *Journal of the American Chemical Society*, **117**, 7035, (1995)
33. Allcock, H.R., Nelson, J.M., Reeves, S.D., Honeyman, C.H., Manners, I., *Macromolecules*, **30**, 50, (1997)
34. Niecke, E., Bitter, W., *Inorg. Nucl. Chem. Lett.*, **9**, 127, (1973)
35. Honeyman, C.H., Lough, A.J., Manners, I., *Inorg. Chem.*, **33**, 2988, (1994)
36. Allcock, H.R., Crane, C.A., Morrissey, C.T., Olshavsky, M.A. *Inorganic Chemistry*, **30**, 50, (1997)
37. Allcock, H.R., Crane, C.A., Morrissey, C.T., Nelson, J.M., Reeves, S.D., Honeyman, C.H., Manners, I., *Macromolecules*, **29**, 7740, (1996)
38. Allcock, H.R., Reeves, S.D., Nelson, J.M., Crane, C.A., Manners, I., *Macromolecules*, **30(7)**, 2213, (1997)

39. Nelson, J.M., Allcock, H.R., *Macromolecules*, **30**, 1854, (1997)
40. Allcock, H.R., Reeves, S.D., Nelson, J.M., Manners, I., *Macromolecules*, **33**, 3999, (2000)
41. Hornbaker, E.D., Li, H.M., 'Process for preparing low molecular weight linear phosphonitrilic chloride oligomers', *US Patent*, US4198381 (1980)
42. Allcock, H.R., Fuller, T.J., Mack, D.P., Matsumura, K., Smeltz, K.M., *Macromolecules*, **10**, 824, (1977)
43. Allcock, H.R., Pucher, S.R., Scopelianos, A.G. *Biomaterials* **15(8)**, 563, (1994)
44. Crommen, J., Vandorpe, J., Schacht, E., *Journal of Controlled Release*, **24**, 167, (1993)
45. Ruiz, E.M., Ramirez, C.A., Aponte, M.A., Barbosa-Canovas, G.V., *Biomaterials*, **14(7)**, 491, (1993)
46. Goedemoed, J.H., De Groot, K., *Makromol. Chem., Macromol. Symp.*, **19**, 341, (1988)
47. Veronese, F.M., Marsilio, F., Lora, S., Caliceti, P., Passi, P., Orsolini, P., *Biomaterials*, **20(1)**, 91, (1999)
48. Vandorpe, J., Schacht, E., Dunn, S., Hawley, A., Stolik, S., Davis, S.S., Garnett, M.C., Davies, M.C., Illum, L., *Biomaterials*, **18**, 1147, (1997)
49. De Scheerder, I.K., Wilczek, K.L., Verbeken, E.V., Vandorpe, J., Lan, P.N., Schacht, E., Degeest, H., Piessens, J., *Atherosclerosis*, **114**, 105, (1995)
50. Ibim, S.M., Ambrosio, A.A., Larrier, D., Allcock, H.R., Laurencin, C.T., *Journal of Controlled Release*, **40**, 31, (1996)
51. Vandorpe, J., Schacht, E., Stolnik, S., Garnett, M.C., Davies, M.C., Illum, L., Davis, S.S., *Biotechnology and Bioengineering*, **52**, 89, (1996)
52. Song, S.C., Lee, S.B., Jin, J.I., Sohn, Y.S., *Macromolecules*, **32**, 2188, (1999)
53. Lee, S.B., Song, S.C., Jin, J.I., Sohn, Y.S., *Macromolecules*, **32(23)**, 7820, (1999)
54. Lee, S.B., Song, S.C., Jin, J.I., Sohn, Y.S., *Polymer Bulletin*, **45**, 389, (2000)
55. Crommen, J.H.L., Schacht, E.H., Mense, E.H.G., *Biomaterials*, **13(9)**, 601, (1992)
56. Allcock, H.R., Fuller, T.J., Matsumura, K., *Inorganic Chemistry*, **21**, 515, (1982)
57. Crommen, J.H.L., Schacht, E.H., Mense, E.H.G., *Biomaterials*, **13(8)**, 511, (1992)
58. Vandorpe, J., Schacht, E., *Polymer*, **37(14)**, 3141, (1996)
59. Allcock, H.R., Fuller, T.J., *J. Am. Chem. Soc.*, **103**, 2250, (1981)
60. Laurencin, C.T., Koh, H.J., Neenan, T.X., Allcock, H.R., Langer, R., *J. Biomed. Mater. Res.*, **21**, 1231, (1987)
61. Laurencin, C.T., Norman, M.E., Elgendy, H.M., El-Amin, S.F., Allcock, H.R., Pucher, S.R., Ambrosio, A.A., *J. of Biomed. Mat. Res.*, **27**, 963, (1993)
62. Caliceti, P., Lora, S., Marsilio, F., Guiotto, A., Veronese, F.M., *Il Farmaco*, **49(1)**, 69, (1994)
63. Caliceti, P., Lora, S., Marsilio, F., Veronese, F.M., *Il Farmaco*, **50(12)**, 867, (1995)
64. Conforti, A., Bertani, S., Lussignoli, S., Grigolini, L., Terzi, M., Lora, S., Caliceti, P., Marsilio, F., Veronese, F.M., *J. Pharm. Pharmacol.*, **48**, 468, (1996)
65. Langone, F., Lora, S., Veronese, F.M., Caliceti, P., Parnigotto, P.P., Valenti, F., Palma, G., *Biomaterials*, **16(5)**, 347, (1995)
66. Allcock, H.R., Kwon, S., *Macromolecules*, **21(7)**, 1980, (1988)
67. Allcock, H.R., Scopelianos, A.G., **16(4)**, 715, (1983)
68. Allcock, H.R., Pucher, S.R., *Macromolecules*, **24**, 23, (1991)
69. Allcock, H.R., Pucher, S.R., Scopelianos, A.G., *Macromolecules*, **27(1)**, 1, (1994)

70. Ibim, S.E.M., Ambrosio, A.M.A., Kwon, M.S., El-Amin, S.F., Allcock, H.R., Laurencin, C.T., *Biomaterials*, **18(23)**, 1565, (1997)
71. Qiu, L.Y., Zhu, K.J., *Polymer International*, **49(11)**, 1283, (2000)
72. Andrianov, A.K., Payne, L.G., Visscher, K.B., Allcock, H.R., Langer, R., *Journal of Applied Polymer Science*, **53(12)**, 1573, (1994)
73. Grossesommer, A., Prudhomme, R.K., *Journal of Controlled Release*, **40(3)**, 261, (1996)
74. Allcock, H.R., Fuller, T.J., *Macromolecules*, **13**, 1338, (1980)
75. Allcock, H.R., Austin, P.E., Neenan, T.X., *Macromolecules*, **15(3)**, 689, (1982)
76. Allcock, H.R., Austin, P.E., *Macromolecules*, **14**, 1616, (1981)
77. Allcock, H.R., Kwon, S., *Macromolecules*, **19**, 1502, (1986)
78. Allcock, H.R., Neenan, T.X., Kossa, W.C., *Macromolecules*, **15**, 693, (1982)
79. Allcock, H.R., Hymer, W.C., Austin, P.E., *Macromolecules*, **16(9)**, 1401, (1983)
80. Neenan, T.X., Allcock, H.R., *Biomaterials*, **3**, 78, (1982)
81. Grolleman, C.W.J., De Visser, A.C., Wolke, J.G.C., Van Der Groot, H., Timmerman, H., *Journal of Controlled Release*, **3**, 143, (1986)
82. Grolleman, C.W.J., De Visser, A.C., Wolke, J.G.C., Van Der Groot, H., Timmerman, H., *Journal of Controlled Release*, **4**, 119, (1986)
83. Grolleman, C.W.J., De Visser, A.C., Wolke, J.G.C., Klein, C.P.A.T., Van Der Groot, H., Timmerman, H., *Journal of Controlled Release*, **4**, 133, (1986)
84. Allcock, H.R., Allen, R.W., O'brien J.P., *J. Am. Chem. Soc.*, **99**, 3984, (1977)
85. Sohn, Y.S., Baek, H., Cho, Y.H., Lee, Y.-A., Jung, O.-S., Lee, C.O., Kim, Y.S., *International Journal of Pharmaceutics*, **153**, 79, (1997)
86. Song, S.-C., Sohn, Y.S., *Journal of Controlled Release*, **55**, 161, (1998)
87. Song, S.C., Lee, C.O., Sohn, Y.S., *Polymer International*, **48(8)**, 627, (1999)
88. Goedemoed, J.H., De Groot, K., Claessen, A.M.E., Scheper, R.J., *Journal of Controlled Release*, **17**, 235, (1991)
89. Goedemoed, J.H., Mense, E.H.G., De Groot, K., Claessen, A.M.E., Scheper, R.J., *Journal of Controlled Release*, **17(3)**, 245, (1991)
90. Schacht, E., Crommen, J., 'Bioerodible sustained release implants', *US Patent*, US4975280 (1990)
91. Andrianov, A.K., Cohen, S.C., Visscher, K.B., Payne, L.G., Allcock, H.R., Langer, R., *Journal of Controlled Release*, **27**, 69, (1993)
92. Andrianov, A.K., Payne, L.G., *Advanced Drug Delivery Reviews*, **31**, 185, (1998)
93. Allcock, H.R., Ambrosio, A.M.A., *Biomaterials*, **17**, 2295, (1996)
94. Schacht, E., Vandorpe, J., Dejardin, S., Lemmouchi, Y., Seymour, L., *Biotechnology and Bioengineering*, **52(1)**, 102, (1996)
95. De Scheerder, I., Wang, K., Wilczek, K., Vandorpe, K., Verbeken, E., Desmet, W., Schacht, E., Piessens, J., *Coronary Artery Disease*, **7(2)**, 161, (1996)
96. Veronese, F.M., Marsilio, F., Caliceti, P., De Fillippis, F., Giunchedi, P., Lora, S., *Journal of Controlled Release*, **52(3)**, 227, (1998)
97. Caliceti, P., Veronese, F.M., Lora, S., *International Journal of Pharmaceutics*, **211**, 57, (2000)
98. Caliceti, P., Aldini, N.N., Fini, M., Rocca, M., Gnudi S., Lora, S., Giavaresi, G., Monfardini, C., Giardino, R., Veronese, F.M., *Il Farmaco*, **52(11)**, 697, (1997)

99. Aldini, N.N., Caliceti, P., Lora, S., Fini, M., Giavaresi, G., Rocca, M., Torricelli, P., Giardino, R., Veronese, F.M., *Journal of Orthopaedic Research*, **19**, 955, (2001)
100. Chang, Y.C, Lee, S.C., Kim, K.T., Kim, C., Reeves, S.D., Allcock, H.R., *Macromolecules*, **34**, 269, (2001)
101. Allcock, H.R., Pucher, S.R., Turner, M.L., Fitzpatrick, R.J., *Macromolecules*, **25**, 5573, (1992)
102. Qiu, L.Y., Zhu, K.J., *Journal of Applied Polymer Science,* **77(13)**, 2987, (2000)
103. Qiu, L.Y., Zhu, K.J., *International Journal of Pharmaceutics*, **219**, 151, (2001)
104. Bonassar, L.J., Vacanti, L., *Journal of Cellular Biochemistry Supplements*, **30/31**, 297, (1998)
105. Tabata, Y., *Pharmaceutical Science and Technology Today*, **3**, 80, (2000)
106. Cohen, S., Baño, M.C., Cima, L.G., Allcock, H.R., Vacanti, J.P., Vacanti, C.A., Langer, R., *Clinical Materials*, **13(1-4)**, 3, (1993)
107. Aldini, N.N., Fini, M., Rocca, M., Martini, L., Giardino, R., Caliceti, P., Veronese, F.M., Lora, S., Maltarello, M.C., *Journal of Bioactive and Compatible Polymers*, **12**, 3, (1997)
108. Aldini, N.N., Fini, M., Rocca, M., Giavaresi, G., Giardino, R., *International Orhopaedics,* **24**, 121, (2000)
109. Veronese, F.M., Marsilio, F., Lora, S., Caliceti, P., Passi, P., Orsolini, P., *Biomaterials*, **20**, 91, (1999)
110. Passi, P., Marsilio, Z.A., Lora, S., Caliceti, P., Veronese, F.M., *Journal of Materials Science-Materials in Medicine*, **11(10)**, 643, (2000)

In: Applicative Aspects of Poly(organophosphazenes)
Editors: M. Gleria and R. De Jaeger, pp. 33-47
ISBN 1-59454-025-X

Chapter 2

USES OF POLYPHOSPHAZENE IN DENTISTRY

Lawrence Gettleman
Professor of Prosthodontics & Biomaterials
School of Dentistry
University of Louisville
Louisville, Kentucky 40292-0001 USA

MATERIALS SCIENCE AND DENTISTRY

The use of manmade materials in dentistry has had a long tradition (1). Dental artifacts made of noble metals have been recovered from ancient Egyptian, Etruscan, and Inca tombs. This attests to the skill of their craftsmen and the wise choice of inert materials to serve in the hostile environment of the human body. If less durable materials were used in ancient times, they have not survived.

The teeth and the jawbones that support the teeth are not replaced by the body when destroyed by bacterial action in dental caries (tooth decay) and periodontal (gum) disease. Accordingly, ancient and modern dentistry has made extensive use of manmade materials to replace lost structures in order to restore chewing and speech function, esthetics, and psychological well-being. Many of today's medical materials used for orthopedic and cardiovascular surgery have had their origins in dental materials.

TOOTH LOSS AND PROSTHESES

Bone from the maxilla and mandible resorbs after the functioning teeth are lost, sometimes to an astounding degree. The partially or totally edentulous patient can then be treated with a partial or complete denture made to replace the teeth, using metal (CoCrMo, NiCrMn, or AuCuAg alloys) or more often, plastic [poly(methyl methacrylate) (PMMA) and copolymers, or vinyl polymers] for the base (2). Individual plastic or occasionally ceramic teeth are set to occlude with the remaining natural teeth or with the opposing denture. Complete (or full) dentures serve to replace all the teeth in the jaw and are retained by accuracy of fit and selective compression of the border tissues, with the presence of saliva as a sealant (3).

The tightly adherent soft tissues (attached gingivae), which once covered the bone supporting the teeth, shrinks drastically as the bone resorbs. The ligaments which attach the facial muscles of expression, the muscles of mastication (chewing), and the muscles of the floor of the mouth and tongue remain attached to the maxilla and mandible in their original positions, and may appear to encroach on the territory of the denture base as bone is lost. Coupled with decreasing saliva due to aging, the xerostomic effects of multiple medications, or effects of radiation treatment to the head and neck, the ability to wear dentures becomes a serious problem for many dental patients as they age.

Today, people are retaining their teeth longer, due to better hygiene and the use of fluoride in drinking water and toothpaste. But lifespan has also been extended so there are more elderly people alive today than ever before, many of whom will need future denture service in their old age. So the cohort needing dentures has been displaced but not lessened. Approximately 33% of people over 65 in the U.S. are edentulous and most wear dentures (4). Millions more need them but go without, due to cost, fear, or dissatisfaction with dentures in the past.

Resilient or Soft Denture Liners

There is clearly a need for resilient liners in removable partial and complete dentures. Prior to the use of acrylic denture bases in the 1930s, complete dentures were made of vulcanized natural rubber and were somewhat resilient to patients during chewing. The acrylics are much harder and inflexible, making dentures more difficult to wear for some patients, pinching the delicate tissues of the mouth between hard acrylic resin and the underlying residual bone. Resilient lining materials that have been used against acrylic denture bases include a) phthlate- or alcohol-plasticized acrylics, which bond well to the denture but harden in days or weeks by the loss of plasticizer; b) methyl silicone rubbers which bond less well, remain resilient, but have a low mechanical storage modulus and will grow fungus within pores in the bulk material in many patients (after antifungal agents have leached away); or c) polyurethane or vinyl polymers, which require more complex processing conditions, special equipment, have higher hardness, absorb environmental colorants, and also promote fungal growth (5).

The National Institute of Dental and Craniofacial Research, one of the U.S. National Institutes of Health, listed the development of improved resilient lining materials as a goal for many years. Following a conference in May 1975 on the potentials of polyphosphazenes, sponsored by the U.S. Army at the Watertown Arsenal, Watertown, Massachusetts, my predecessors at Gulf South Research Institute in New Orleans, Elias Klein, Paul D. May, H. Ralph Rawls, and Luis R. Guerra, determined that polyphosphazene (6) would be a good candidate as a resilient soft liner for dentures.

Candidate Elastomers

Preliminary work was done at the same time that another NIH-funded project was underway in 1976-79, using thermoplastic chlorinated polyethylene (CPE®, Dow) for maxillofacial prosthetics (7). A proposal to the N.I.H. submitted in 1979 was funded in 1980 to examine several candidate materials for resilient denture liners. Thermoplastic CPE, an isoprene-styrene-isoprene block copolymer (Kraton®, Shell) (8), and the polyphosphazene material made in the 1970s by Horizon Research and in the 1980s by Firestone (PNF-200®) were compounded with help from the respective manufacturers. The following criteria were followed: a) use of additives common to other dental materials (to ease U.S. Food & Drug Administration approval); and b) not require processing equipment or processing conditions beyond those already in use in the dental laboratory industry (to gain acceptance by users) (9-11).

CPE received favorable use as a maxillofacial prosthetic material owing to its ability to be remolded until fully dense, reparability, laminated coloring procedure, excellent edge strength, compatibility with all skin adhesives, and low cost. But CPE formulations were too hard (H_{DA} 45-50) as a denture liner. The isoprene-styrene-isoprene material failed a UV-exposure test and was also too hard (12).

Choice of Polyphosphazene

The polyphosphazene was found to be nearly ideal, owing to its reported biocompatibility, acceptance of fillers and pigments, compatibility with an interpenetrating network of di-and tri-functional crosslinking acrylic monomers, and ability to be compounded at various hardnesses from H_{DA} 10 to 90 by varying the filler and acrylic components. It promised to be reasonable in cost as Firestone quoted $44-50/pound ($97-110/ kilogram) in 1984 (13-15).

Between 1980 and 1985, Firestone's polyphosphazene (PNF®-200) was developed by systematically blending with a number of additives: acrylic monomers and copolymers, fillers, and pigments commonly used in other dental products. Purification of the PNF®-200 was performed by dissolving in acetone, filtering, and reprecipitation of the polymer in agitated deionized water.

Final Denture Liner Composition

Advice from Firestone was received during the closeout of their production in 1984-85; Ethyl Corporation resumed production of the identical polymer in 1987, under the trade name Eypel®-F.

A number of reports were published (16-18), and 4 U.S. patents were received and assigned to Gulf South Research Institute, the first with Paul D. May and Luis R. Guerra in 1981 (19), two documenting intermediate steps which were eventually superceded (20-21), and the final formulation by Lawrence Gettleman and Paul Gebert in 1987 (22). The last patent reveals the final composition of the material. Of particular importance is the 1:9 ratio of di-functional (ethyleneglycol dimethacrylate) to tri-functional acrylic monomers (trimethylolpropane trimethacrylate), respectively. Twenty parts per hundred rubber (phr) of acrylic monomer yields a usable rubber for a denture liner after incorporating the filler and pigment. Barium sulfate filler was chosen to impart radiopacity to the product (13, 23), and because the optical opacity is of little importance when used as a denture liner (see Figs. 1a & 1b).

Although the acrylics are optically clear and the phosphazene is translucent yellow, the 2 blended substances, produce a milky white rubber due to the differences in refractive indices between acrylic and phosphazene compartments.

The one-part material is designed to be cured without mixing at temperatures common to a dental laboratory, such as compression at 20.7 MPa (3000 psi) and heating up to boiling water temperature (100°C). Initially, lauroyl peroxide (LP) was used as an activator because its long-chain residue would act as a plasticizer in the final product. LP's relatively low activation temperature, however, demanded that the final product be kept cold to prevent premature curing. Later, benzoyl peroxide (BP) was used because it is more commonly used in dental resins, leaves a smaller residue, and has a higher activation temperature, providing longer shelf life for the product during shipment or storage. Still, refrigeration or freezing of the final product is recommended.

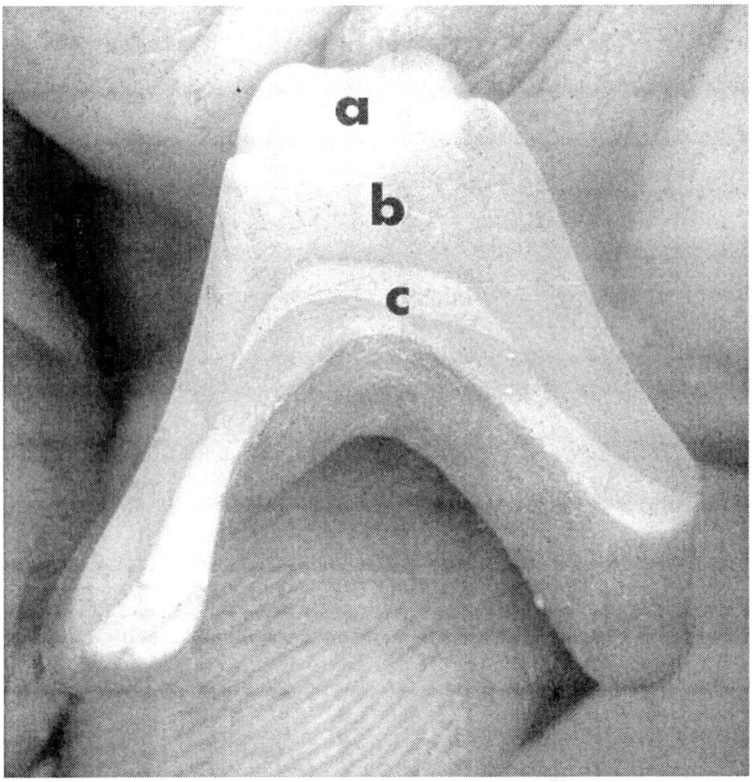

Figure 1a. Cross-section of a mandibular denture showing a) porcelain denture tooth, b) poly(methyl methacrylate) denture base, c) 2 layers of Novus™ resilient denture liner which rests against the soft tissues overlying the mandible of the patient.

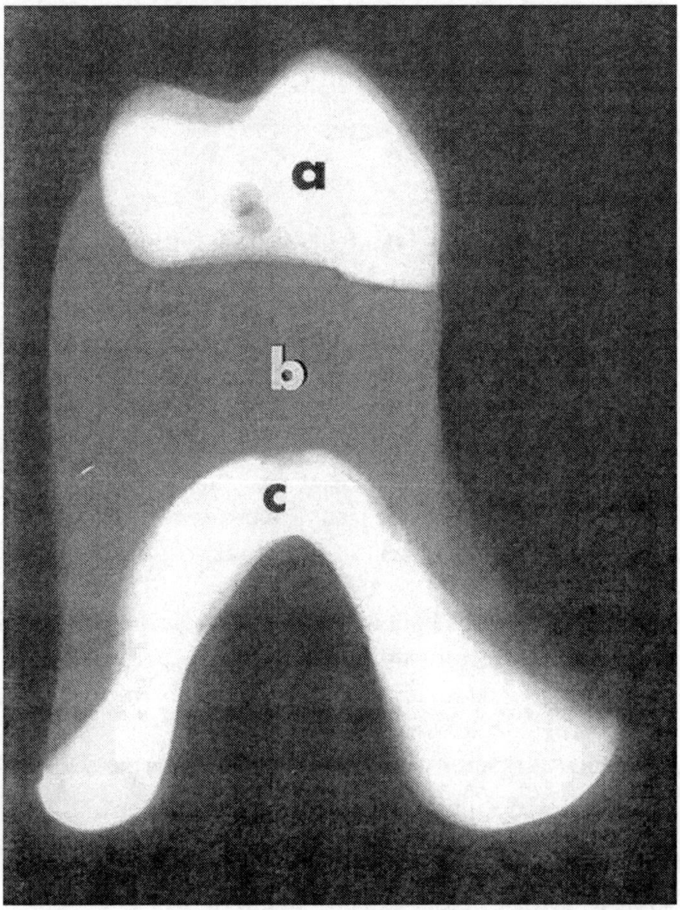

Figure 1b. Radiograph of the specimen in Figure 1a, showing the relative radiodensity of the porcelain denture tooth, acrylic denture base, and Novus™ denture liner. Radiopacity is important if the denture were to be swallowed or fractured in a traumatic incident. The fragments could become lost or lodged in the perioral tissues, the esophagus, or the lung. Radiopacity of all medical materials would help locate the fragments in tissues and assure that they have been completely removed.

INTERPENETRATING NETWORKS

Firestone's PNF®-200 (and later, Ethyl's Eypel®-F) is a fluoroalkoxy-substituted polyphosphazene of approximately 50,000 —(–N=P$^{=}$–)— units, substituted with C3 to C9 fluoroalkoxy units derived from fluoroalcohols. Commercial cross-linking is achieved with small quantities of phenol groups on the chain whose activation temperature is above the highest temperature to be used in the denture curing process (100°C). It is believed that curing of the multifunctional acrylics with BP at 65°C does not establish crosslinks between the acrylic components and the phosphazene rubber chains (24, 25). An interpenetration of acrylics (20 phr) with the fluoroalkoxy-substituted phosphazene is the result, achieving useful rubber properties.

Physical Properties

As compounded, H_{DA} is 35, tensile strength is 1.65 MPa, elongation is 18%, bond strength to PMMA is 3.4 kN/m, and tan δ is 1.4 (26, 27).

Biological Testing

Resistance to fungal growth was tested by incubating *Candida albicans* on nutrient agar after placing discs of soft liner and maxillofacial prosthetic materials. Colonies growing up to but not within or under the phosphazene was observed, unlike silicone rubber where colonies were observed under the specimens. Hyphae of the fungus were observed microscopically deep within the silicone's substance, making permanent eradication nearly impossible (28).

Biocompatibility (29) was also tested, first at the University of Tennessee's toxicology laboratories, named as a national center for testing biomaterials. Hard and soft variations of the polyphosphazene denture liner were sent to Drs. John Autian and Elwood Dillingham in Memphis. Fifteen tests were performed, including samples made from disks, crumbled polymer, and extracts in soybean oil and polyethylene glycol. Combined toxicity scores of 56 and 65 (out of 1500) were reported for the hard and soft materials, respectively, indicating "the materials present a very limited toxic potential" (10, 13).

Subsequent biological evaluations at the University of Texas Health Sciences Center at San Antonio tested the final version of the polyphosphazene formula for hemolysis (the only test that was positive from the previous toxicology tests) and mucous membrane irritation (in a hamster cheek pouch). Both tests, under the direction of Dr. Ramon Baez, revealed very low levels of toxicity for the polyphosphazene, with the most irritation due to the presence of the sutures used to hold the specimens in the pouch. This fortunate result came about despite the use of cadmium sulfoselenide (Ferro) as the red pigment, which was, at the time, in common use for pink-colored dental products (16, 17).

Laboratory Processing

The material is applied to a precured or pre-existing (relined) acrylic denture base by a compression technique in a custom-made gypsum mold against the acrylic resin denture base, which carries the individual acrylic denture teeth. A spacer piece in the shape of the denture liner is made of silicone rubber or other suitable material, against which the acrylic resin is pressed and then hardened by initial curing at 74°C (30). A patty of preformulated phosphazene denture liner is then cut from the cake of refrigerated material and pressed in the mold after removing the spacer. Trial packs are done several times, separated from the acrylic denture base resin with a polyethylene sheet, removing excess material or adding fresh material to deficient areas (31).

The polyphosphazene component is a fully polymerized rubber material. As such, it has an elastic memory that requires time to relax. The material appears to regain its original shape if the mold is opened quickly; compression for 10-12 minutes is recommended before initial opening for the material to lose its memory (32).

If fresh, partially cured acrylic is in place, a small amount of methyl methacrylate monomer is used as a wetting agent between the acrylic denture base and the phosphazene liner. If older acrylic is used, as in a reline procedure, a special wetting agent is used which contains methyl methacrylate monomer containing dissolved poly(methyl methacrylate).

Recommended curing time for the liner is 1.5 hours in 74°C water, followed by boiling for 0.5 hours, typical for most heat-curing dental acrylic resins. Since the mono-, di-, and tri-functional monomer molecules are asymmetric, they absorb energy in an oscillating magnetic field, so the material is curable in a microwave oven, which shortens the overall time of curing from 2+ hours to 15-20 minutes. The labor required is approximately the same, however, between conventional hot water curing and microwave curing. Recovery and removal of the denture with a resilient liner preserves the master cast of the patient's tissues, but the tooth side of the gypsum flask is usually lost unless it is made of an elastic, recoverable material.

Clinical Trials

Forty edentulous patients were recruited between 1985 and 1987 at Charity Hospital, part of the Louisiana State University system in New Orleans, and at the Dental School at Northwestern University in Chicago. A randomized single-blind multiple crossover clinical trial was designed by making complete dentures with the phosphazene liner in one denture and the control silicone rubber denture liner material (Molloplast B, Buffalo Dental, Brooklyn, NY) in the other. Maxillary and mandibular sites were alternated from one patient to the next so that each material would be used in either dental arch an equal number of times. Patients wore the dentures up to 29 months in the trial and then were allowed to keep the dentures. A few were followed up to 74 months (33-40).

Following analysis of the clinical trial, patients were found to prefer the phosphazene-lined denture to the silicone-lined denture, especially when used on the mandibular arch, due in part to the high mechanical modulus of the phosphazene elastomer (41, 42). This was unexpected but made sense because most patients expressed a preference for the chewing ability of the "dead rubber" compared to the "live rubber" in the silicone-lined dentures (26, 43).

A second area of superiority of phosphazenes was resistance to the overgrowth of fungus. Regardless of location in the mouth, the phosphazene-lined dentures resisted the growth of *Candida albicans* in almost all patients, while more than half of the silicone-lined dentures became fouled. As noted earlier, fungal hyphae were shown to reside within the pores of dental silicone rubber materials but not within the phosphazene (28, 44).

Both materials resisted debonding well and maintained their durometer hardness throughout the course of the clinical trial. Patients who soaked their dentures in bleaching solutions saw the color of the liner brighten from deep pink to light pink or white, without a change in mechanical properties. These were mostly patients who had fungal problems with the opposing silicone rubber dentures, and so they soaked both dentures in concentrated bleaching solutions.

Resilient liners are especially effective for denture patients whose anatomy has deep undercuts (see Fig. 2).

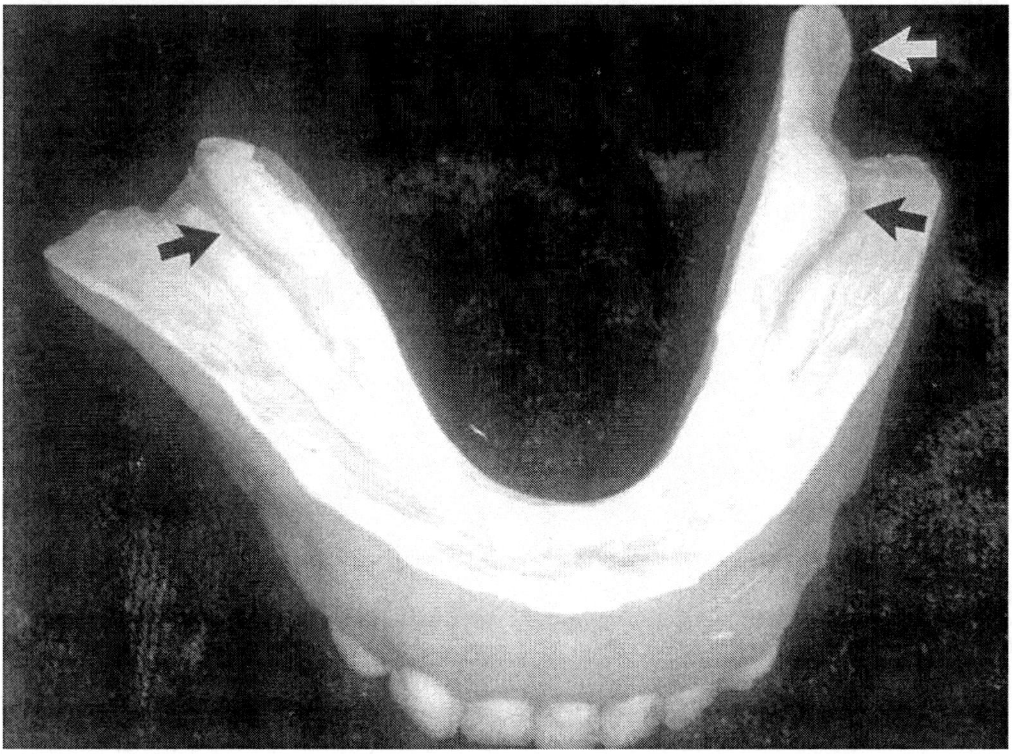

Figure 2. Denture of a patient with significant pre- and post-mylohyoid undercuts in the mandible (see arrows). Resilient liners penetrate deeper into these undercuts and provide better retention than an unlined hard acrylic denture.

U.S. Food & Drug Administration Approval

Presentation of evidence of safety (toxicity tests) and effectiveness (results of the clinical trial), and the reversible nature of use in removable dentures resulted in the FDA's approval using the 510(k) process to market the phosphazene material in interstate commerce in 1987. This was the first time that a phosphazene material had been approved for a biomedical product in human subjects.

Marketing

A search was made to find a dental manufacturer to compound and market the denture liner. An agreement was concluded between Gulf South Research Institute and LouisianA Research & Development, Inc. (LARD), a small business established to market the product. GSRI ceased operations in late 1987. LARD then subcontracted in 1988 with The Hygenic Corporation of Akron, Ohio, a company familiar with rubber products in dentistry, which marketed a line of denture acrylic resins but no resilient soft liner material. The liner became the jewel in the crown of their dental laboratory product line after its introduction in January, 1989 under the trade name Novus™.

OBTAINING ETHYL'S EYPEL®-F

The phosphazene was purchased from Ethyl Corporation's Specialty Polymers Division, with the polymer made in its Orangeberg, South Carolina plant. Ethyl synthesized the fluoroalkoxy-phosphazene that was used by the U.S. military for gaskets, seals, O-rings, osmotic membranes, plenums, and non-burning rubber products. As Ethyl engineers pointed out, if the rubber O-rings used in the solid rocket boosters of the U.S. space shuttle *Challenger* had been made of polyphosphazene rubber, the tragic explosion in 1986 would not have occurred because a) the glass transition temperature of the rubber is -40°C and would not have hardened in the cold weather at Cape Canaveral, and b) the rubber would not burn. An arylalkoxy-phosphazene, Eypel®-A was also produced and marketed by Ethyl as a fireproof structural foam to replace polyurethane foams which burn and give off cyanide gas.

Ethyl was at first unwilling to sell one of their polymer products for use in a biomedical device. They cited the unfortunate experience of A.H. Robbins Co., which had recently declared bankruptcy after litigation over the Dalkon Shield intrauterine device (Robbins and Ethyl are both headquartered in Richmond, Virginia). They were persuaded to allow the use of Eypel®-F as a denture liner because a) the device was removable, not implanted, b) there was no evidence of lawsuits against U.S. dental manufacturers for denture liner products in the past, c) the toxicity of the phosphazene was lower than other commercial products on the market in our tests, d) FDA approval had already been received, e) Ethyl was already a bulk producer of drugs like penicillin and ibuprofen, and f) we discovered that the son of the chairman of the board of Ethyl was a physician, looking for new uses for Ethyl's products in the health care field. Since then, U.S. Public Law 105-230, "Biomedical Access Assurance Act," August 13, 1998, has been enacted, exempting originating biomedical materials manufacturers from liability except in cases of gross negligence.

COMPOUNDING

Under the direction of chemist John Riazi at Hygenic Corporation, the process of compounding the phosphazene was scaled up to 25 kg batches in a Brabender 2-arm blender. Small changes in the composition were made: a) substitution of benzoyl peroxide (present in small quantities on acrylic fillers used previously) for the lauroyl peroxide due to its higher activation temperature; and b) change of pigment from red cadmium sulfoselenide (Ferro) to an iron-based pink pigment, due to concurrent bans on the use of cadmium-containing components. Hygenic developed methods to press the material into 60 g sheets and package it in Saran. Novus™ was marketed initially at $143 per package in early 1989. At the time, it was the most expensive denture liner product on the U.S. market, and it remained that way.

After an introductory campaign, sales increased yearly to approximately 700 kg of product per year through 1994, mostly in the United States, representing treatment for several hundred thousand-denture wearers. Some sales were recorded in Italy, Germany, and the United Kingdom, but were limited due to Hygenic's small marketing presence outside the USA. Fewer than ten adverse reports were received from dentists, dental laboratories, or patients, all due to errors in processing.

WITHDRAWAL OF ETHYL'S EYPEL®-F

Citing loss of military contracts at the end of the Cold War, Ethyl announced that it would cease manufacturing phosphazene materials by the end of 1994. Hygenic purchased sufficient material for several years of production but sold it shortly thereafter to an outside client. Sales of Novus™ ceased early in 1995 despite continuing demand from members of the dental profession.

FUTURE USES OF POLYPHOSPHAZENES

Our good experience with the polyphosphazenes has led to a few tantalizing speculations and small steps toward developing other biomedical uses of the material. FDA approvals should also be easier after successful use of the material as a denture liner.

Concurrent research in New Orleans dealt with **maxillofacial prosthetics**, replacement of parts of the face lost due to cancer, trauma, and birth defects. This is a sub-specialty of prosthodontics and solves a medical problem with a dental solution. Typically, ears, noses, contents of the orbit, and intraoral prostheses are fabricated from silicone rubber elastomers. Compounding the phosphazene with little or no filler and decreasing the ratio of acrylic to rubber yields a softer rubber, with a H_{DA} of 25, similar to human skin (24, 45, 46). The rubber is compounded with pigments appropriate to the patient's skin color. Cost would be high, however, limiting it to patients with the ability to pay, and limiting its use in developing countries where the need is great.

Creative uses in prosthetic dentistry include **implant denture retention** and stability by processing the rubber over implant heads (see Figs. 3a, 3b, & 3c).

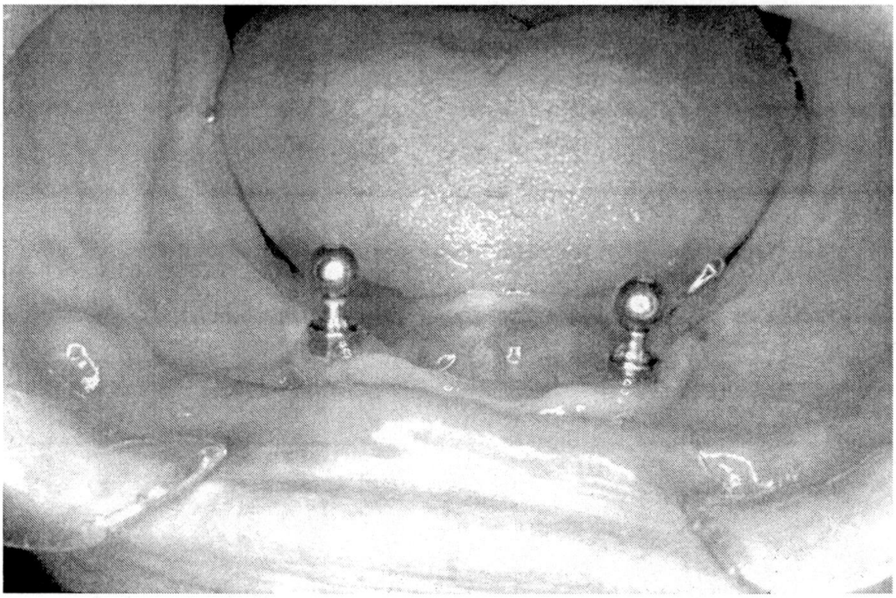

Figure 3a. Patient with 2 titanium endosseous implants in place in the mandibular bone. Large ball-headed attachments are threaded into the embedded implants, which are united with the bone surrounding the implants.

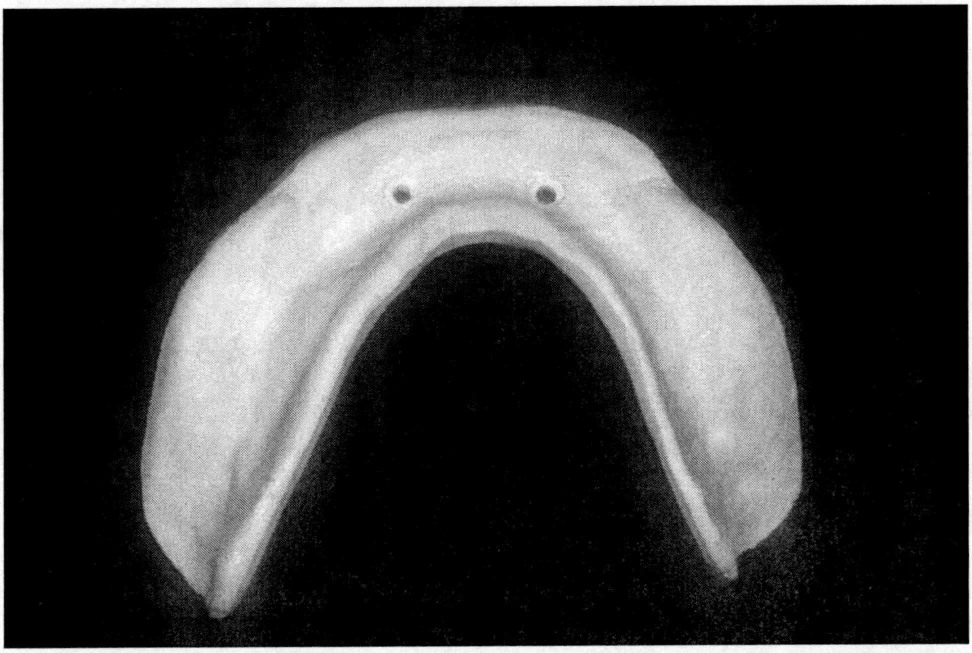

Figure 3b. The denture is processed with Novus™, which provides retention and stability and allows movement of the denture towards the tissues, similar to movement of natural teeth suspended in the bone by the periodontal ligament. Rigid anchorage of the denture to the implant does not allow for this physiologic motion.

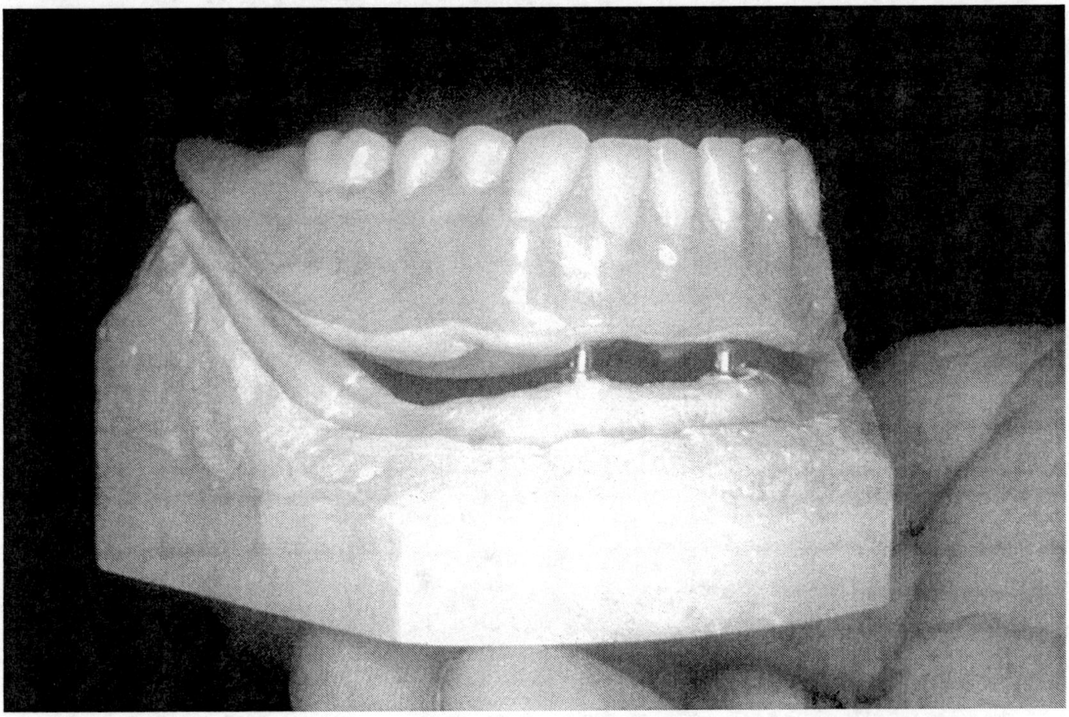

Figure 3c. Denture partially inserted over the implant heads on a cast of the patient's anatomy.

The advantage over mechanical clips and attachments is the freedom of movement, using the rubber to envelop the implant head which is rigidly fixed to bone, providing retention and also movement of the denture toward the tissues, simulating the periodontal membrane around natural teeth. (47-52).

Another potential use would be in load-bearing areas of **artificial limb prostheses**. Below- or above-the-knee artificial prostheses are similar to dentures, supporting cyclic uniaxial and rotary muscular loads, with the bone pinching the soft tissues of the stump in the same way that dentures fit the maxillae or mandible. The environment is not as wet, however. Selective use of the phosphazene rubber to load-bearing areas only will limit the cost and weight of the prosthesis, and greatly extend the benefits of phosphazine polymers (53).

Ballerinas carry their entire weight on two or three toes. The toe box of ballet shoes are filled with a variety of materials, such as lambs wool, papier maché, or even beefsteak. A characteristic of these materials is their ability to absorb shock while dancing *en pointe*. Professional dancers consume shoes rapidly and would benefit from **custom-molded phosphazene toe pieces** that could be transferred from shoe to shoe. The high storage modulus and tan δ of phosphazene polymers will be of immense help. Other beneficial uses would also be in all forms of **athletic shoes**, such as the heelpieces in running shoes. Foamed rubber has been the material of choice but suffers from collapse and water entrapment.

The prospects for polyphosphazenes in the health sciences and elsewhere are bright. Success of the material in dentistry should open the door for other medical uses. Nearly everyone who has worked with polyphosphazene has become enamored with it, and has been equally frustrated with the difficulties obtaining this "forbidden fruit" of the polymer field (54). Manufacturing by a medium-sized company may be best: Horizon Research in the 1970s was too small, and Firestone and Ethyl were too large. When a stable and reasonably priced source of the polymer becomes available, I predict that the phosphazene system will become the silicone rubber of the 21st century.

References

1. Craig, R.G.: *Restorative Dental Materials*, 9th ed. St. Louis: Mosby, 1993. Chapter 1.
2. Anusavice, K.J.: *Phillips' Science of Dental Materials*, 10th ed. Philadelphia: Saunders, 1996. Chapter 11.
3. Zarb, G.A., Bolender, C.L., and Carlsson, G.E.: *Boucher's Prosthodontic Treatment for Edentulous Patients*. St. Louis: Mosby, 1997.
4. Department of Health & Human Services, *Oral Health in America. A Report of the Surgeon General.* Rockville, MD, U.S. Department of Health & Human Services, National Institute for Dental & Craniofacial Research, National Institutes of Health, 2000.
5. Craig, R.G.: *Restorative Dental Materials*, 9th ed. St. Louis: Mosby, 1993. Chapter 19.
6. Allcock, H.R.: Polyphosphazenes: New polymers with inorganic backbone atoms. *Science*, **193**: 1214-1219, 1976.
7. May, P.D.: Maxillofacial prostheses of chlorinated polyethylene. *J. Biomed. Mater. Res.* **12**: 421-431, 1978.
8. Sefton, M.V., and Merrill, E.W.: Surface hydroxylation of styrene-butadiene-styrene block copolymers for biomaterials. *J. Biomed. Mater. Res.* **10**: 33-45, 1976.

9. Gettleman, L., Farris, C.L., and Ruiz, S.B.: Properties of experimental elastomers for permanent soft denture liners. *J. Dent. Res.* **61**: 262 (No. 753), 1981.
10. Gettleman, L. Farris, C.L., LeBoeuf, R.J., Rawls, H.R., and Dillingham, E.O.: Improvement of experimental elastomers for permanent soft denture liners: Physical and biological properties. *Biomat., Med. Dev., Artif. Organs,* **9**: 345-347, 1981.
11. P.D. May, C.L. Farris, L. Gettleman, F.F. Holland, and I. Cabasso: New elastomers for soft denture liners. *J. Dent. Res.* **60 A**: 437 (No. 507), 1981.
12. Gettleman, L., Ross-Bertrand, L., Gebert, P.H., and Guerra, L.R.: Novel elastomers for denture and maxillofacial prostheses. In *Biomed. Engin. IV, Recent Devel.* B. Sauer, ed. New York: Pergamon, pp. 141-144, 1985.
13. Gettleman, L., LeBoeuf, R.J. Jr., and Rawls, H.R.: Soft and firm PNF denture liners. *J. Dent. Res.* **62**:250 (No. 725), 1983.
14. Gettleman, L., LeBoeuf, R.J., Jr., and Rawls, H.R.: Polyphosphazene fluoroelastomers as a permanent soft liner for removable dentures – Improvements in radiopacity and soft/firm properties. *Soc. for Biomat. Trans.* No. **71**, 1983.
15. Gettleman, L., Vargo, J.M, Gebert, P.,H., Farris, C.L., LeBoeuf, R.,J., Jr., and Rawls, H.R.: Polyphosphazene fluoroelastomer (PNF) as a permanent soft liner for removable dentures. In *Advances in Biomed. Polymers*, Polymer Sci. & Tech. Series, Vol.35: sr C.G. Gebelein, ed. New York: Plenum, pp. 55-61, 1987.
16. Gebert, P.H., Capezza, M.R., Rawls, H.R., Smith, J.K., and Gettleman, L.: The effect methacrylate on the physical properties of poly(fluoroaloxy) phosphazene elastomers. *2nd World Biomat. Congr. Trans.,* Washington, No. 61, 1984.
17. Gettleman, L., and Ross, L.M.: Polyphosphazene fluoroelastomer (PNF) as a permanent soft liner for removable dentures. American Chemical Society Symposium on Advances in Biomedical Polymers. *Proc. ACS Div. of Polymeric Mater. Sci. & Engin.*, **53**:772-774, 1985.
18. Gettleman, L., Ross, L.M., Gebert, P.H.,and Guerra, L.R.: Novel elastomers for denture and maxillofacial prostheses. *4th Southern Biomedical Engineering Conf. Proc.*, Jackson, Mississippi, Oct. 12, 1985.
19. U.S. Patent No. 4,251,215, *Phosphonitrilic fluoroelatomer lined denture.* Paul D. May, and Luis R. Guerra, February 17, 1981. Assigned to Gulf South Research Institute, Baton Rouge, Louisiana.
20. U.S. Patent No. 4,432,730, *Soft and Firm Denture Liner for a Composite Denture and Method of Fabricating.* L. Gettleman, C.L. Farris, H.R. Rawls, and R.J. LeBoeuf, Jr. awarded February 21, 1984. Assigned to Gulf South Research Institute, Baton Rouge, Louisiana.
21. U.S. Patent No. 4,543,379, *Soft and Firm Denture Liner for a Composite Denture and Method of Fabricating.* L. Gettleman, C.L. Farris, H.R. Rawls, and R.J. LeBoeuf, Jr. awarded September 24, 1985. Assigned to Gulf South Research Institute, Baton Rouge, Louisiana.
22. U.S. Patent No. 4,661, 065, *Soft Denture Liner.* L. Gettleman and P.H. Gebert. Awarded April 28, 1987. Assigned to Gulf South Research Institute, Baton Rouge, Louisiana.
23. Goshima, T., Gettleman, L., Goshima, Y., and Yamamoto, A. Evaluation of a radiopaque denture liner. *Oral Surg., Oral Med., Oral Radiol.* **74**:379-82, 1992.

24. Gettleman, L., and Riazi, J.:Interpenetrating acrylics used with Eypel®-F for biomedical devices, especially resilient denture liners. (Novus™). *Proc. Conf. on Status and Future of Polyphosphazene*. Raleigh, NC, March 16, 1992.
25. Gettleman, L., Polyphosphazene fluoroelastomer for denture liners and facial prosthetics. *Phosphorus, Sulfur, and Silicon*, **144-146**:205-208, 1999.
26. Selvaraj, R.: *Dynamic Mechanical Analysis of Polyphosphazenes*. Master of Science thesis, Speed Scientific School, University of Louisville, 1992.
27. Selvaraj, R., and Gettleman, L.: Dynamic mechanical analysis of polyphosphazene resilient denture liners. *J. Dent. Res.* **82**: Spec. Iss. A. (No. 221), 2003.
28. Gettleman, L. Fisher, D.L., and Farris, C.: Self-sanitizing soft denture liners: Paradoxical results. *J. Biomed. Mater. Res.* **17**:731-734, 1983
29. Stanley, H.R.: *Toxicity Testing of Dental Materials*. Boca Raton, FL: CRC Press, 1985
30. Maginnis, M.J., and Gaubert, G.T.: Soft Liners, Chapter 16. In R.M. Morrow, K.D. Rudd, and J.E. Rhoades: *Dental Laboratory Procedures: Complete Dentures*, Vol. 1. St. Louis: Mosby, 1986.
31. Gettleman, L., Huband, M., Leatherbury, E., Vargo, J.M., Sherry, D., Guerra, L.R., and Jameson, L.M.: Comparative properties of GSRI-PNF and Molloplast B soft denture liners for clinical trials. *J. Dent. Res.* **65**:284 (No. 1036), 1986.
32. Gettleman, L., Guerra, L.R., Jameson, L.M., Vargo, J.M., Sherry, D.E., Huband, M., and Leatherbury, E.: Design of clinical trials for soft denture liners: Early lab comparisons. *Soc. for Biomat. Trans.* No. 72, 1986.
33. Gettleman, L., Guerra, L.R., Jameson, L.M., Finger, I.M., Agarwal, A., Larsen, H., Salib, M., McDonald, G., and Vargo, J.M.: Baseline results from clinical comparison of two permanent soft denture liner materials. *J. Dent. Res.* **65**:728 (No. 27), 1986.
34. Gettleman, L., Jameson, L.M., Guerra, L.R., Agarwal, A.K., Finger, I.M., Salib, M.M., Larsen, H.D., McDonald, G.T., Vargo, J.M., and Goist, K.C.: Interim results of clinical trial of two permanent resilient denture liners. *J. Dent. Res.* **66**:134 (No. 220), 1987.
35. Gettleman, L., Vargo, J.M., Goist, K.C., Guerra, L.R., Finger, I.M., McDonald, G.T., Jameson, L.M., Salib, M.M., and Agarwal, A.K.: Update of clinical trial results at two medical centers of a polyphosphazene vs. a silicone rubber resilient denture liner. *Trans. Acad. of Dent. Mater.,* Chicago, No. 11, Feb. 19, 1988.
36. Gettleman, L., Vargo, J.M., Goist, K.C., Guerra, L.R., Finger, I.M., McDonald, G.T., Jameson, L.M., Salib, M.M., and Agarwal, A.K.: Patient evaluation of resilient denture liners: GSRI-PNF vs. Molloplast B. *J. Dent. Res.* **67**:375 (No. 2100), 1988.
37. Gettleman, L., Guerra, L.R., Finger, I.M., McDonald, G.T., Jameson, L.M., Salib, M.M., Agarwal, A.K., Vargo, J.M., and Goist, K.C.: Results of clinician's evaluation of a polyphosphazene resilient denture liner and a silicone rubber liner. *3rd World Biomater. Congr. Trans.* Kyoto, No. 2P-32, p. 162, 1988.
38. Gettleman, L., Vargo, J.M., Goist, K.C., Jameson, L.M., Salib, M.M., Agarwal, A.K., Guerra, L.R., Fnger, I.M., and McDonald, G.T.: Laboratory and clinical testing of a polyphosphazene resilient denture liner vs. a silicone rubber liner. *Proc. 3rd Int'l. Congr. of Implantol. and Biomater. in Stomatol.* Osaka, No. A-3, p. 22, April 28, 1988.
39. Gettleman, L., Guerra, L.R., and Jameson, L.M.: Properties and clinical efficacy of a polyphosphazene rubber soft denture liner (Novus™) vs. Molloplast B. *Trans. Int'l. Congr. of Dent. Mater.* Honolulu, No. 023, pp. 254-255, 1989.

40. Gettleman, L., Vargo, J.M., Goist, K.C., Jameson, L.M., Salib, M.M., Agarwal, A.K., Guerra, L.R., Finger, I.M., and McDonald, G.T.: Laboratory and clinical testing of a polyphosphazene resilient denture liner vs. a silicone rubber liner. In *Oral Implantol. and Biomater. Progress in Biomed. Engin. Series*, H. Kawahara, ed. Amsterdam: Elsevier, **7**, 15-20, 1989.
41. Gettleman, L., Guerra, L.R., and Jameson, L.M.: Clinical trial of Novus soft denture liner vs. Molloplast B: Final results and four-year follow-up. *J. Dent. Res.* **69**: 166 (No. 458), 1990.
42. Gettleman, L.: Clinical trial of Novus™ soft denture liner (polyphosphazene) vs. Molloplast B: Final results and four-year follow-up. *Proc. 8th Southern Biomed. Engineering Conf.*, Miami, 1991.
43. Aksit, K.S., Kutay, O, Gettleman, L., and Bayli, M.S.: Effect of a soft denture liner on masticatory performance and efficiency. *J. Dent. Res.* **73**: 122 (No. 167), 1994.
44. Hazelrigg, D.E., Selvaraj, R., Watters, J.C., and Gettleman, L.: Polyphosphazenes for maxillofacial prosthetics. *J. Dent. Res.* **72**, 128 (No. 200), 1993.
45. Segal, M.A., Haeberle, C.B., Windchy, A., Shay, J., Goldsmith, L.J., Khan, Z., and Gettleman, L.: Soft and hard denture liners for xerostomic patients. *J. Dent. Res.* **76**: 62 (No. 385), 1997.
46. Gettleman, L., Rappoport, L., Marx, S.A., Watters, J.C., Selvaraj, R., Hazelrigg, D., and Khan, Z.: Doughy polyphosphazene elastomers for denture liners and maxillofacial prosthetics. *J. Dent. Res.* **73**: 274 (No. 1382), 1994.
47. Shaygan, F., Khan, Z., and Gettleman, L.: Retention and longevity of resilient denture liner in implant overdentures. *J. Dent. Res.* **72**: 272 (No. 1354), 1993.
48. Esfahani, A.R., Haeberle, C.B., and Gettleman, L.: Overdenture retention of an implant bar comparing resilient liners and Hader clips. *J. Dent. Res.* **76**: 179 (No. 1321), 1997.
49. Kiat-amnuay, S., Khan, Z., and Gettleman, L.: Overdenture retention of four resilient denture liners over an implant bar. *J. Dent. Res.* **77**: 937 (No. 2442), 1998.
50. Kiat-amnuay, S., Khan, Z., and Gettleman, L.: Overdenture retention of four resilient denture liners over an implant bar. *J. Prosthet. Dent.* **81**: 568-573, 1999.
51. Kiat-amnuay, S., Gettleman, L., Makayarajjananonth, T., and Khan, Z.: Hardness of resilient denture liners over time. *J. Dent. Res.* **79**: 157 (No. 106), 2000.
52. Kiat-amnuay, S., Mekayarajjananonth, T., Khan, Z., and Gettleman, L.: Simplified methods for fabricating tissue-supported implant-retained overdentures with retention from a resilient liner. *J. Prosthet. Dent.* **82**: 242-0245, 1999.
53. Gettleman, L. Chlorinated polyethylene and polyphosphazenes, In Materials Res. in Maxillofac. Prosthet., *Trans. Acad. Dent. Mater.* **5(1)**: 156-172, 1992.
54. Gettleman, L.: Polyphosphazene fluoroelastomers for denture liners and facial prosthetics. *ACS 4th Int'l. Conf. on Phosphorus Chem.*, No. LT3-3, Cincinnati, July 14, 1998.

In: Applicative Aspects of Poly(organophosphazenes)
Editors: M. Gleria and R. De Jaeger, pp. 49-83
ISBN 1-59454-025-X

Chapter 3

PHOSPHAZENES AND LIGHT

Mario Gleria[1], Pietro Bortolus[2] and Francesco Minto[2]
1-Istituto di Scienze e Tecnologie Molecolari del C.N.R., c/o Dipartimento C.I.M.A. dell'Università di Padova, Via Marzolo 1, 35131 Padova, Italy, and Istituto per la Sintesi Organica e la Fotoreattività, Consiglio Nazionale delle Ricerche, Viale dell'Università, 2 – 35020 Legnaro, Padova, Italy
2-Istituto per la Sintesi Organica e la Fotoreattività del C.N.,R., Sezione di Legnaro, Viale Dell'Università 2, 35020 Legnaro, Padova, Italy

INTRODUCTION

Research on the interactions among different types of electromagnetic radiations and phosphazene substrates started relatively early in the past century, mostly with the aim of finding out new ways to induce the ring-opening polymerization of cyclic phosphazene oligomers (hexachlorocyclotri- and/or octachlorocyclotetra-phosphazene) to high molecular weight linear polymers both in solution and in solid state. These investigations were partially successful, the polymerization of $(NPCl_2)_3$ to $(NPCl_2)_n$ being obtained only when high energy radiations (X-ray[1,2], γ-rays[3,4] or electrons[5]) were exploited. On the contrary, the original attempt by Dishon[6] of polymerizing $(NPCl_2)_3$ in solid state and/or in decaline solution upon UV irradiation was unsuccessful. Hexachlorocyclophosphazene was completely unaffected when irradiated as solid, whereas only radical substitution processes of the chlorine atoms of the phosphazene substrates with solvent residues were observed during the photochemistry of these compounds in decaline. The attempts successively reported by Konecny[7,8] and Allcock[9] were similarly unsuccessful.

These observations, extremely important from a photochemical point of view, can be rationalized in terms of the intrinsic spectroscopic characteristics of the -P=N- units of both cyclo- and poly- phosphazene substrates[10-16] (see Figure 1).

These groups are completely transparent to radiations of wavelengths longer than 220-230 nm, thus photons of $\lambda > 230$ nm are not absorbed by phosphazene substrates, and are therefore completely unable to induce any type of photochemical event in them.

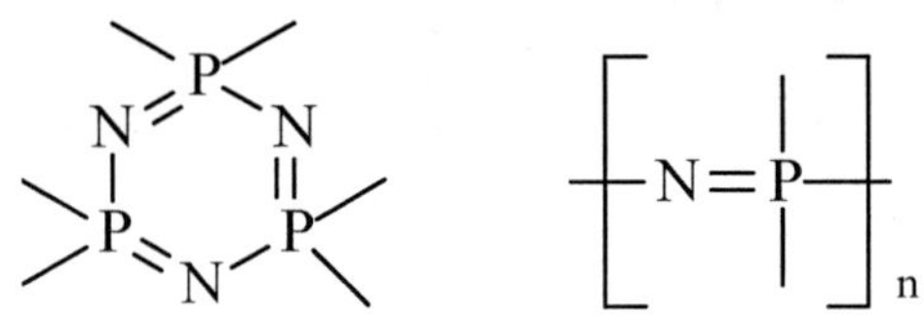

Figure 1

Indeed, all the successive photochemical investigations carried out on phosphazenes during the last 30 years clearly showed that the great variability of photochemical phenomena undergone by these substrates (e.g. degradation[17,18], crosslinking[19], oxidation[20,21], fragmentation[22,23], photochromisms[24-26]) is mostly dependent on the spectroscopic and photochemical characteristics of the substituent groups attached to the phosphorus atoms of the phosphazene materials[14-16].

It became evident that bimolecular[15] interactions between different types of substrates also play an important role in determining the overall photoreactivity of these sytems. Thus energy[27,28], and electron[29-32] transfer processes and hydrogen abstraction phenomena among excited chromophores attached to the phosphazene skeleton and external quenchers[19,33-36], or *viceversa*[16,35-44], are equally possible for these substrates.

A second very important feature of the polyphosphazene skeleton in determining the photochemical behavior of these materials, is the presence of two substituent groups on each phosphorus of the phosphazene skeleton. In fact, when taking the solid state structure of phosphazene macromolecules into consideration, it can be seen that most of these substrates show the *cis-trans* planar configuration (see Figure 2), reported below[45],

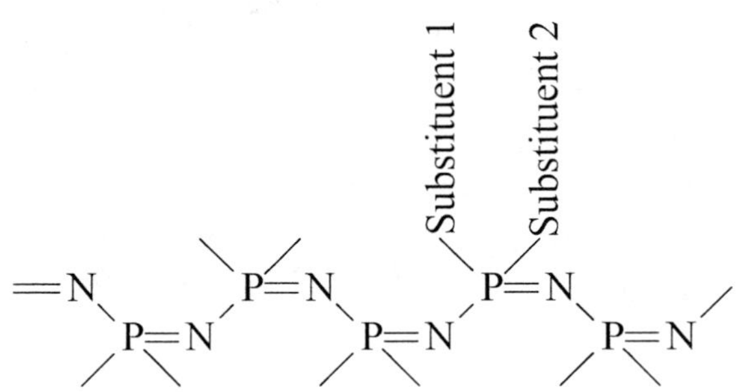

Figure 2

where the two phosphorus substituents lie on the same side of the phosphazene skeleton. On the basis of molecular models it could be estimated that the two substituent groups geminally attached to the same phosphorus of the phosphazene chain are very close to each others (2.5-3 Å), at a distance which is perfectly compatible with the possibility of generating intramolecular interactions between the two groups. This was first ascertained by the discovery of the existence of intramolecular excimers in aryloxy-substituted cyclophosphazenes[46] and polyphosphazenes[46-51] by means of fluorescence spectroscopy experiments, and was successively confirmed by research carried out in our laboratory on the photochemical behavior of phosphazene copolymers geminally substituted with benzophenone/β-naphthoxy[52], benzophenone/4-isopropylphenoxy[53], benzophenone/3-N,N'-dimethylphenoxy[30-32] and benzophenone/methoxyethoxyethoxy[36,54] couples. The overall photochemical behavior of these materials depends, in fact, on the type of intramolecular interactions existing between the substituent couples in phosphazenes (energy-[52], electron-transfer[30-32], and hydrogen abstraction[36,53] processes). The efficiency of these phenomena can easily be regulated and finely tuned[52].

On this basis, we will summarize the research on the photochemical behavior of cyclo- and polyphosphazenes carried out in several laboratories over more than 30 years. We will describe the basic discoveries on the intimate mechanism of the reactions explored, possible practical applications of the compounds investigated in the photochemical domain, utilization of photochemical techniques to modify polyphosphazenes and transform them into "useful materials"[55,56], and the future perspectives of phosphazene substrates in photochemistry.

LIGHT-INDUCED DEGRADATION PHENOMENA IN POLY(ORGANOPHOSPHAZENES)

When arylamino-[17,18,29], aryloxy-[18,20,37], and/or alkyl/aryl-[57,58] substituted polyphosphazenes are dissolved in air-saturated THF or in other solvents of low electron affinity, such as CH_2Cl_2, and irradiated with UV light the main process observed is the degradation of the macromolecules with decrease in both weight and number average molecular weights and intrinsic viscosity.

Photosensitization experiments carried out with poly[bis(4-methylanilino)phosphazene] (PMAP)[17,18,29] and with poly[bis(β-naphthoxy)phosphazene] (PβNP)[18] with benzophenone or anthraquinone sensitizers clearly suggested that the first excited singlet state of the chromophore attached to the polyphosphazene chain is responsible for the photoreactivity of these polymers. Similar observations hold true for poly(phenyl/methyl)lphosphazene[57,58], whose photochemistry is not influenced by the presence in solution of piperylene, a typical triplet state quencher. Moreover, flash photolysis experiments[18], together with mass spectrometric determinations of the low molecular weight products formed upon irradiation[17,57,58] indicated that the primary photochemical step in this process is the homolytic scission of the P—YChromophore and, to a far less extent, of PY—Chromophore bonds with formation of phosphazene macroradicals, according to the following Scheme (see Figure 3):

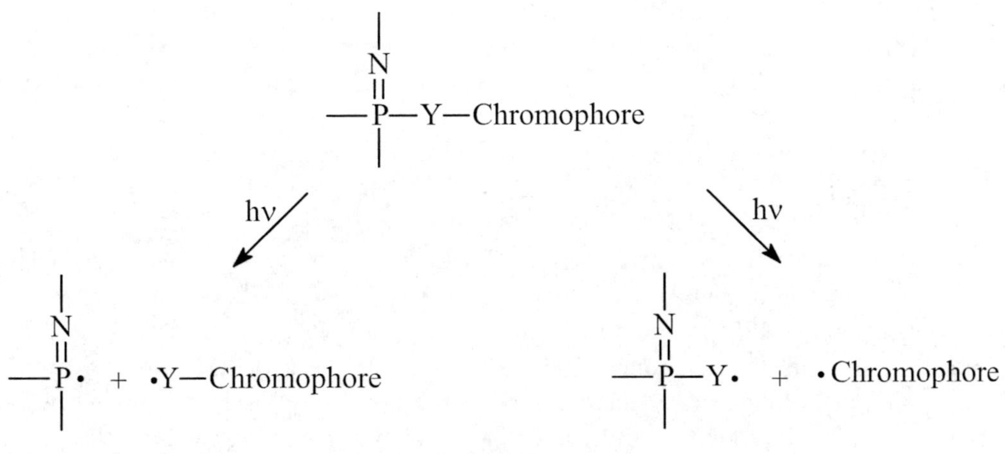

Figure 3

These species easily stabilize to

Chromophore—H or Chromophore—Y-H

by hydrogen abstraction from the solvent, while phosphazene macroradicals form hydroperoxidic species by oxygen uptake (as demonstrated by analytical techniques), whose cleavage eventually leads to chain scission and degradation (see Figure 4).

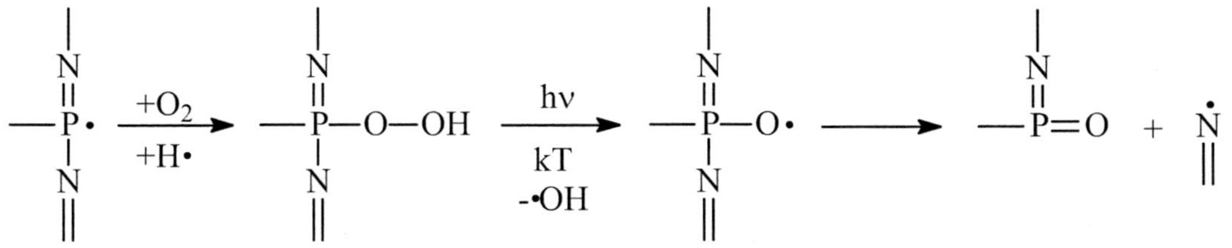

Figure 4

Similar processes are also operative when other types of phosphazene macroradicals are considered (see Figure 5).

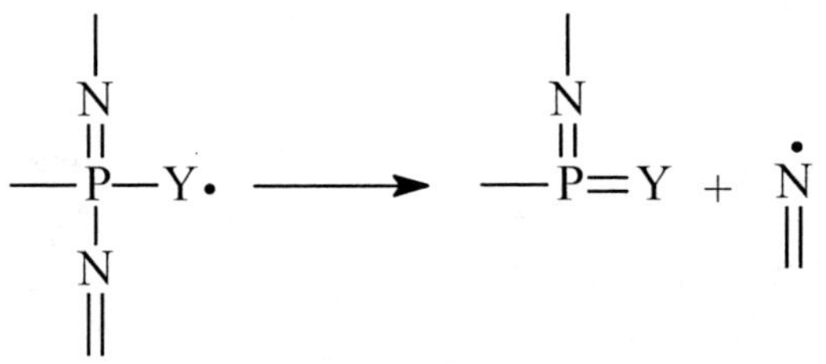

Figure 5

When the same processes are carried out in air-saturated solvents possessing higher electron affinities, such as $CHCl_3$ and CCl_4 (+ 1% w/w ethanol)[29], photodegradation is still the dominant process, and the efficiency of the phenomenon appears related to the electron affinity of the reaction medium. In this case, charge-transfer processes between the arylamino substituent of PMAP and the solvent itself are operative, as demonstrated by the UV spectra of the polymers dissolved in halogenated solvents[29], and by flash photolysis experiments[29] that indicated the formation, under illumination, of amine radical cations attached to the polyphosphazene skeleton, according to the following Scheme (see Scheme 6).

Once formed, phosphazene macroradicals follow the same degradation pathway illustrated in Figure 4, the charge transfer mechanism being considerably more efficient in inducing degradation phenomena in amino-substituted polyphosphazenes.

There are two polymers, however, that do not conform to this general behavior of phosphazene polymers irradiated in aerated THF or CH_2Cl_2 solutions: poly[bis(4-isopropyl-phenoxy)phosphazene (PiPP)[37] and poly[bis(4-benzylphenoxy)phosphazene] (PBPP)[20]. These polymers possess substituent groups that are particularly reactive and prone to undergo photooxidation and/or photofragmentation processes instead of phosphorus-substituent bond cleavage reactions, with formation of hydroperoxides on the side groups and/or release of low molecular weight photolysis products (*e.g.* benzaldehyde and, to a far less extent, benzene in the case of PBPP[20]).

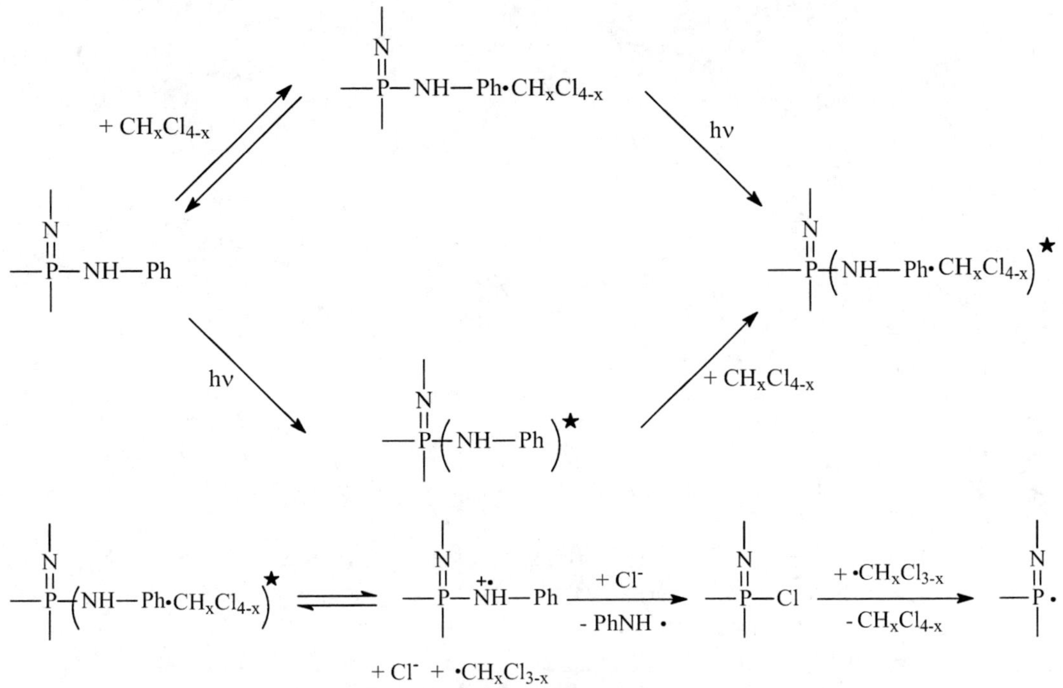

Figure 6

When the photochemistry of poly[bis(4-benzoylphenoxy)phosphazene] (BzPOP) is considered[19], the overall reaction mechanism is dominated by the reactivity of the first excited triplet state of the benzophenone group. This state easily abstracts hydrogen atoms from the solvent, producing highly reactive ketyl radicals (see Scheme in Figure 7 below)

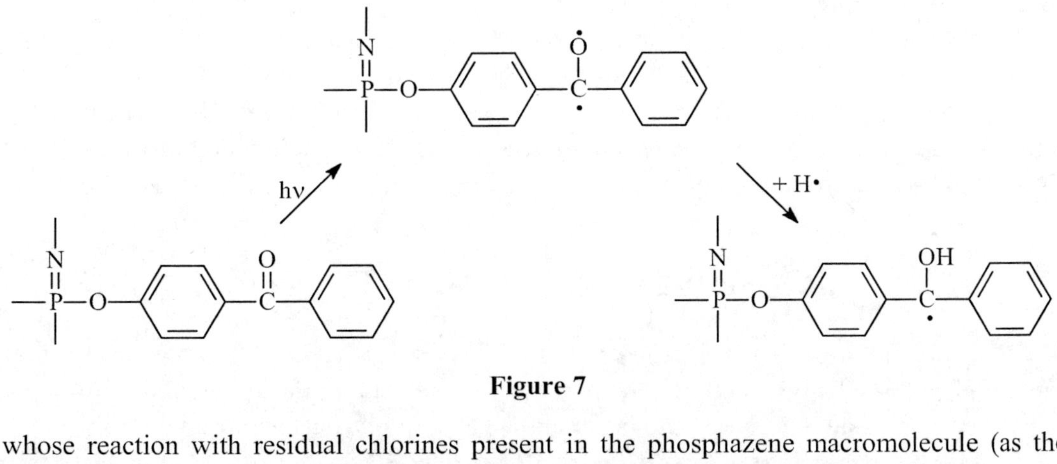

Figure 7

whose reaction with residual chlorines present in the phosphazene macromolecule (as the consequence of an incomplete substitution process of polydichlorophosphazene), leads to the formation of phosphazene macroradicals (see Figure 8)

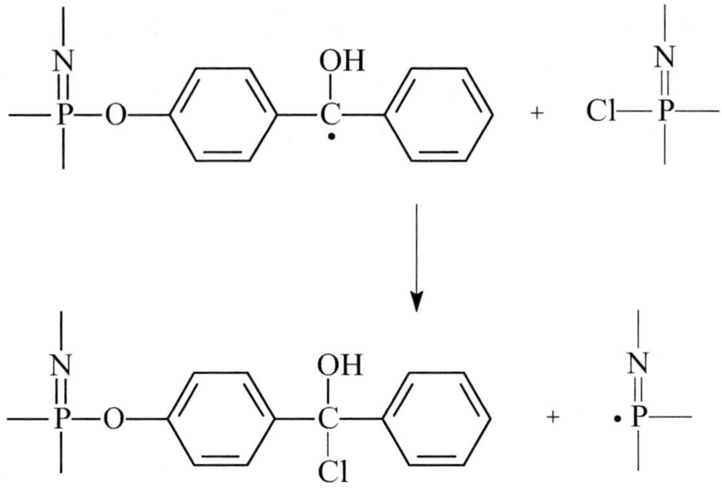

Figure 8

that degrade to low molecular weight polymers according to the mechanism reported in Figure 4.

In conclusion, from these investigations it clearly emerged that the light-induced degradation of poly(organophosphazenes) always takes place when these polymers are irradiated in air-saturated solutions. Furthermore, when the first excited singlet state of the chromophore attached to the polyphosphazene skeleton is the photoactive state[17,18,29], the overall photochemical reaction pathway seems to be dominated by four factors, i.e. the presence of molecular oxygen (to form hydroperoxide and start the photooxidation mechanism), the possibility of the solvent supplying hydrogen atoms (to stabilize the radical species by hydrogen abstraction processes), the electron affinity of the solvent used for the photolysis (which introduces into the degradation mechanism a very efficient charge transfer pathway to chain scission of the macromolecules) and the photolability of the polyphosphazene substituent which preferentially photofragments or photooxidizes instead of detaching from the polymer chain.

On the contrary, when the first excited triplet state is involved[19] the most important process is the abstraction reaction of hydrogen atoms from the solvent.

In both cases the presence of residual chlorines in the irradiated polymers due to an incomplete substitution process seems to play a decisive role in determining the efficiency of the light-induced degradation processes [19].

PHOTOCROSSLINKING PROCESSES IN PHOSPHAZENES

Photocrosslinking of phosphazene derivatives is a very interesting and important process capable of substantially modifying both the chemical and physical properties of these substrates[55,56], transforming them into materials suitable for practical applications.

A) Poly(organophosphazenes)

When phosphazene polymers are considered for light-induced reticulation processes, a number of previously inaccessible practical applications become possible, for example in: microelectronics[59-63], ionic conductivity[64-68], membranes[69-73], thermally resistant coating compositions[74,75] biomaterials for microencapsulation processes[25,56,76-80], and so on. These processes can be obtained by a variety of methods that utilize bimolecular reactions[15].

Thus, aryloxy-[18] and arylamino-[17,18,29] substituted polyphosphazenes can be crosslinked by irradiation in oxygen free solutions. In fact the radical species primarily formed upon illumination (see Scheme in Figure 3) combine very efficiently with each other, inducing the reticulation of the polymer chains and precipitation, according to the following Scheme (see Figure 9):

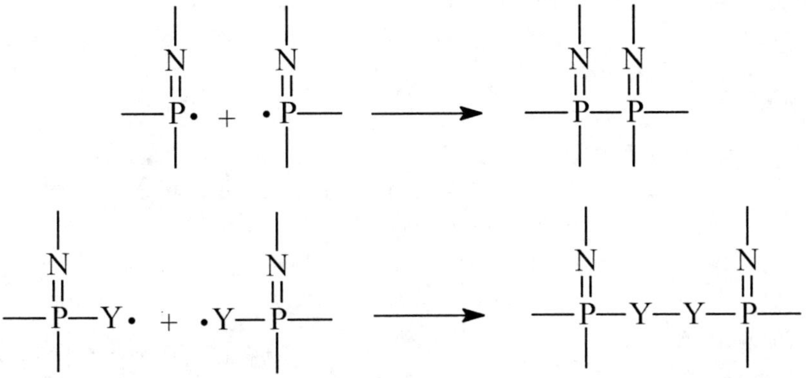

Figure 9

The same mechanism is operative when the solvent has a high electron affinity (as CCl_4)[29], or when poly[bis(phenoxy)phosphazene] (PPP)[81,82], poly[bis(trifluoroethoxy)-phosphazene (PTFEP)[81-83], poly[bis(ethoxy)phosphazene] (PEP)[81,82], or poly(phenyl/methyl)-phosphazene[57,58] are irradiated in film with UV light.

In all the above reported examples, the light-induced crosslinking phenomena of polyphosphazenes take place by recombination reactions of radical species localized mainly on the skeleton or in its own strict proximity (see Scheme in Figure 9).

Many reactions exist, however, where the bimolecular processes leading to crosslinking involve mostly polyphosphazene substituents. PiPP[37,84] and PBPP[20,21], for instance, show a feeble crosslinking when irradiated in CH_2Cl_2 or THF solutions, while their treatment in film leads to rapid insolubilization of the materials due to the formation of interchain linkages. Furthermore, in the case of PiPP the irradiation of films induces deep surface modifications[85] with the introduction of hydroperoxide and carbonyl functions on the surface.

When BzPOP is irradiated in a solution of hydrogen-supplying solvents (e.g. THF)[19] the crosslinking reactions occurs by recombination reaction of the ketyl radicals on the benzophenone chromophores formed by hydrogen abstraction reaction (see Scheme in Figure 7), as reported in the following reaction Scheme (see Figure 10).

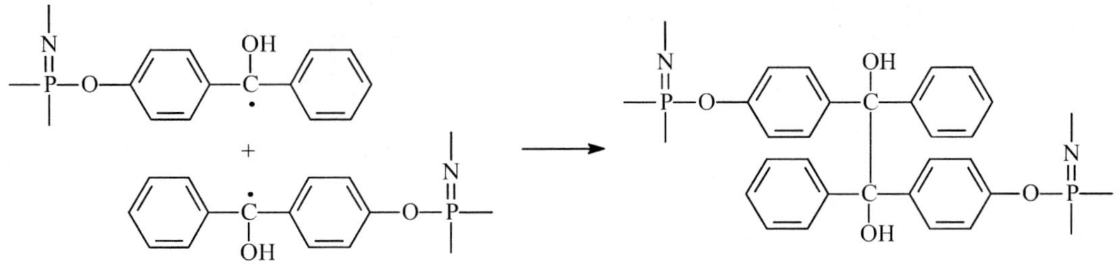

Figure 10

Similar reactions should also be operative when the same polymer is irradiated in the solid state, although the reticulation mechanism is not presently fully understood[52,53]. In fact, the light excitation of BzPOP films[19,52] induces a crosslinking phenomenon that starts a few minutes after the beginning of irradiation and is almost complete (≈100% of gel formed) within 3 h of illumination. The extent of this process could be finely tuned by using naphthalene quenchers (see Figure 11).

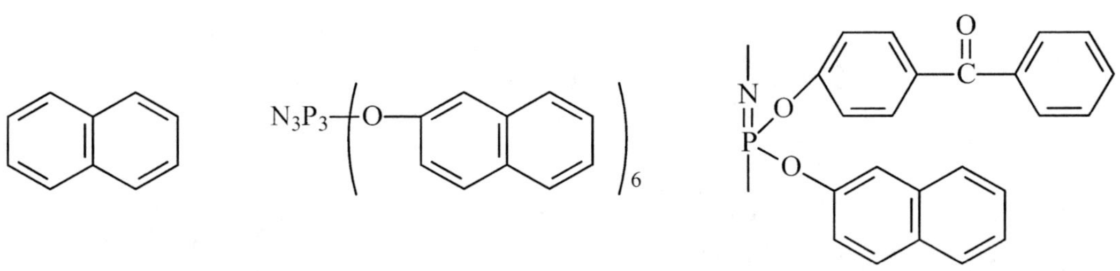

Figure 11

either free in solutions[28], attached to the cyclophosphazene ring[28] or geminally substituted on the same phosphorus atoms of the phosphazene skeleton already bearing 4-hydroxybenzophenone groups[52]. In all cases, in fact, the photoreactivity of the excited benzophenone triplet attached to the polyphosphazene skeleton can be quenched by energy transfer processes, thus reducing the extent of the reticulation process itself and considerably prolonging the induction period necessary for observing reticulation and gel formation processes.

The discovery that the photochemical properties of free molecules are preserved when attached to the phosphazene skeleton allowed BzPOP to be utilized as a heterogeneous phase triplet photosensitizer[27], to efficiently induce the trans-cis isomerization of stilbene and piperylene, the cycloaddition of indene and the valence isomerization of norbornadiene to quadricyclane by energy transfer processes.

The preparation of phosphazene copolymers led to the synthesis of phosphazenes with variable percentages of the monomeric units shown in Figure 12.

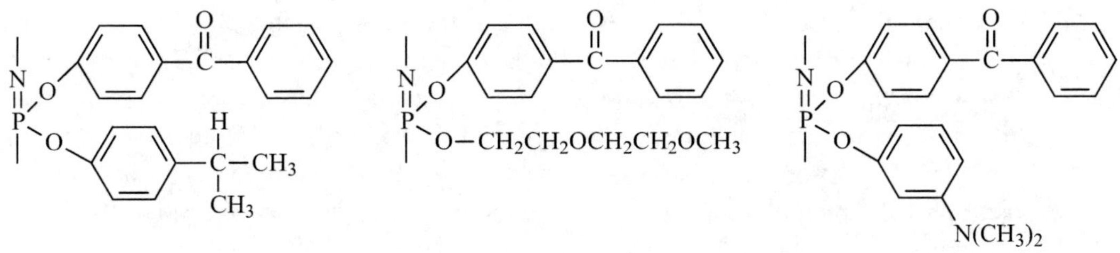

Figure 12

In these polymers, the selective excitation of the benzophenone group leads to the generation of excited triplets with unitary efficiency, and, to the onset of extremely efficient crosslinking phenomena by hydrogen abstraction reactions from the geminally substituted 4-isopropylphenoxy[53] and methoxyethoxyethoxy[36,54] moieties, or by electron transfer processes in the case of 3-N,N'-dimethylaminophenoxy groups[30-32]. For these phosphazene copolymers one could envisage possible applications as negative photoresists[13,15] or ionic conductive materials[36].

It is important to observe that benzophenone works equally well to induce crosslinking phenomena as an external photosensitizer free in solution or dispersed in the polyphosphazene films. A few examples of these processes are provided below:

(a) free benzophenone[37] and benzophenone moieties attached to linear and/or cyclic phosphazene supports[38] sensitize the photochemistry of PiPP in both solution and film (see Figure 13) so inducing a relevant photocrosslinking of this polymer together with surface modification and functionalization reactions;

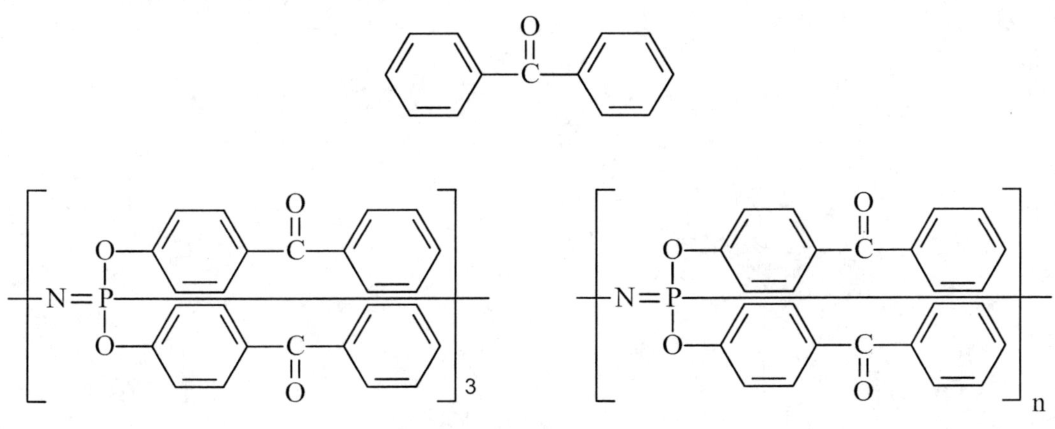

Figure 13

(b) spectacular reticulation effects, in both solution and films, can be obtained in the case of the photochemistry of PBPP carried out in the presence of free benzophenone[86];

(c) the benzophenone sensitized photochemistry of poly[bis(methoxyethoxyethoxy)-phosphazene] (MEEP)[64,65,67,68] led to the controlled crosslinking of this polymer and the remarkable improvement of its dimensional stability without detrimental for its ionic conductivity;

(d) the same phenomenon is observed in the case of the benzophenone sensitized photochemistry of phosphazene copolymers containing almost 50% phenoxy and 50% 4-ethylphenoxy, 4-methylphenoxy or 4-isopropylphenoxy moieties[71-73]. The triplets of the aromatic ketone easily induces the crosslinking of these materials by hydrogen abstraction reactions from the labile hydrogen atoms present in the para position of the phenoxy substituents, eventually generating new phosphazene-based membranes;

(e) fluoroalkoxy-substituted polyphosphazenes also containing acrylate or methacrylate groups as side phosphorus substituents[87-89], having the structure shown in Figure 14 are easily reticulated by selective irradiation of benzophenone in the presence of triethanolamine. The substrates obtained are usually exploited as insoluble coatings.

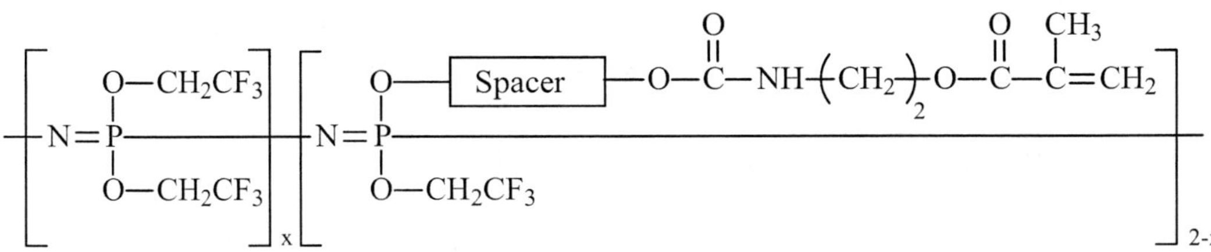

Figure 14

(f) fluoroalkoxy-substituted phosphazene polymers also containing 20-27 mol % of 4-allyloxyphenol or 4-[4'-(allyloxy)phenyl]phenol having the structure shown in Figure 15 readily crosslink upon irradiation in the presence of benzophenone as a photoinitiator[90].

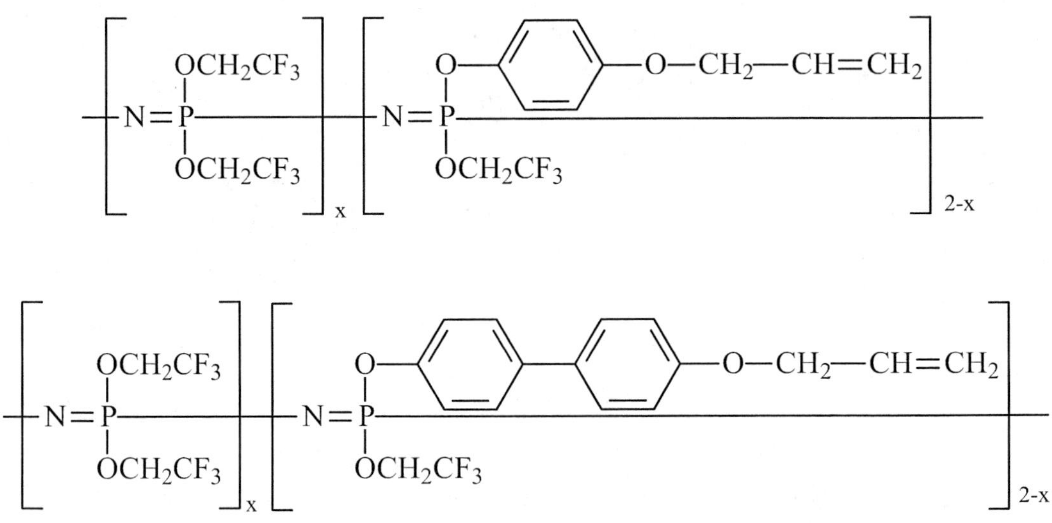

Figure 15

The corresponding crosslinked polymers are usually exploited as low-temperature elastomers.

Lastly, it is interesting to observe that in the case of both the direct[20] or sensitized[86] photochemistry of PBPP in the presence of oxygen and during photo-oxidation of this polymer under accelerated conditions[21] benzophenone moieties are always generated by oxidative processes that takes place on the 4-benzylphenoxy substituents of the polymer, according to the reaction Scheme reported in Figure 16

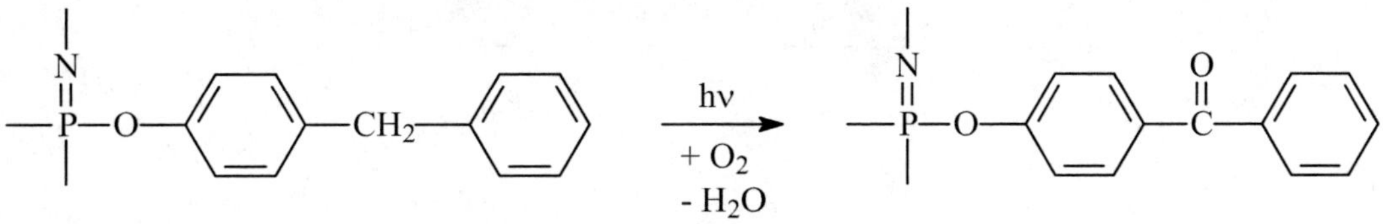

Figure 16

Once formed, these species are themselves able to absorb light and participate in the overall photochemical processes of PBPP macromolecules.

The following polyphosphazenes have been reported to undergo photocrosslinking by excitation of their side substituents:

a) unsaturated polyphosphazene containing allyl groups have been exploited as photocurable systems because of the ability of these polymers to reticulate when irradiated in the presence of multithiolic compounds. The overall reaction is shown in Figure 17:

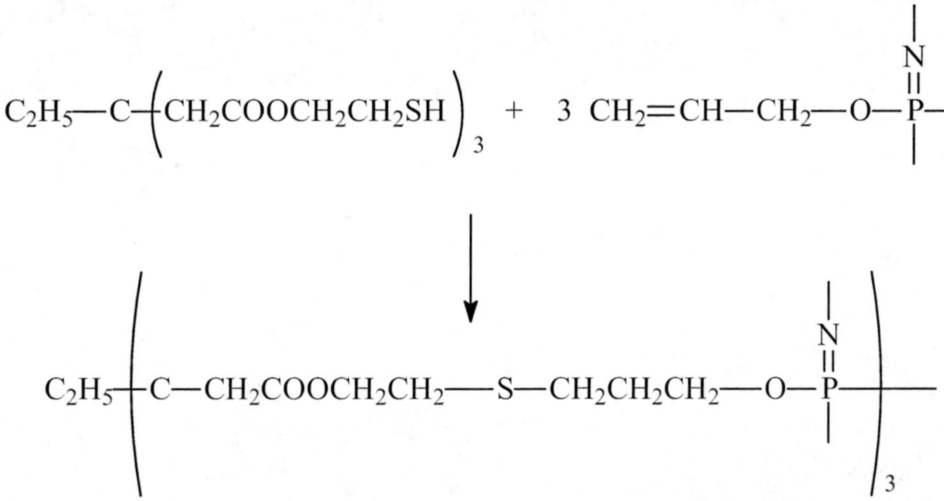

Figure 17

Suitable polymers for these processes are poly[bis(allylamino)phosphazene][91] or phosphazene copolymers partially substituted with 2-allylphenoxy groups[74,75]. In the example given in Figure 17 trimethylolpropane tri(3-mercaptopropionate) was used. This process was claimed to reduce significantly the negative influence of molecular oxygen on the speed and efficiency of the curing process[91], and to be greatly facilitated by the presence of a photoinitiating species like benzoin ethyl ether.

b) phosphazene macromolecules containing allyl unsaturation in their chemical structure (see figure 15)[90] and other macromolecules containing ethyleneoxide-allyl substituents[92-94], as shown in Figure 18

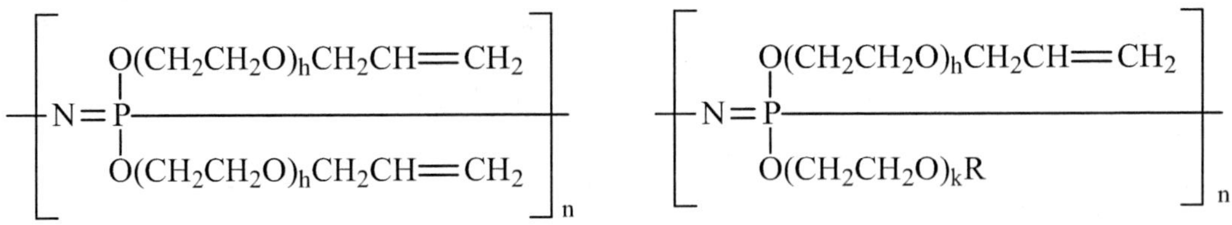

where "h" and "k" are integer between 0 and 15, and R = alkyl groups

Figure 18

can easily be crosslinked when irradiated at room temperature[90], even in the absence of photoinitiators, to produce crosslinked, low-temperature elastomers[90] or membranes which can retain their shapes without being washed away with water; these are suitable for the preparation of secondary batteries[92-94].

c) the [2+2] photoaddition reaction of cinnamate and chalcone residues attached to the phosphazene skeleton to form cyclobutane rings and crosslinked phosphazene materials (Figure 19) have been described by Allcock[76-80,95]; these have been used as biomaterials for microencapsulation purposes.

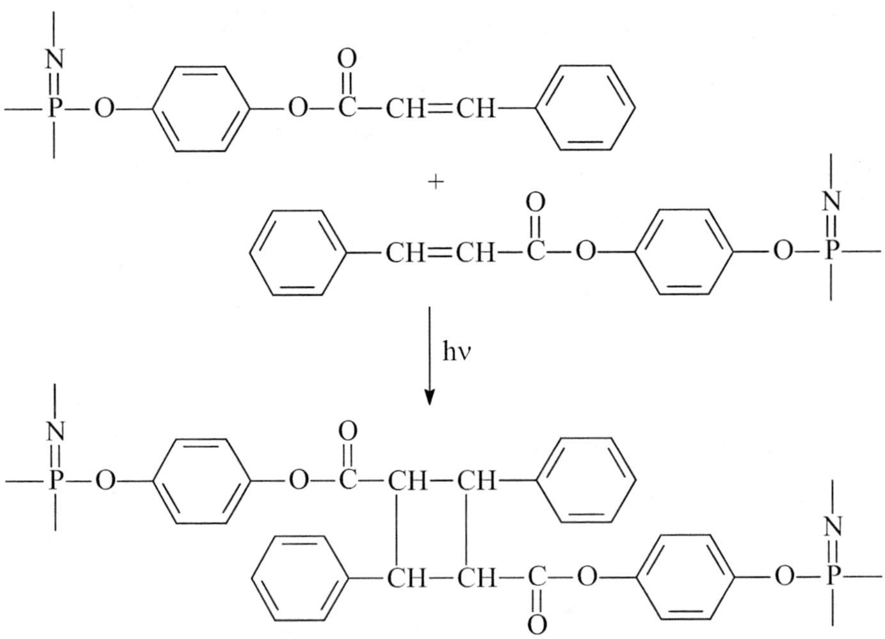

Figure 19

B) Cyclophosphazenes

Although crosslinking reactions involving cyclophosphazenes to produce cyclomatrix polymers is one of the most important topics in cyclophosphazene chemistry[96], the use of photochemical reactions to induce these processes started relatively late.

At the beginning of the seventies, the photocrosslinking of cyclophosphazenes substituted with allyl residues (*e.g.* allyl alcohol or 2-allylphenol) (see Figure 20)

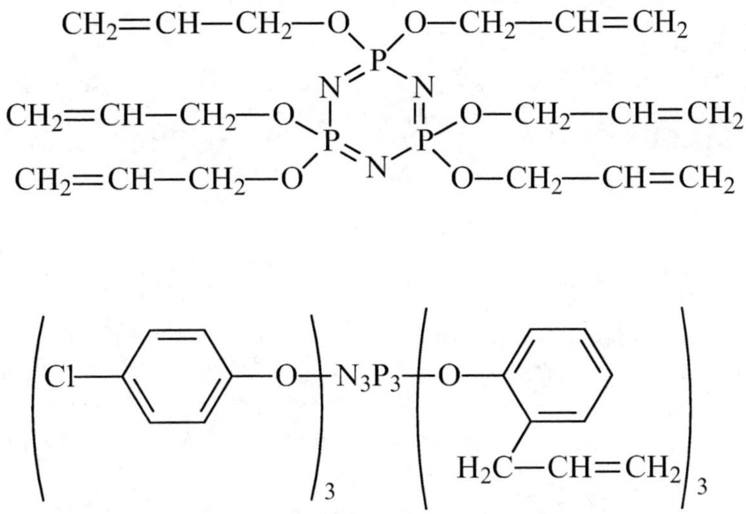

Figure 20

could be reached by irradiating these cyclophosphazenes with multithiol derivatives (sometimes in the presence of suitable photoinitiating agents[91]) as described in Figure 17, to obtain burn-resistant compositions that also showed improved thermal stability.

Following this strategy, Quella[97,98] reported the synthesis of a spirocyclophosphazene having the chemical structure shown in Figure 21:

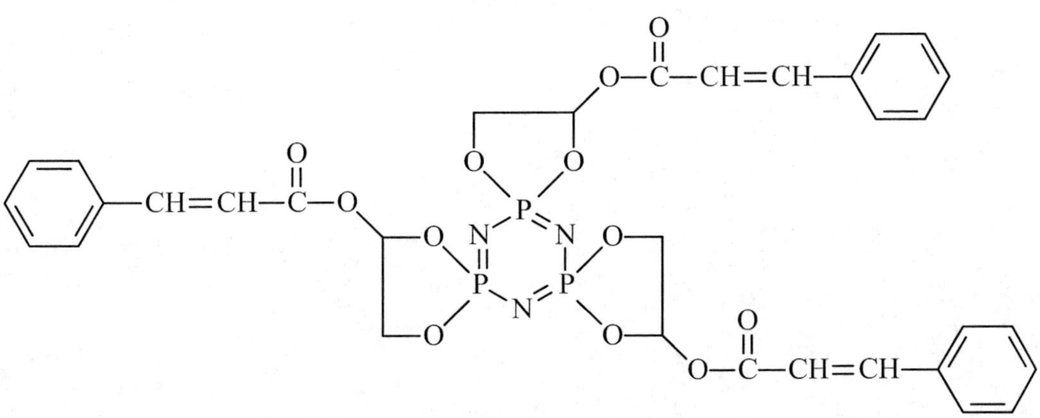

Figure 21

in which three cinnamate residues are present. The crosslinking of this material proceeds by UV irradiation of these groups and formation of cyclobutane rings by [2+2] cycloaddition reactions. The resulting crosslinked materials are low-shrinkage coatings, resist to α-ray irradiation, and are suitable for use as a cover layer for microelectronic components.

The most important photoreaction in this field, however, is the photoreticulation process of 1,1,3,3,5,5-hexakis(methacryloylethylenedioxy) cyclophosphazene (see Figure 22)

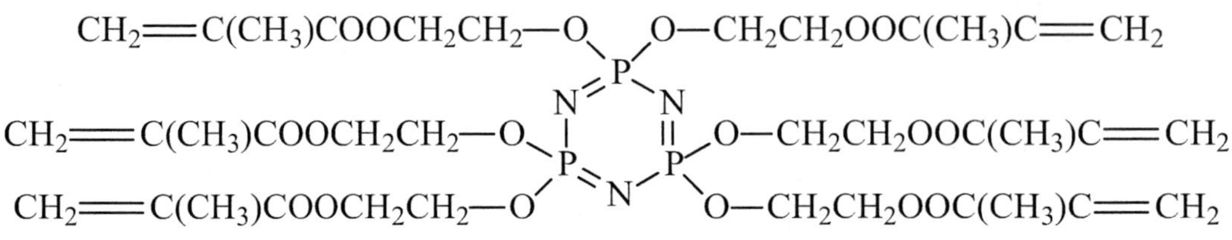

Figure 22

in the presence of photoinitiating species, to obtain new substrates with potential practical applications in printed circuits[99], heat-[100-103] and flame-[101,102] resistant compositions, surface coatings[101,102,104-120], dental resins[121,122], optical decoration[123-127], abrasion-[103,104,107,109,112,114,120,128-149], weather-[132], chemical-[109,114,131,132,143,148], heat and/or flame-[109,126,130-132], or solvent-[130] resistant hard substrates, water repellant products[107,145], lenses[150-154], adhesives[120,132,135,138,141,155,156], resin compositions[112,134,149,157], photochromic products[158], liquid crystal devices[159], UV transparent derivatives[103,112,114,130,134,135,137,138,141,142,144], image receptors[137,160-163], electroconductive compositions[164], recording display[165], leather substitutes[166], antistatic[141] or antifogging[112] agents, touch panel films[108,118], low shrinkage materials[114], and styrene-based resin composites[167].

Similar reactions have also been reported by Kimura[168] for dichlorophosphinoyliminotrichlorophosphorane substituted with five HEMA residues, whose structure is reported in Figure 23

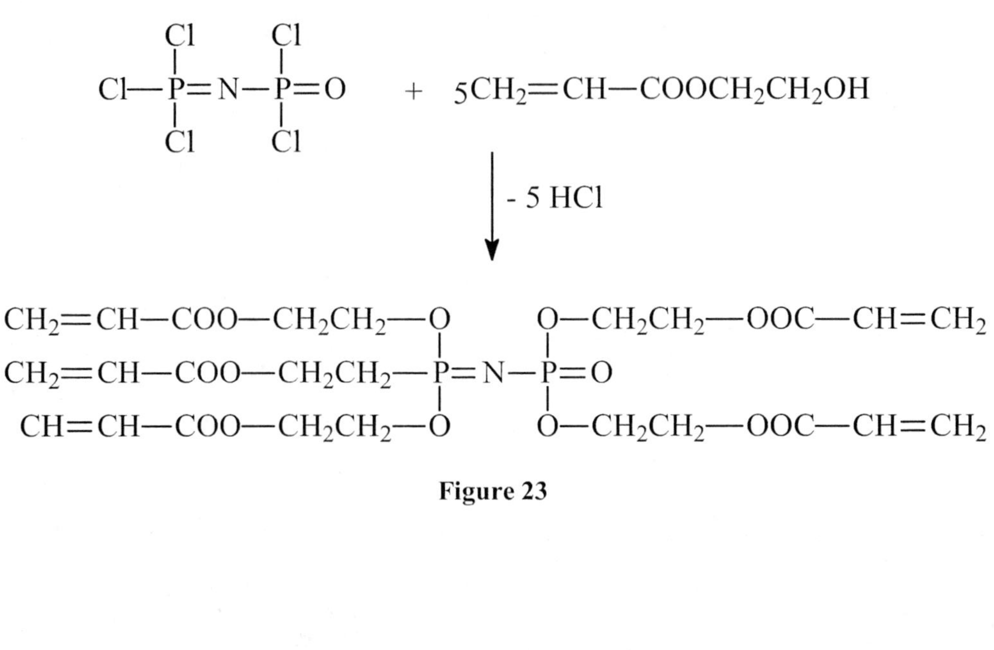

Figure 23

LIGHT-INDUCED GRAFTING REACTIONS IN PHOSPHAZENES

Grafting organic substrates onto poly(organophosphazenes)[16,42] is one of the simplest ways to modify their reactivity[169,170], surface properties[41], and bulk features[171]. This process is usually obtained in a rather simple way by introducing a reactive site in the phosphazene compound that is able to start grafting reactions[172]. This site could be a carbon anion (by proton extraction reactions from the polyphosphazenes by strong bases) according to Wisian-Neilson[173-177], or a radical species (by hydrogen abstraction processes) following Gleria[16,42,172], Lora[178-181] and Allcock[169,182]. From both these species the grafting copolymerization reactions of styrene[171], methylmethacrylate[39,40], 4-vinylpyridine[169,170], glycidyl methacrylate[169], acrylic and methacrylic acids[170], acrylates containing HALS groups[43,183], vinyl acetate and vinyl alcohol[44,184] monomers onto polyphosphazenes could be obtained. In some cases, maleic anhydride[185,186] and various maleates[187] have been exploited as unsaturated substrates in the grafting processes.

Among the different experimental strategies exploited to produce radical sites in phosphazenes the use of light in the presence of suitable sensitizing species is very important[16,42].

In fact the selective excitation of suitable photoinitiators[39,41] produces radical species by photofragmentation[40,42-44,170,172,184,188], hydrogen abstraction reactions[39,40,42,43,169,170,172,183,188] or electron- transfer processes[41], according to the Scheme in Figure 24

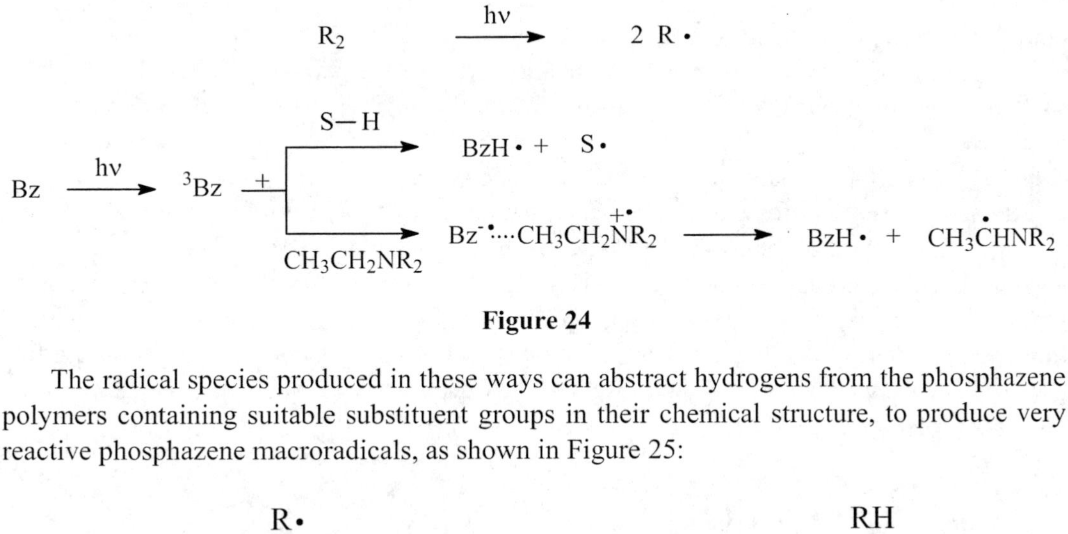

Figure 24

The radical species produced in these ways can abstract hydrogens from the phosphazene polymers containing suitable substituent groups in their chemical structure, to produce very reactive phosphazene macroradicals, as shown in Figure 25:

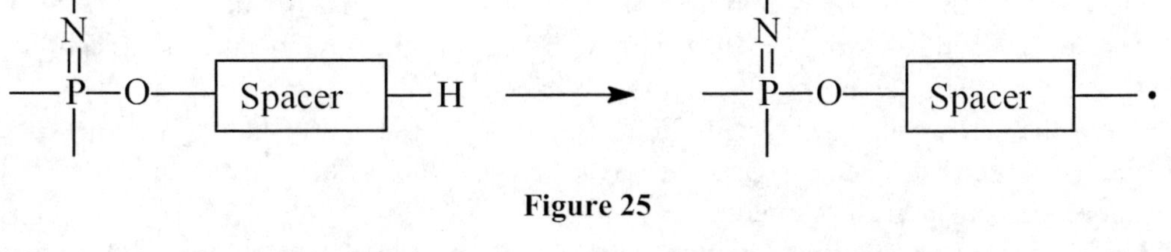

Figure 25

from which the successive grafting reaction takes place with unsaturated organic monomers leading to comb-like phosphazene copolymers (Figure 26).

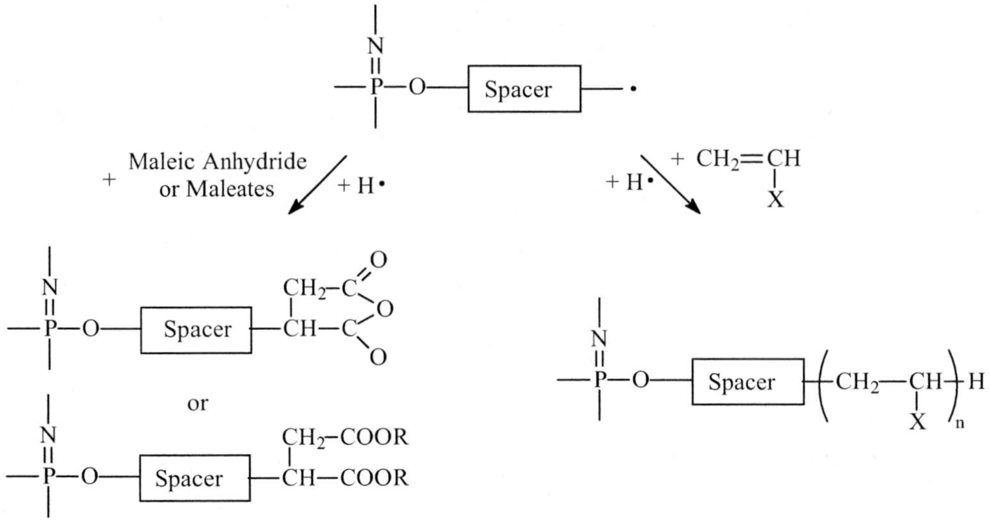

Figure 26

In the case of maleic anhydride or maleates, however, the grafting process stops at monomer level due to the intrinsic inability of these compounds to undergo homopolymerization processes[189].

Obviously, this process is always accompanied by the crosslinking of the polyphosphazene due to macroradical coupling reactions and by simultaneous homopolymerization of the unsaturated monomers[16,42].

Possible substituents for phosphazene macromolecules are para-substituted alkyl phenols, such as 4-methylphenol[44,169,184,190,191], 4-ethylphenol[44,184,190,191], 4-isopropylphenol[39,40,190,191], 4-secbutylphenol[40,44,184,190,191] and 4-benzylphenol[40,43,169,170,183,190], or trifluoroethoxy[40] residues, all of which contain abstractable hydrogen atoms.

It may also be stressed that glycidylmethacrylate[169] and 4-vinylpyridine[169,170] are bifunctional unsaturated monomers, with double bonds suitable for grafting, while the oxirane and the pyridine groups, unmodified during this process, easily undergo additional post-grafting functionalizations. The resulting grafted phosphazene copolymers are species with an improved reactivity[169,170], higher hydrophilicity or hydrophobicity compared with the starting products[41], enhanced photo-[43,190] and thermo-[183] oxidative stability, and have potential use in blending processes[192].

PHOSPHAZENE-BASED PHOTORESISTS

Electromagnetic radiations impinging upon polymeric films for microlithography may induce a decrease in solubility for these substrates, for instance by crosslinking reactions, an enhancement of their solubility, for example by degradation phenomena, or a change in the polarity of the macromolecular system. In the first case the polymeric material may be used as a negative photoresist, in the second case as a positive photoresist; in the third case a dual tone photoresists may be developed[193].

As far as phosphazene macromolecules are concerned, negative photoresists have been developed mostly by exploiting electron beams[59,194] or γ-rays[194-200], while the use of UV or visible light seems to be less common. A few examples, however, are reported in literature, such as the use of the poly[bis(phenoxy)phosphazene]/CBr_4 system[59]; moreover, phosphazene homopolymers substituted with cinnamate[78-80,95] or chalcone[76,77,80,95] residues, or phosphazene copolymers partially substituted with benzophenone groups, and with 4-isopropylphenol[53], 3-N,N'-dimethylphenol[30-32], or methoxyethoxyethanol[36], respectively, have been considered as possible candidates for negative resist technology due to their capability to crosslink rapidly and efficiently upon irradiation with UV or visible light.

The same situation holds true for positive photoresists, as only a few papers are known on the use of phosphazene macromolecules as positive resist systems. For example the solubility of poly(methyl/phenyl)phosphazene functionalized with various percentages of carboxylic groups[201] is greatly increased by irradiating substituted 1,2-naphthoquinone diazide photosensitizers[202] with light of 405 nm. The development of this composition could be obtained by the action of 0.05 M tetrabutylammonium carbonate solutions, showing positive features down to 4 μm. In the field of positive photoresists, moreover, Kim[203] recently reported the synthesis of hexakis[(1,2-naphthoquine-2-diazide-5-sulfonyl)phenoxy]cyclophosphazene photosensitizer (see Figure 27)

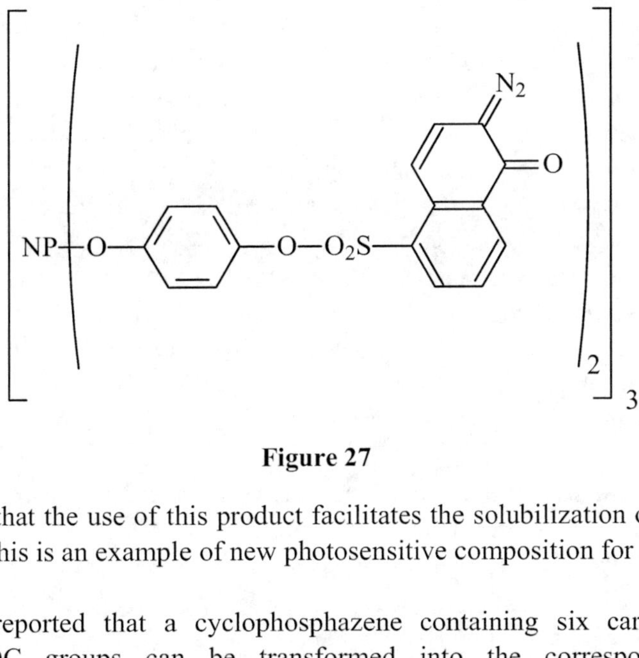

Figure 27

and demonstrated that the use of this product facilitates the solubilization of Novolak Resins upon irradiation. This is an example of new photosensitive composition for traditional organic photoresists.

It has been reported that a cyclophosphazene containing six carboxylic functions protected by BOC groups can be transformed into the corresponding hexakis(4-carboxylatephenoxy)cyclophosphazene by irradiating a suitable photoacid generator (usually triphenylsolfonium hexafluoroantimoniate)[204]. The reaction is shown in Figure 28:

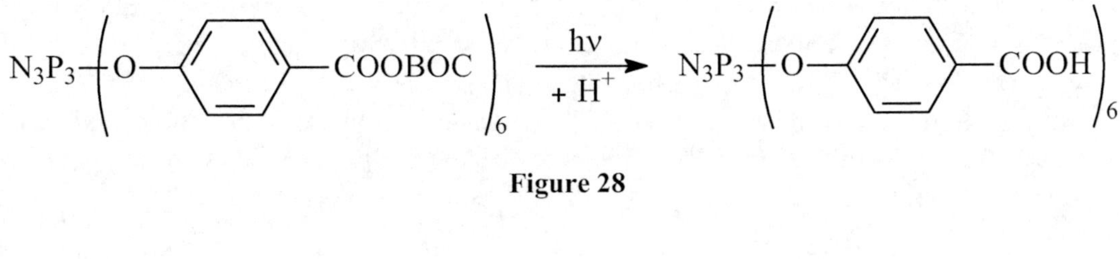

Figure 28

During this process isobutylene is released. When present in the Novolak composition as a protected trimer, the cyclophosphazene acts as a dissolution inhibitor for the resin, whereas when transformed to its free COOH structure its principal function is to promote Novolak dissolution, thus working as a positive resist system.

Finally, another BOC-protected cyclophosphazene has been described in literature[205], having the structure reported in Figure 29

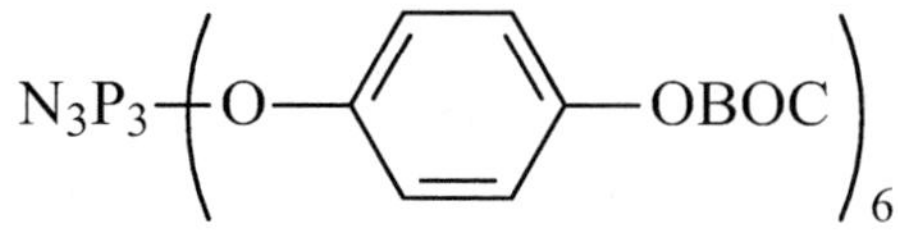

Figure 29

whose potential to act as a dual tone photoresists has been claimed.

PHOTOCONDUCTIVITY IN POLYPHOSPHAZENES

Phosphazene polymers are usually considered as insulating materials, with resistivity ranging between 10^{14} and 10^{16} Ohm x cm[206,207]. They may become conductors when doped with suitable products, such as, iodine[207], tetracyanoquinondimethane[208], tetrachloro-1,4-benzoquinone[208], silver or lithium salts[209]. Sometimes the efficiency of the doping process can be enhanced by illumination with suitable light, as in the case of PMAP doped with iodine after illumination with light of $\lambda > 320$ nm[207]. These polymers, moreover, are not photoconductive.

It has been found, however, that treating PβNP and PMAP with electron acceptors, such as trinitrofluorenone, causes these materials to increase their dark conductivity of two order of magnitude and to show conspicuous photoconductivity phenomena[206,210,211] (photocurrent density of 10^{-6} A/cm^2 with 5 x 10^3 photons/cm^2, at 450 nm, with 105 V/cm as applied field). The origin of this new phosphazene feature is recognizeable in the formation of charge-transfer complexes between the β-naphthoxy (or 4-methylaniline) substituent acting as an electron donor and the trinitrofluorenone considered as an electron acceptor. This induces the shift of the absorption spectrum of the doped polymer to well inside in the visible range of the absorption (~600 nm)[206,210]. The corresponding photocurrent in this spectral region appears to be generated through an Onsager mechanism, in which a large number of charge carriers (electrons or holes) are formed upon irradiation and immediately trapped.

The β-naphthoxy-substituted polyphosphazene has also been investigated as possible photoconductive material for the preparation of electrophotographic photoreceptors, its charge-discharge performance described, and its potential applicability in this field evaluated in view of its photoconductive characteristics and mechanical properties[208]. Subsequent research on the photoconductive features of both PβNP and PMAP and on the possibility of using them in layered photoresponsive imaging member have also been reported[212,213].

Other phosphazene macromolecules investigated as potentially photoconductive materials are polyphosphazenes containing carbazole, naphthoxy and naphthylamino substituents, described mainly in patent literature[214-216] or congress presentations[217].

MISCELLANEOUS PHOSPHAZENE-LIGHT INTERACTIONS

Phosphazene-light interactions not previously described have been collected here to demonstrate the variety of possible applications accessible to these substrates.

Poly[bis(4-methylphenoxy)phosphazene] functionalized with a various percentages of 4-bromomethylene phenoxy moieties[218] was allowed to react with Rose Bengal to prepare an insoluble, deeply colored, material which showed very interesting applicability as a heterogeneous phase photosensitizer for the generation of singlet oxygen[219]. Similar results have been reported for BzPOP and its capability to act as a heterogeneous phase triplet state energy sensitizer also described[27].

A spiropyran unit[25,26] was appended to a diethylenoxy- or triethylenoxy-substituted polyphosphazene chain by an esterification process. Films and solutions of these polymers, and a copolymer containing the trifluoroethoxy-group, displayed the classical, ring-opening photoreaction[220], as shown in Figure 30

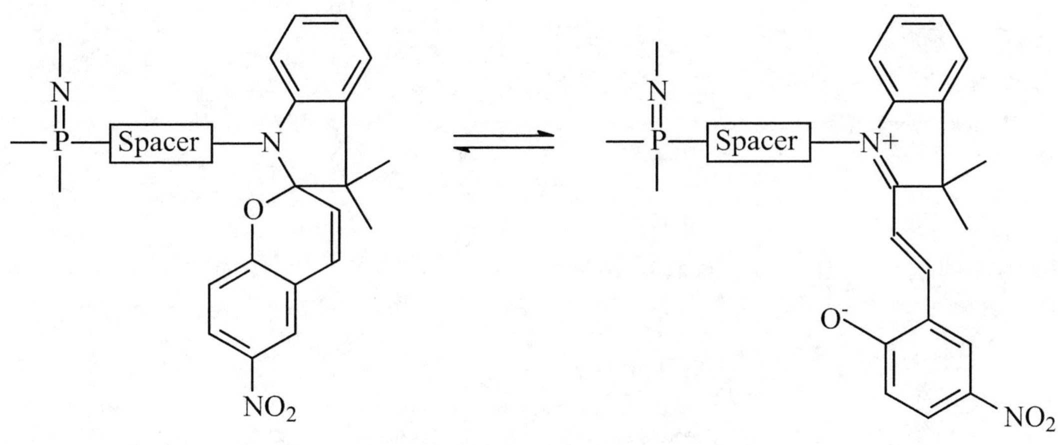

Figure 30

with formation of the coloured merocyanine when exposed to light. The rate of the thermal back-reaction depends on the polarity of the solvent and, in films, on the nature of the spacers and the degree of substitution of the phosphazene chain.

Light induced reversible changes of the UV-visible absorption spectrum have also been observed in films of the copolymer trifluoroethoxy/4-hydroxyazobenzene bearing 14% *trans*-azobenzene residues[24] (see Figure 31)

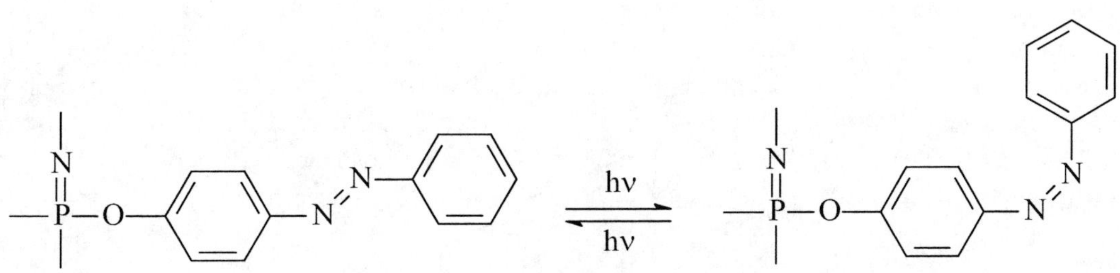

Figure 31

the spectral variation sobserved after irradiation at 313 nm were attributed to the formation of the *cis*-isomer. The recovery of the pristine *trans*-form could be achieved either thermally or by irradiation with light of 435 nm, which is preferentially absorbed by the *cis*-form.

The application of light to promote the synthesis of phosphazene substrates has been exploited in the reaction of tetrachlorophenyl(diphenylphosphine)cyclophosphazene, $N_3P_3Cl_4PhPPh_2$ with chromium hexacarbonyl $Cr(CO)_6$ to obtain $N_3P_3Cl_4PhPPh_2{\bullet}Cr(CO)_5$[221]. Similarly, Majoral and Potin[222-224] succeeded in the preparation of polyphosphazene bearing allyl groups by UV irradiation of a toluene solution of poly(dichlorophosphazene) in the presence of allylbutyltin[222-224]. The polymer obtained contained diallyl- (8%) and monoallyl-substituted (26%) phosphorus atoms. The residual chlorines were successively substituted by thermal reaction with sodium phenoxide.

Furthermore, films of poly[bis(trifluoroethoxy)phosphazene] SiO_x have been exploited as a deformable membrane light modulator with good thermal stability (220°C) and interesting applicative features[225].

PHOTOPHYSICS OF PHOSPHAZENES

Emission spectra of cyclic trimers and poly(phosphazenes) have been reported for aryl-[57,58,226], aryloxy-[46-49], and arylamino-[29,227] substituted compounds. At room temperature, the three classes of phosphazenes display in dilute THF[46-49] or CH_2Cl_2 solutions[29,57,58,226,227], a double emission composed of a monomer-like fluorescence similar to that of the unbound, free aromatic group and a broad structureless band, red-shifted with respect to the monomer-like band, which disappears in rigid glasses at 77 K. The long-wavelength emission was attributed to the formation of excimers, dimeric species arising from the interaction of two aromatic rings, one in the singlet excited state and the other in the ground state[228].

The class of aryloxy-substituted phosphazenes was the most investigated in both solution and film. From the temperature effect on the intensity of the two emission bands of phenoxy- and naphthoxy-substituted cyclophosphazenes an activation energy of about 10 kJ mol^{-1} for the formation of the excimers in solution was found. The excimer stabilization energy (8 kJ mol^{-1} $< \Delta H <$ 24 kJ mol^{-1}) was found to depend on the dimensions of the substituent.

The different effect of temperature on the emission characteristics (excimer bandwidth and band maximum) examined for hexaphenoxy- and hexa-*p*-cresoxy-cyclotriphosphazene and the corresponding polymers disclosed interesting features of the emission properties of cyclic trimers and linear polyphosphazenes[48,49] in dilute solutions. In linear polyphosphazenes, excimer bandwidth and maximum change with the temperature, contrary to what observed for the corresponding cyclo derivatives. Moreover, for polymers, no Arrhenius behaviour was observed for the activation energy of excimer formation contrary to what observed for trimers and, apparently, polymer solutions display excimer emission at low temperatures whereas no such emission was observed for the trimers. These differences were attributed to the fact that excimer formation in poly(aryloxyphosphazenes) involves a number of different possible conformations, in contrast to the situation in the trimers, as schematized in Figure 32.

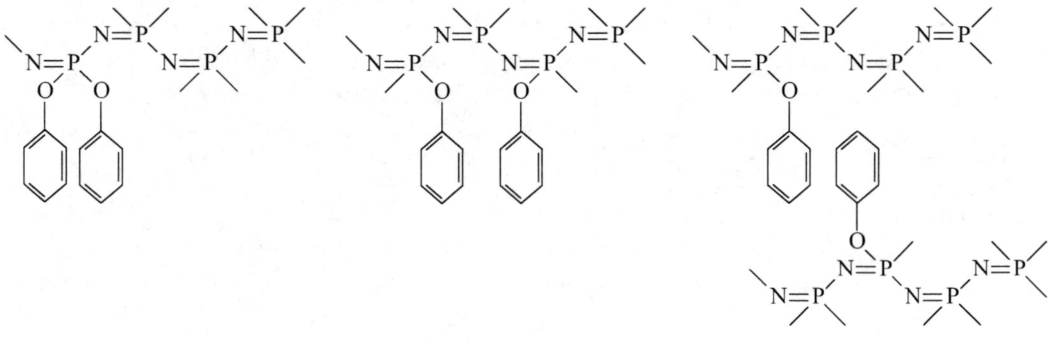

Figure 32

The first possible conformation is the intramolecular excimer between two aryloxy groups bonded to the same phosphorus atom, as found in phosphazene trimers. The second results from an intramolecular interaction between the substituent appended to an adjacent phosphorus and requires a *trans-trans* or *trans-gauche* backbone and eclipsed chain conformation; this can only occur in phosphazene polymers. The third is an intermolecular excimer resulting from the interaction of two aromatic groups bonded to different phosphazene chains or to non adjacent phosphorus atoms of the same chain.

Confirmation that different types of interactions occur in polyphosphazenes and the corresponding cyclic trimers was achieved in a comparative study of the photophysics of solutions of hexakis(β-naphthoxy)cyclophosphazene, the corresponding linear polyphosphazene, and a series of copolymers containing both the β-naphthoxy- and the 4-hydroxybenzophenone- groups in different percentages[229]. It was found that the formation of excimers is much more efficient in the homopolymer than in the trimer, indicating that the deactivation of the singlet excited state of the β-naphthoxy- chromophore occurs in the homopolymer without the formation of the corresponding triplet state. In fact no naphthalenic triplet-triplet absorption or phosphorescence could be detected in the polymer, in contrast to the situation observed for the trimer, whose spectroscopic characteristics strictly resembled those of the free naphthalenic chromophore.

In copolymers containing benzophenone, naphthalenic phosphorescence can be seen as a result of an intramolecular energy transfer from the benzophenone to the naphthalene moiety following the selective excitation of the aromatic ketone group. The phosphorescence of the copolymer with low naphthoxy content is identical to that of the hexakis(β-naphthoxy)cyclophosphazene, while that of copolymers with high naphthoxy content has a red-shifted, broader emission very similar to that detected in films of poly(2-vinylnaphthalene) and attributed to the formation of triplet excimers.

The emission of poly(phenoxyphosphazene) film has also been studied and compared with that in CH_2Cl_2 solution[47]. No substantial difference in behaviour was found for the polymer in the two physical states: at variance with polystyrene for which the monomer component is almost totally quenched in films, the film of poly(phenoxyphosphazene) maintains an appreciable monomer emission. Attempts to carry-out a reliable multi-esponential analysis of the emission decay failed, probably due to the presence of multiple emitting species. An estimate of singlet exciton migration led to the conclusion that the monomer state of the phenoxy-substituent is essentially immobile.

The temperature dependence of the excimer emission has been exploited to probe the dynamics of some aryloxy-substituted polyphosphazene films[48,51]. It was found that the variations of the of the excimer (I_D) to monomer(I_M) intensities well correlate with the glass transition temperature, T_g, and the mesomorphic temperature, T(1), obtained by differential scanning calorimetry. These studies also showed that the intensity and wavelength of the excimer emission are related to the crystal structure of the polymer.

The cyclophosphazene skeleton has been used as support for ligands of lanthanide ions with the goal of ameliorating the emission properties of these cations. In water, hexakis(4-carboxylatophenoxy)cyclotriphosphazene forms a stable complex with two Eu^{3+} and Tb^{3+} ions having luminescence emission significantly higher than that of the corresponding free ions[230]. The effect was attributed to protection of the coordinated ions from the quenching action of water molecules and to an efficient energy transfer from the electronically excited ligand to the metal ions. An increase in the emitting efficiency of Eu^{3+} was also observed when the trimer cyclophosphazene was functionalized with five $-NH-(CH_2)_3-N(CH_3)_2$ or $-O-(CH_2)_2-N(CH_3)_2$ and a 4-vinylphenoxy-group. The increase in efficiency observed with the trimers mentioned above was smaller than that obtained with the carboxylatophenoxy-group as chelating agent, and the polymerization of the vinyl group did not further ameliorate the emitting properties[231].

CYCLOPHOSPHAZENES IN PHOTOCHEMISTRY

Cyclophosphazenes[96] have long been considered as the model compounds of polyphosphazenes[232,233], for their synthesis, reactivity, structure, spectroscopic and kinetic behavior. They assumed, however increasing importance when it was realized that they can be used as important compounds themselves, as anticancer products[234,235] in biomedicine, clathrates[236], substrates for supporting metals[237-239], ion receptors[240,241], and phase transfer catalysts[242,243].

Some of the most interesting recent applications of cyclophosphazenes in photochemistry[16] are their use as photostabilizing additives for commercial organic macromolecules[244-252], photoinitiating species for unsaturated monomers[22,23,246,253], and antioxidants[254-257] to stabilize polyolefins thermally and photochemically. This use developed when it was realized[16,233] that a) these compounds have a molecular weight ranging around 500 Dalton, which allows them to be blended rather easily without segregation from the polymeric matrices and migration toward the surface of the polymer system[251]; b) their reactivity, rather similar to that of the corresponding polyphosphazenes[232,233], allows easy synthesis, functionalization and modification of these substrates according to the final goal they have to fulfil.

Thus, cyclophosphazenes containing o-hydroxybenzophenone[246,249,251,252], benzo-triazole[244,245], and **H**indered **A**mine **L**ight **S**tabilizers (HALS)[247,248,250] (Figure 33) have been developed over time to stabilize polyolefins, polyesters, and so on, and to protect them from damages deriving from light exposure.

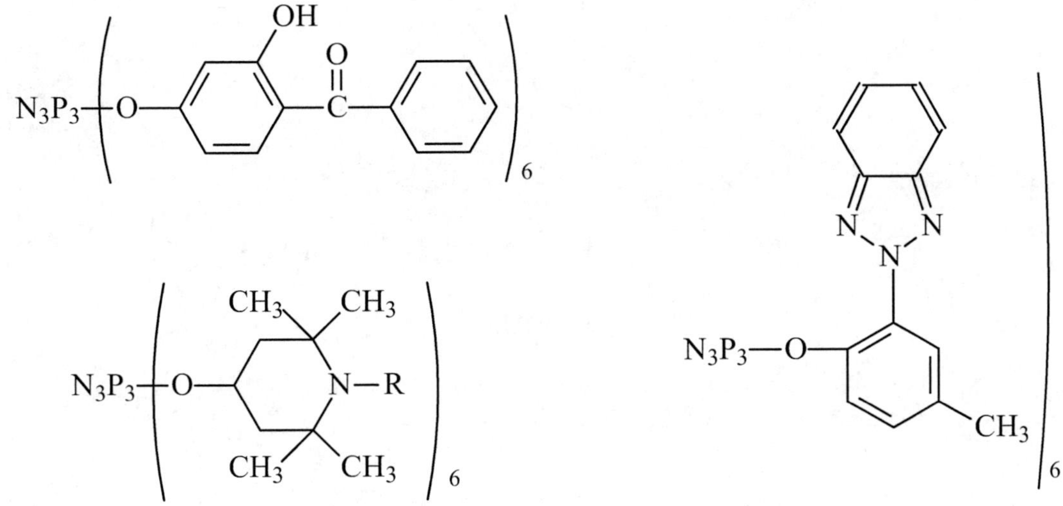

where R=H, CH_3

Figure 33

By modification of the phosphazene substituents, these compounds turn from photostabilizers to photoinitiators, that is photosensitive products that can produce very reactive radicals upon irradiation and initiate free radical polymerization of unsaturated organic monomers. Typical substituents for this application are benzophenone[246], alkoxyacetophenones[23], benzoin[22] and azide[258] derivatives attached to the cyclophosphazene rings (see Figure 34), sometimes made water-soluble by co-substituting oligo polyethylene oxide monomethyl ether residues on the same cyclophosphazene ring[253] (see Figure 35).

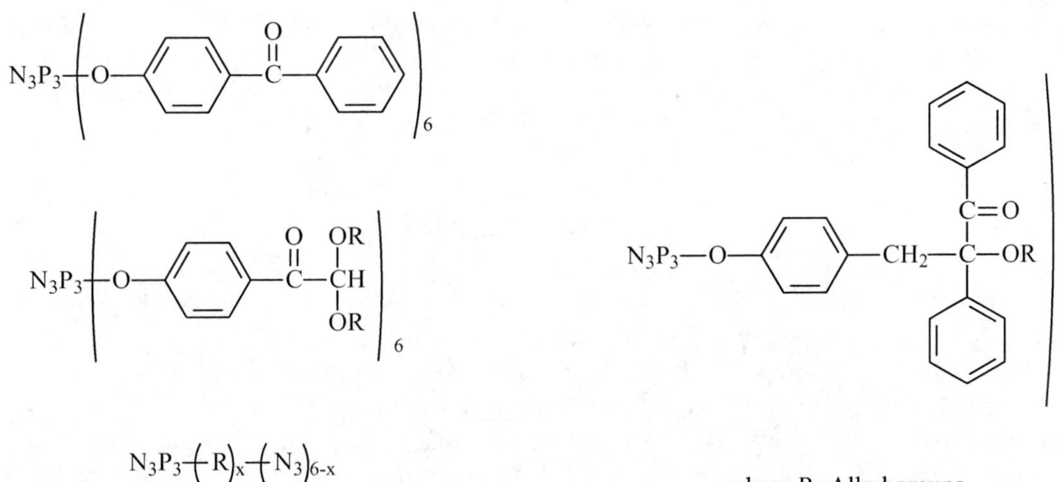

where: R=C_6H_5O, CF_3CH_2O or $(CH_3)_2N$ and x=3,4,5

where R=Alkyl groups

Figure 34

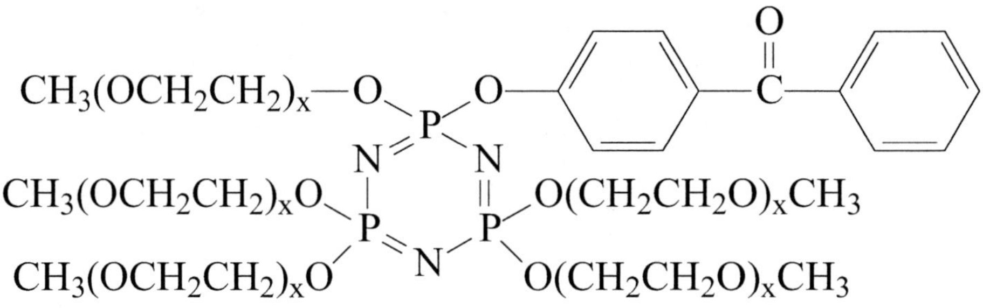

Figure 35

Finally, the synthesis of cyclophosphazenes substituted with hindered phenols[254,255,257], having the general structure shown in Figure 36

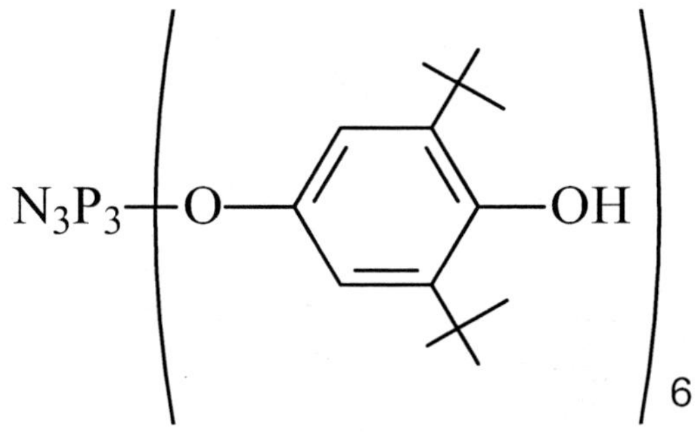

Figure 36

leads to the preparation of antioxidant species for the thermal and/or photochemical treatment of organic macromolecules (e.g. polyolefins). Alternatively, cyclophosphazenes substituted with linear aliphatic alcohols (e.g. butanol) or glycols (e.g. diethylene glycol) in combination with hindered phenols[256] can provide a synergist effect on the thermal and/or photochemical stabilization of polyolefins.

CONCLUSION

We have reviewed the large variety of chemical reactions induced in cyclo and polyphophazenes by absorption of UV or visible light.

It has been shown that the photochemical features of phosphazenes[13-16] above 220-230 nm depend strictly on the spectroscopic and photochemical characteristics of the side phosphorus substituents, in view of the spectral transparency of the -P=N- skeleton[10,11,13-16] and on the possible multiple substitution of the phosphazene system[233]. This latter property allows the fine tuning of both the photochemical reactivity and the physical features of phosphazenes.

Photodegradation and photocrosslinking are important processes which arise from macroradicals formed by light absorption[17-19,30-36,53,58,249]. These macroradicals can be formed in cyclic and polymeric derivatives by both photofragmention[22,23] and hydrogen[19,33-35,53,54] or electron[30-32] abstraction processes, thus allowing the occurrence of photografting[16,35,36,39-42,44,169,170] processes, light-induced polymerization reactions[253] and light-promoted surface modification[41,85] reactions. When charge-transfer complexes are formed by doping phosphazenes with electron-acceptors or donors, electric and/or photo-conductivity is generated in originally insulating phosphazene materials[206,210,211]. The possibility of geminally supporting different substituents on the same phosphorus of the phosphazene chain allowed the onset of intramolecular interactions between chromophores able to enhance[30-32,34,36,53,54] or depress[52] the photoreactivity of these materials or to induce the formation of intramolecular excimers[46-51,226,229]. Finally, cyclophosphazenes are widely utilized as basic substrates for the photochemical production of transparent, flexible and abrasion-resistant cyclomatrix polymers[99,104,122,128-131,140,142,143,150-153,161,259-261], or as additives for the photopolymerization of vinyl monomers[22,23,28,31,32,34,42,75,103,128,138,142,156,246,253,262] or the photostabilization[28,43,110,244-248,250-252,254-257] of polyolefins, polyesters or other commercial organic macromolecules.

It can be concluded, therefore, that the photochemistry of phosphazenes is one of the most important and investigated topics in modern phosphazene research, still susceptible of impressive progress in the future.

REFERENCES

1. Caglioti, V.; Cordischi, D.; Mele, A. *Nature* **1962**, *195*, 491.
2. Caglioti, V.; Cordisci, D.; Mele, A.; Porta, P., *Proceedings of the Ind. Use Large Radiation Sources*, Salzburg, Austria, 1964, Vol. 1, p 233, *Chem. Abstr.* 60, 10798 (1964).
3. Cordisci, D.; Delle Site, A.; Mele, A.; Porta, P. *J. Macromol. Chem.* **1966**, *1*, 219.
4. Liu, H. Q.; Stannett, V. T. *Macromolecules* **1990**, *23*, 140.
5. Spindler, M. W.; Vale, R. L. *Makromol. Chem.* **1961**, *43*, 231.
6. Dishon, B. R.; Hirshberg, Y. *J. Polym. Sci.* **1949**, *4*, 75.
7. Konecny, J. O.; Douglas, C. M. *J. Polym. Sci.* **1959**, *36*, 195.
8. Konecny, J. O.; Douglas, C. M.; Gray, M. Y. *J. Polym. Sci.* **1960**, *42*, 383.
9. Allcock, H. R. *Phosphorus-Nitrogen Compounds. Cyclic, Linear, and High Polymeric Systems*; Academic Press: New York, 1972, p 296.
10. Allcock, H. R. ibidem p 368.
11. Allcock, H. R. *Angew. Chem. Int. Ed. Engl.* **1977**, *16*, 147.
12. Allcock, H. R., In *Inorganic and Organometallic Polymers with Special Properties*; Laine, R. M., Ed.; Nato Asi Series E: Applied Sciences, 1992; Vol. 206; p 43.
13. Gleria, M.; Bortolus, P.; Minto, F.; Flamigni, L., ibidem p 375.
14. Bortolus, P.; Gleria, M. *J. Inorg. Organomet. Polym.* **1994**, *4*, 1.
15. Bortolus, P.; Gleria, M. *J. Inorg. Organomet. Polym.* **1994**, *4*, 95.
16. Bortolus, P.; Gleria, M. *J. Inorg. Organomet. Polym.* **1994**, *4*, 205.
17. Bortolus, P.; Minto, F.; Beggiato, G.; Lora, S. *J. Appl. Polym. Sci.* **1979**, *24*, 285.
18. Gleria, M.; Minto, F.; Lora, S.; Bortolus, P. *Eur. Polym. J.* **1979**, *15*, 671.
19. Gleria, M.; Minto, F.; Flamigni, L.; Bortolus, P. *Macromolecules* **1987**, *20*, 1766.
20. Minto, F.; Gleria, M.; Scoponi, M.; Pradella, F.; Bortolus, P. *J. Inorg. Organomet. Polym.* **1992**, *2*, 405.

21. Scoponi, M.; Pradella, F.; Carassiti, V.; Gleria, M.; Minto, F. *Makromol. Chem.* **1993**, *194*, 3047.
22. Gleria, M.; Facchin, G.; Bertani, R.; Minto, F. Italian Patent, 1,270,949 (1997), assigned to Consiglio Nazionale delle Ricerche.
23. Fantin, G.; Medici, A.; Fogagnolo, M.; Gleria, M.; Minto, F. Italian Patent, 1,265,090 (1996), assigned to Consiglio Nazionale delle Ricerche.
24. Gleria, M.; Minto, F. Italian Patent, 1,240,644 (1993), assigned to Consiglio Nazionale delle Ricerche.
25. Allcock, H. R.; Kim, C. *Macromolecules* **1991**, *24*, 2846.
26. Allcock, H. R.; Kim, C. *Department of Chemistry, Pennsylvania State University, University Park, PA*, Report No AFOSR-TR-91-0649, availble NTIS AD-A238 719 **1991**, Chem. Abtr. 117, 234861t (1992).
27. Gleria, M.; Minto, F.; Lora, S.; Busulini, L.; Bortolus, P. *Macromolecules* **1986**, *19*, 574.
28. Gleria, M.; Minto, F.; Flamigni, L.; Bortolus, P. *Polym. Degrad. Stab.* **1988**, *22*, 125.
29. Gleria, M.; Minto, F.; Lora, S.; Bortolus, P.; Ballardini, R. *Macromolecules* **1981**, *14*, 687.
30. Gleria, M.; Minto, F.; Flamigni, L. Italian Patent, 1,231,179 (1991), assigned to Consiglio Nazionale delle Ricerche.
31. Gleria, M.; Minto, F.; Flamigni, L. Eur. Pat. Appl. EP, 410,287 (1991), *Chem. Abstr.* 115, 123861g (1991), assigned to Consiglio Nazionale delle Ricerche.
32. Gleria, M.; Minto, F.; Flamigni, L. U.S. US, 5,260,103 (1993), *Chem. Abstr.* 115, 123861g (1991), assigned to Consiglio Nazionale delle Ricerche.
33. Gleria, M.; Minto, F.; Flamigni, L. Eur. Pat. Appl. EP, 369,398 (1990), *Chem. Abstr.* 113, 181450s (1990), assigned to Consiglio Nazionale delle Ricerche.
34. Gleria, M.; Minto, F.; Flamigni, L. U.S. US, 5,041,524 (1991), *Chem. Abstr.* 113, 181450s (1990), assigned to Consiglio Nazionale delle Ricerche.
35. Minto, F.; Pegoretti, A.; Fambri, L.; Gleria, M. *Phosphorus Res. Bull.* **1999**, *10*, 752.
36. Minto, F.; Gleria, M.; Pegoretti, A.; Fambri, L. *Macromolecules* **2000**, *33*, 1173.
37. Gleria, M.; Minto, F.; Bortolus, P.; Porzio, W.; Bolognesi, A. *Eur. Polym. J.* **1989**, *25*, 1039.
38. Gleria, M.; Minto, F.; Bortolus, P.; Porzio, W.; Meille, S. V. *Eur. Polym. J.* **1990**, *26*, 315.
39. Minto, F.; Scoponi, M.; Fambri, L.; Gleria, M.; Bortolus, P.; Pradella, F. *Eur. Polym. J.* **1992**, *28*, 167.
40. Minto, F.; Gleria, M.; Bortolus, P.; Scoponi, M.; Pradella, F., *Proceedings of the 3rd International Congress on Polymer Photochemistry*, Sestri Levante (Genoa), Italy, 5-10th September, 1993, p 162.
41. Minto, F.; Scoponi, M.; Gleria, M.; Pradella, F.; Bortolus, P. *Eur. Polym. J.* **1994**, *30*, 375.
42. Minto, F.; Gleria, M.; Bortolus, P.; Scoponi, M.; Pradella, F.; Fambri, L. In *Current Trends in Polymer Photochemistry*; Allen, N. S., Edge, M., Bellobono, I. R., Selli, E., Eds.; Printice Hall, UK, 1995; Chapt. 11, p 165.
43. Gleria, M.; Minto, F.; Scrima, R.; Borzatta, V. *J. Appl. Polym. Sci.* **1996**, *61*, 1675.
44. Minto, F.; Fambri, L.; Gleria, M. *Macromol. Chem. Phys.* **1996**, *197*, 3099.
45. Allcock, H. R. *Phosphorus-Nitrogen Compounds. Cyclic, Linear, and High Polymeric Systems*; Academic Press: New York, USA, 1972.

46. Gleria, M.; Barigelletti, F.; Dellonte, S.; Lora, S.; Minto, F.; Bortolus, P. *Chem. Phys. Lett.* **1981**, *83*, 559.
47. Hargreaves, J. S.; Webber, S. E. *Polym. Photochem.* **1982**, *2*, 359.
48. Yeung, A. S.; Frank, C. W.; Singler, R. E. *ACS Polym. Prep.* **1986**, *27(2)*, 214.
49. Yeung, A. S.; Frank, C. W.; Singler, R. E. *Polymer* **1990**, *31*, 1092.
50. Hoyle, C. E.; Wisian-Neilson, P.; Chatterton, P. M.; Trapp, M. A. *Macromolecules* **1991**, *24*, 2194.
51. Yeung, A. S.; Frank, C. W.; Singler, R. E. *Macromolecules* **1991**, *24*, 5539.
52. Minto, F.; Flamigni, L.; Bortolus, P.; Gleria, M. *J. Inorg. Organomet. Polym.* **1991**, *1*, 53.
53. Gleria, M.; Minto, F.; Flamigni, L.; Bortolus, P. *J. Inorg. Organomet. Polym.* **1992**, *2*, 329.
54. Minto, F.; Gleria, M.; Bertani, R.; Di Noto, V.; Vidali, M. *J. Inorg. Organomet. Polym.* **1998**, *8*, 67.
55. Allcock, H. R. *Chem. Mater.* **1994**, *6*, 1476.
56. Allcock, H. R. *Department of Chemistry, Pennsylvania State University, University Park, PA*, TR-23, available NTIS AD-A279696 **1994**, *Chem. Abstr.* 123, 341925n (1995).
57. Hoyle, C. E.; Creed, D.; Subramanian, P.; Chatterton, P.; Rufus, I. B.; Wisian-Neilson, P. *ACS Polym. Prep.* **1993**, *34(1)*, 276.
58. Hoyle, C. E.; Creed, D.; Subramanian, P.; Rufus, I. B.; Chatterton, P.; Bahadur, M.; Wisian-Neilson, P. In *Inorganic and Organometallic Polymers II. Advanced Materials and Intermerdiates*; USA ed.; Wisian-Neilson, P., Allcock, H. R., Wynne, K. J., Eds.; ACS Symposium Series: Washington, DC, 1994; Vol. 572, pp 324.
59. Hiraoka, H.; Chiong, K. N. *J. Vac. Sci. Technol.* **1987**, *B5*, 386.
60. Yamashita, Y.; Ito, T.; Kajiwara, N. Jpn. Kokai Tokkyo Koho JP, 01 44,927 [89 44,927] (1989), *Chem. Abstr.* 111, 48164u (1989), assigned to Oki Electric Industry Co., Ltd.
61. Yamashita, Y.; Ito, T.; Kajiwara, N. Jpn. Kokai Tokkyo Koho JP, 01 44,935 [95 44,935] (1989), *Chem. Abstr.* 111, 48165v (1989), assigned to Oki Electric Industry Co., Ltd.
62. Hashimoto, K.; Koizumi, T.; Kitigawa, K.; Nomura, N. Jpn. Kokai Tokkyo Koho JP, 02 10,357 [90 10,357] (1990), *Chem. Abstr.* 113, 14832m (1990), assigned to Matsushita Electric Industrial Co., Ltd.
63. Choung, Y. S.; Lee, K. H.; Lee, D. C. *Polym. Eng. Sci* **1999**, *39*, 1153.
64. Nelson, C. J.; Coggio, W. D.; Allcock, H. R. *Chem. Mater.* **1991**, *3*, 786.
65. Nelson, C. J.; Coggio, W. D.; Allcock, H. R. *Department of Chemistry, Pennsylvania State University, University Park, PA*, TR-3, Available NTIS AD-A241746 **1992**, *Chem. Abstr.* 118, 255465c (1993).
66. Shinohara, H.; Sakuta, N. Jpn. Kokai Tokkyo Koho JP, 04 145,143 [92 145,143] (1992), *Chem. Abstr.* 118, 23216u (1993), assigned to Tokuyama Soda Co., Ltd.
67. Allcock, H. R.; Nelson, C. J.; Coggio, W. D. *Polym. Mater. Sci. Eng.* **1993**, *68*, 76.
68. Allcock, H. R.; Nelson, C. J.; Coggio, W. D. U.S. US, 5,414,025 (1995), *Chem. Abstr.* 123, 61305h (1995), assigned to The Penn State Research Foundation.
69. Kajiwara, M.; Kimura, T. *Int. Conf. Compos. Mater. Energy* **1995**, 26, *Chem. Abstr.* 124, 119365s (1996).
70. Kajiwara, M.; Kimura, T. *J. Mater. Sci.* **1996**, *31*, 5457.

71. Wycisk, R.; Pintauro, P. N.; Wang, W.; O'Connor, S. *J. Appl. Polym. Sci.* **1996**, *59*, 1607.
72. Wycisk, R.; Pintauro, P. N.; Wang, W.; O'Connor, S. *Tulane University, New Orleans, LA, USA*, available NTIS AD-A305441 **1996**, *Chem. Abstr.* 125, 197150c (1996).
73. Graves, R.; Pintauro, P. N. *J. Appl. Polym. Sci.* **1998**, *68*, 827.
74. Adams, B. E.; Quinn, E. J.; Rodgers, W. R. Ger. Offen., 2,948,127 (**1980**), *Chem. Abstr.* 94, 32391k (1981), assigned to Armstrong Cork Co.
75. Adams, B. E.; Quinn, E. J.; Rodgers, W. R. Brit. UK Pat. Appl., 2,046,627 (1980), *Chem. Abstr.* 94, 210466y (1981), assigned to Armstrong World Industries, Inc.
76. Allcock, H. R.; Cameron, T. S. *Department of Chemistry, Pennsylvania State University, University Park, PA*, Report No. TR-13, available NTIS AD-A264915 **1993**, *Chem. Abstr.* 122, 56734x (1995).
77. Allcock, H. R.; Cameron, C. G. *Macromolecules* **1994**, *27*, 3131.
78. Allcock, H. R.; Cameron, C. G. *Macromolecules* **1994**, *27*, 3125.
79. Wynne, K. J. *Department of Chemistry, Pittsburgh University, PA*, TR-21, available NTIS AD-A279782 **1994**, *Chem. Abstr.* 123, 341128e (1995).
80. Allcock, H. R.; Cameron, C. G.; Smith, D. E. PTC Int. Appl. WO, 95 28,150 (1995), *Chem. Abstr.* 124, 89025m (1995), assigned to The Penn State Research Foundation.
81. Ferrar, W. T.; O'Brien, J. P.; Allcock, H. R. *ACS Polym. Prep.* **1978**, *19(2)*, 557.
82. O'Brien, J. P.; Ferrar, W. T.; Allcock, H. R. *Macromolecules* **1979**, *12*, 108.
83. Hiraoka, H.; Lee, W. Y.; Welsh, L. W.; Allen, R. W. *Macromolecules* **1979**, *12*, 753.
84. Scoponi, M.; Pradella, F.; Gleria, M.; Minto, F. *J. Inorg. Organomet. Polym.* **1996**, *6*, 325.
85. Minto, F.; Gleria, M.; Bortolus, P.; Daolio, S.; Facchin, B.; Pagura, C.; Bolognesi, A. *Eur. Polym. J.* **1989**, *25*, 49.
86. Gleria, M.; Minto, F.; Fambri, L.; Pegoretti, A. *Eur. Polym. J.* **1995**, *31*, 791.
87. Chang, S. C.; De Poortere, M. Eur. Pat. Appl. EP, 334,090 A2 (**1989**), *Chem. Abstr.* 112, 99561z (1990), assigned to Ethyl Corporation.
88. De Poortere, M.; Chang, S. C. U.S. US, 5,001,203 (**1991**), *Chem. Abstr.* 111, 78886y (1989), assigned to Ethyl Corp.
89. Kolich, C. H.; Chang, S. C. U.S. US, 5,102,963 (**1992**), *Chem. Abstr.* 117, 49478k (1992), assigned to Ethyl Corp.
90. Allcock, H. R.; Smith, D. E.; Kim, Y. B.; Fitzgerald, J. J. *Macromolecules* **1994**, *27*, 5206.
91. Ahn, K. D.; Kim, U. Y.; Kim, C. H. *J. Macromol. Sci., Chem.* **1986**, *A23*, 169.
92. Nakacho, Y.; Tada, Y.; Yamada, S. Eur. Pat. Appl. EP, 0 294 495 B1 (**1987**), *Chem. Abstr.* 110, 96027c (1989), assigned to Otsuka Kagaku Yakuhin.
93. Nakacho, Y.; Tada, Y.; Yamada, S. PCT Int. Appl. WO, 88 05,064 (**1989**), *Chem. Abstr.* 110, 96027c (1989), assigned to Otsuka Chemical Co., Ltd.
94. Nakacho, Y.; Tada, Y.; Yamada, S. U.S. US, 4,840,856 (**1989**), *Chem. Abstr.* 110, 96027c (1989), assigned to Otsuka Kagaku Kabushiki Kaisha.
95. Allcock, H. R.; Cameron, C. G.; Smith, D. E. U.S. US, 5,464,932 (**1995**), *Chem. Abstr.* 110, 89025m (1996), assigned to The Penn State Research Foundation.
96. De Jaeger, R.; Gleria, M. *Prog. Polym. Sci.* **1998**, *23*, 179.
97. Quella, F.; Nuyken, O.; Budde, K.; Suefke, T. U.S. US, 4,925,772 (**1990**), *Chem. Abstr.* 114, 153928f (1991), assigned to Siemens Aktiengesellschaft.

98. Nuyken, O.; Suefke, T.; Budde, K.; Quella, F. Eur. Pat. Appl. EP, 318,020 (**1990**), *Chem. Abstr.* 114, 153928f (1991), assigned to Siemens Aktiengesellschaft.
99. Mori, S.; Yaguchi, A. Jpn. Kokai Tokkyo Koho JP, 01 207,986 [89 207,986] (**1989**), *Chem. Abstr.* 112, 120262q (1990), assigned to Idemitsu Petrochemical Co. L.t.d.
100. Frank, V. S.; Stahly, E. E.; Rice, R. G. US Patent, 3,676,311 (**1972**), *Chem. Abstr.* 77, 102633r (1972), assigned to Grace W.R., and Co.
101. Chen-Chang, Y. W.; Chuang, J. R.; Yang, Y. C. *ACS Polym. Prep.* **1996**, *37(2)*, 340.
102. Chen-Yang, Y. W.; Chuang, J. R.; Yang, Y. C.; Li, C. Y.; Chiu, Y. S. *J. Appl. Polym. Sci.* **1998**, *69*, 115.
103. Kurachi, I.; Shibue, T.; Nagayasu, K. Jpn. Kokai Tokkyo Koho JP, 06 220,223 [94 220,223] (**1994**), *Chem. Abstr.* 122, 136263h (1995), assigned to Konishiroku Photo Ind.
104. Mori, S.; Yaguchi, A. Eur. Pat. Appl. EP, 307,861 (**1989**), *Chem. Abstr.* 111, 116946c (1989), assigned to Idemitsu Petrchemical Co. L.t.d.
105. Ando, H. Jpn. Kokai Tokkyo Koho JP, 02 284,925 [90 284,925] (**1990**), *Chem. Abstr.* 114, 187703u (1991), assigned to Idemitsu Petrochemical Co. L.t.d.
106. Ando, H.; Kurahashi, A. Jpn. Kokai Tokkyo Koho JP, 02 284,904 [90 284,904] (1990), *Chem. Abstr.* 114, 187671g (1991), assigned to Idemitsu Petrochemical Co. L.t.d.
107. Kurahashi, A.; Kitayama, M. Eur. Pat. Appl. EP, 394,918 (1990), *Chem. Abstr.* 114, 166457b (1991), assigned to Idemitsu Petrochemical Co. L.t.d.
108. Yaguchi, A. Eur. Pat. Appl. EP, 400,625 (1990), *Chem. Abstr.* 114, 165916p (1991), assigned to Idemitsu Petrochemical Co. L.t.d.
109. Yaguchi, A. Jpn. Kokai Tokkyo Koho JP, 02 133,447 [90 133,447] (1990), *Chem. Abstr.* 113, 117227x (1990), assigned to Idemitsu Petrochemical Co. L.t.d.
110. Yaguchi, A.; Funaki, K. Eur. Pat. Appl. EP, 376,021 (1990), *Chem. Abstr.* 114, 208825c (1991), assigned to Idemitsu Petrochemical Co. L.t.d.
111. Yaguchi, A. Jpn. Kokai Tokkyo Koho JP, 03 26,536 [91 26,546] (1991), *Chem. Abstr.* 115, 10485u (1991), assigned to Idemitsu Petrochemical Co. L.t.d.
112. Yaguchi, A. Jpn. Kokai Tokkyo Koho JP, 03 115,464 [91 115,464] (1991), *Chem. Abstr.* 115, 209437k (1991), assigned to Idemitsu Petrochemical Co. L.t.d.
113. Kurahashi, A.; Kageyama, A. Eur. Pat. Appl. EP, 443,626 (1991), *Chem. Abstr.* 117, 100917g (1992), assigned to Idemitsu Petrochemical Co. L.t.d.
114. Ando, H. Eur. Pat. Appl. EP, 557,943 (1993), *Chem. Abstr.* 120, 194166y (1993), assigned to Idemitsu Petrochemical Co. L.t.d.
115. Yaguchi, A. Jpn. Kokai Tokkyo Koho JP, 04 132,772 [92 132,772] (1992*), Chem. Abstr.* 118, 40920m (1993), assigned to Idemitsu Petrochemical Co. L.t.d.
116. Kurahashi, A. PCT Int. Appl. WO, 93 07,540 (1993), *Chem. Abstr.* 120, 65829t (1994), assigned to Idemitsu Petrochemical Co. L.t.d.
117. Kurahashi, A. Eur. Pat. Appl. EP, 564,655 (1993), *Chem. Abstr.* 120,65829t (1994), assigned to Idemitsu Petrochemical Co., Ltd.
118. Yaguchi, A. U.S. US, 5,213,880 (1993), *Chem. Abstr.* 114, 165916p (1991), assigned to Idemitsu Petrochemical Co., Ltd.
119. Hosono, H.; Kurasaki, S.; Taniguchi, T. Jpn. Kokai Tokkyo Koho JP, 05 51,471 [93 51,471] (1993), *Chem. Abstr.* 119, 97605n (1993), assigned to Toray Industries, Inc.
120. Ootomo, R.; Yamamoto, S.; Umezawa, M.; Kikuchi, A. Jpn. Kokai Tokkyo Koho JP, 08 143,755 [96 143,755] (1996), *Chem. Abstr.* 125, 171119z (1996), assigned to Toyo Ink Mfg Co., Japan.

121. Ishigami, K.; Aoyama, Y.; Umi, T.; Saito, A.; Miura, M.; Maeda, M.; Takeda, T.; Hayakawa, J.; Anzai, M.; Ohki, K. *J. Nihon Univ. Sch. Dent.* **1987**, *29*, 221, *Chem. Abstr.* 108, 118908d (1988).
122. Ishigami, K.; Maeda, R.; Maeda, M.; Hamada, H.; Shou, K.; Shimada, A.; Morii, H.; Fukushi, K.; Takeda, T.; Ohki, K. *J. Nihon Univ. Sch. Dent.* **1993**, *35*, 36, *Chem. Abstr.* 119, 103262s (1993).
123. Hosoya, I. Jpn. Kokai Tokkyo Koho JP, 63 266,703 [88 266,703] (1988), *Chem. Abstr.* 111, 25033m (1989), assigned to Idemitsu Petrochemical Co. L.t.d.
124. Kozuka, A.; Ninomyia, Y.; Murakami, K. Jpn. Kokai Tokkyo Koho JP, 04 12,851 [92 12,851] (1992), *Chem. Abstr.* 117, 92553t (1992), assigned to Fuji Kobunshi Co., Ltd.
125. Shiino, M. Jpn. Kokai Tokkyo Koho JP, 04 156,331 [92 156,331] (1992), *Chem. Abstr.* 117, 253517b (1992).
126. Suzuki, S.; Ito, S. Jpn. Kokai Tokkyo Koho JP, 06 126,868 [94 126,898] (1994), *Chem. Abstr.* 121, 233361a (1994), assigned to Hitachi Chemical Co., Ltd.
127. Suzuki, S.; Ebisawa, F.; Ito, S. Jpn. Kokai Tokkyo Koho JP, 07 246,630 [95 246,630] (1995), *Chem. Abstr.* 124, 10639h (1996), assigned to Hitachi Chemical Co., Ltd.
128. Mori, S.; Yaguchi, A. Jpn. Kokai Tokkyo Koho JP, 63 284,268 [88 284,268] (1988), *Chem. Abstr.* 110, 156263p (1989), assigned to Idemitsu Petrochemical Co. L.t.d.
129. Mori, S.; Yaguchi, A.; Kitayama, M. Jpn. Kokai Tokkyo Koho JP, 63 241,075 [88 241,075] (1988), *Chem. Abstr.* 111, 41464h (1989), assigned to Idemitsu Petrchemical Co. L.t.d.
130. Mori, S.; Kitayama, M.; Kurahashi, A. Jpn. Kokai Tokkyo Koho JP, 01 167,338 [89 167,338] (1989), *Chem. Abstr.* 112, 99504h (1990), assigned to Idemitsu Petrochemical Co. L.t.d.
131. Mori, S.; Kitayama, M.; Kurahashi, A. Jpn. Kokai Tokkyo Koho JP, 01 167,337 [89 167,337] (1989), *Chem. Abstr.* 112, 36725k (1990), assigned to Idemitsu Petrochemical Co. L.t.d.
132. Yaguchi, A. Jpn. Kokai Tokkyo Koho JP, 01 256,609 [89 256,609] (1989), *Chem. Abstr.* 112, 181607c (1990), assigned to Idemitsu Petrochemical Co. L.t.d.
133. Yaguchi, A. Jpn. Kokai Tokkyo Koho JP, 01 315,423 [89 315,423] (1989), *Chem. Abstr.* 112, 236079q (1990), assigned to Idemitsu Petrochemical Co. L.t.d.
134. Yaguchi, A. Eur. Pat. Appl. EP, 368,165 (1990), *Chem. Abstr.* 113, 117200h (1990), assigned to Idemitsu Petrochemical Co. L.t.d.
135. Ando, H. Jpn. Kokai Tokkyo Koho JP, 02 142,832 [90 142,832] (1990), *Chem. Abstr.* 113, 153323j (1990), assigned to Idemitsu Petrochemical Co. L.t.d.
136. Ando, H. Jpn. Kokai Tokkyo Koho JP, 03 227,463 [91 227,463] (1991), *Chem. Abstr.* 117, 9928a (1992), assigned to Idemitsu Petrochemical Co. L.t.d.
137. Yaguchi, A. Jpn. Kokai Tokkyo Koho JP, 03 131,658 [91 131,658] (1991), *Chem. Abstr.* 115, 257862s (1991), assigned to Idemitsu Petrochemical Co. L.t.d.
138. Wada, M.; Yokoshima, M. Jpn. Kokai Tokkyo Koho JP, 04 225,026 [92 225,036] (1992), *Chem. Abstr.* 118, 82993t (1993), assigned to Nippon Kayaku Co., Ltd.
139. Tokaji, H.; Matsuda, M.; Komatsu, M.; Koyanagi, T. Jpn. Kokai Tokkyo Koho JP, 05 08,350 [93 08,350] (1993), *Chem. Abstr.* 119, 119046k (1993), assigned to Catalysis and Chemicals Industries Co., Ltd.
140. Mori, S.; Kitayama, M.; Ando, H. Jpn. Kokai Tokkyo Koho JP, 05 202,209 [93 202,209] (1993), *Chem. Abstr.* 120, 32578g (1994), assigned to Idemitsu Petrochemical Co. L.t.d.

141. Kitayama, M. Jpn. Kokai Tokkyo Koho JP, 05 230,442 [93 230,442] (1993), *Chem. Abstr.* 120, 136251m (1994), assigned to Idemitsu Petrochemical Co. L.t.d.
142. Namiki, M.; Mori, S. Jpn. Kokai Tokkyo Koho JP, 05 306,342 [93 306,342] (1993), *Chem. Abstr.* 120, 194194f (1994), assigned to Idemitsu Petrochemical Co. L.t.d.
143. Mori, S.; Namiki, M. Jpn. Kokai Tokkyo Koho JP, 05 286,988 [93 286,988] (1993), *Chem. Abstr.* 121, 11866e (1994), assigned to Idemitsu Petrochemical Co. L.t.d.
144. Yaguchi, A. U.S. US, 5,239,029 (1993), *Chem. Abstr.* 113, 117200h (1990), assigned to Idemitsu Petrochemical Co. L.t.d.
145. Kurahashi, A.; Kitayama, M. U.S. US, 5,280,090 (1994), *Chem. Abstr.* 114, 166457b (1991), assigned to Idemitsu Petrochemical Co. L.t.d.
146. Kurachi, I.; Shibue, T.; Nagayasu, K. Jpn. Kokai Tokkyo Koho JP, 06 220, 224 [94 220,224] (1994), *Chem. Abstr.* 122, 108847h (1995), assigned to Konishiroku Photo Ind.
147. Suzuki, T.; Yoshida, M.; Suzuki, M. Jpn. Kokai Tokkyo Koho JP, 08 118,807 [97 118,807] (1996), *Chem. Abstr.* 125, 127841e (1996), assigned to Tomoegawa Paper Co., Ltd., Japan.
148. Matsui, A.; Hyodo, K.; Seko, M. Jpn. Kokai Tokkyo Koho JP, 08 183,144 [96 183,144] (1996), *Chem. Abstr.* 125, 224150m (1996), assigned to Takiron Co, Japan.
149. Yoshida, T.; Murata, K. Jpn. Kokai Tokkyo Koho JP, 08 239,395 [96 239,395] (1996), *Chem. Abstr.* 125, 331790s (1996), assigned to Nippon Kayaku Kk, Japan.
150. Kitayama, M.; Mori, S. Jpn. Kokai Tokkyo Koho JP, 01 74,229 [89 74,229] (1989), *Chem. Abstr.* 111, 196098r (1989), assigned to Idemitsu Petrochemical Co. L.t.d.
151. Kitayama, M.; Mori, S. Jpn. Kokai Tokkyo Koho JP, 01 74,230 [89 74,230] (1989), *Chem. Abstr.* 111, 196099s (1989), assigned to Idemitsu Petrochemical Co. L.t.d.
152. Mori, S.; Kitayama, M. Jpn. Kokai Tokkyo Koho JP, 01 14,239 [89 14,239] (1989), *Chem. Abstr.* 111, 78785q (1989), assigned to Idemitsu Petrochemical Co. L.t.d.
153. Mori, S.; Kitayama, M. Jpn. Kokai Tokkyo Koho JP, 01 14,240 [89 14,240] (1989), *Chem. Abstr.* 111, 78839k (1989), assigned to Idemitsu Petrchemical Co. L.t.d.
154. Haga, S.; Kizawa, Y.; Shimizu, K. Jpn. Kokai Tokkyo Koho JP, 09 230,109 (1997), *Chem. Abstr.* 127, 294304 (1997), assigned to Daiseru Amihoshi Sangyo K. K., Japan.
155. Yaguchi, A. Jpn. Kokai Tokkyo Koho JP, 02 32,187 [90 32,187] (1990), *Chem. Abstr.* 112, 218469n (1990), assigned to Idemistu Kosan Co., Ltd.
156. Nishide, R.; Shinoda, M. Jpn. Kokai Tokkyo Koho JP, 04 263,931 [92 263,931] (1992), *Chem. Abstr.* 118, 170626f (1993), assigned to Nissan Motor Co., Ltd.
157. Yaguchi, A. Jpn. Kokai Tokkyo Koho JP, 02 160,872 [90 160,872] (1990), *Chem. Abstr.* 114, 45132h (1991), assigned to Idemitsu Petrochemical Co. L.t.d.
158. Kurahashi, A. Jpn. Kokai Tokkyo Koho JP, 04 25,564 [92 25,564] (1992), *Chem. Abstr.* 117, 9964j (1992), assigned to Idemitsu Petrochemical Co. L.t.d.
159. Ogawa, H.; Kuryama, T. Jpn. Kokai Tokkyo Koho JP, 07 294,893 [95 294893] (1995), *Chem. Abstr.* 124, 216259t (1996), assigned to Dainippon Ink & Chemicals Japan.
160. Kagami, K.; Suzuki, M.; Fukui, T.; Kondo, Y. Jpn. Kokai Tokkyo Koho JP, 03 282,469 [91 282,469] (1991), *Chem. Abstr.* 117, 140644x (1992), assigned to Oriental Photo Industrial Co., Ltd.; Canon K.K.
161. Morishita, K. Jpn. Kokai Tokkyo Koho JP, 03 65,390 [91 65,390] (1991), *Chem. Abstr.* 115, 171012c (1991), assigned to Nisshimbo Industries, Inc.

162. Hashimoto, Y.; Koyama, T.; Maruyama, A.; Nagahara, S. Jpn. Kokai Tokkyo Koho JP, 04 353,860 [92 353,860] (1992), *Chem. Abstr.* 119, 82844r (1993), assigned to Canon K.K.
163. Maruyama, K.; Kosho, N. Jpn. Kokai Tokkyo Koho JP, 07 160,031 (1995), *Chem. Abstr.* 123, 270744q (1995),assigned to Fuji Electric Co., Ltd.
164. Shinohara, H.; Sakuta, N. Jpn. Kokai Tokkyo Koho JP, 03 130,297 [91 130,297] (1991), *Chem. Abstr.* 115, 208253k (1991), assigned to Tokuyama Soda Co., Ltd.
165. Sekya, T.; Nishioka, Y.; Okabe, Y.; Okada, Y. Jpn. Kokai Tokkyo Koho JP, 07 179,058 [95 179,058] (1995), *Chem. Abstr.* 123, 325813u (1995), assigned to Oki Electric Industry Co., Ltd.
166. Suzura, Y. Jpn. Kokai Tokkyo Koho JP, 03 275,316 [91 275,316] (1991), *Chem. Abstr.* 116, 153466v (1992), assigned to Dainippon Printing Co., Ltd.
167. Yaguchi, A.; Funaki, K. U.S. US, 5,082,717 (1992), *Chem. Abstr.* 114, 208825c (1991), assigned to Idemitsu Petrochemical Co. L.t.d.
168. Kimura, T.; Kajiwara, M. *Polymer* **1995**, *36*, 713.
169. Allcock, H. R.; Nelson, C. J.; Coggio, W. D. *Chem. Mater.* **1994**, *6*, 516.
170. Minto, F.; Gleria, M.; Bortolus, P.; Fambri, L.; Pegoretti, A. *J. Appl. Polym. Sci.* **1995**, *56*, 747.
171. Gleria, M.; Bolognesi, A.; Porzio, W.; Catellani, M.; Destri, S.; Audisio, G. *Macromolecules* **1987**, *20*, 469.
172. Gleria, M.; Minto, F.; Bortolus, P.; Facchin, G.; Bertani, R. In *Inorganic and Organometallic Polymers II. Advanced Materials and Intermediates*; Wisian-Neilson, P., Allcock, H. R., Wynne, K. J., Eds.; ACS Symposium Series: Washington DC, USA, 1994; Chapt. 22, Vol. 572, p 279.
173. Wisian-Neilson, P.; Schaefer, M. A.; Islam, M. S. *ACS Polym. Prep.* **1989**, *30(1)*, 185.
174. Wisian-Neilson, P.; Islam, M. S.; Schaefer, M. A. *Phosphorus, Sulfur, and Silicon* **1989**, *41*, 145.
175. Wisian-Neilson, P.; Islam, M. S. *Macromolecules* **1989**, *22*, 2026.
176. Garcia-Alonso, F. J.; Wisian-Neilson, P., *Proceedings of the Workshop on the Status and Future of Polyphosphazenes*, Raleigh, NC, USA, March 15-18th, 1992, Poster No. 10.
177. Bailey, L.; Bahadur, M.; Wisian-Neilson, P. *ACS Polym. Prep.* **1993**, *34(1)*, 318.
178. Lora, S.; Carenza, M.; Palma, G.; Pezzin, G.; Caliceti, P.; Battaglia, P.; Lora, A. *Biomaterials* **1991**, *12*, 275.
179. Lora, S.; Carenza, M.; Palma, G.; Caliceti, P. *Radiat. Phys. Chem.* **1990**, *35*, 117.
180. Lora, S.; Palma, G.; Carenza, M.; Caliceti, P.; Pezzin, G. *Biomaterials* **1994**, *15*, 937.
181. Pezzin, G.; Palma, G.; Lora, S.; Carenza, M. *Chem. Ind. (Milan)* **1991**, *73*, 185.
182. Allcock, H. R.; Fitzpatrick, R. J.; Visscher, K.; Salvati, L. *Chem. Mater.* **1992**, *4*, 775.
183. Minto, F.; Borzatta, V.; Gleria, M. *J. Inorg. Organomet. Polym.* **1996**, *6*, 171.
184. Fambri, L.; Minto, F.; Gleria, M. *J. Inorg. Organomet. Polym.* **1996**, *6*, 195.
185. Gleria, M.; Minto, F.; Scoponi, M.; Pradella, F.; Carassiti, V. *Chem. Mater.* **1992**, *4*, 1027.
186. Gleria, M.; Minto, F.; Fontana, G.; Bertani, R.; Facchin, G. *Macromolecules* **1995**, *28*, 4399.
187. Gleria, M.; Minto, F.; Doriguzzi, F.; Bertani, R.; Facchin, G.; Tondello, E. *Macromolecules* **1997**, *30*, 4310.

188. Gleria, M.; Minto, F.; Bortolus, P.; Facchin, G.; Bertani, R.; Scoponi, M.; Pradella, F. *ACS Polym. Prep.* **1993**, *34(1)*, 270.
189. Trivedi, B. C.; Culberstone, B. M. *Maleic Anhydride*; Plenum Press: New York, USA, 1982.
190. Minto, F.; Borzatta, V.; Bertani, R.; Gleria, M. *J. Appl. Polym. Sci.* **1997**, *65*, 217.
191. Minto, F.; Gleria, M.; Favero, G. *J. Macromol. Sci., Pure Appl. Chem.* **1997**, *A34*, 2177.
192. Gleria, M.; Minto, F.; Bertani, R.; Facchin, G.; Tondello, E. *Gazz. Chim. Ital.* **1997**, *127*, 343.
193. Soane, D. S.; Martynenko, Z. *Polymers in Microelectronics*; Elsevier: Amsterdam, Netherland, T., 1989.
194. Grune, G. L.; Greer, R. W.; Chern, R. T.; Stannett, V. T. *J. Macromol. Sci., Pure Appl. Chem.* **1994**, *A31*, 1193.
195. Welker, M. F.; Allcock, H. R.; Grune, G. L.; Chern, R. T.; Stannett, V. T. *Polym. Mater. Sci. Eng.* **1992**, *66*, 259.
196. Grune, G. L.; Stannett, V. T.; Chern, R. T.; Harada, J. *Polym. Adv. Technol.* **1993**, *4*, 341.
197. Welker, M. F.; Allcock, H. R.; Grune, G. L.; Chern, R. T.; Stannett, V. T. *Department of Chemistry, Pennsylvania State University, University Park, PA*, Report No. TR-9, available NTIS, AD-A264914 **1993**, *Chem. Abstr.* 122, 57992u (1995).
198. Stannett, V. T.; Grune, G. L.; Chern, R. T.; Allcock, H. R. *ACS Polym. Prep.* **1994**, *35(2)*, 876.
199. Welker, M. F.; Allcock, H. R.; Grune, G. L.; Chern, R. T.; Stannett, V. T. *ACS Symp. Ser.* **1994**, *537*, 293.
200. Welker, M. F.; Allcock, H. R.; Grune, G. L.; Chern, R. T.; Stannett, V. T. *Department of Chemistry, Pennsylvania State University, University Park, PA*, TR-18, available NTIS No. AD-A279694 **1994**, *Chem. Abstr.* 123, 325423s (1995).
201. Wisian-Neilson, P.; Bailey, L.; Bahadur, M. *Macromolecules* **1994**, *27*, 7713.
202. Lynch, T. J.; Durham, D.; Sobodacha, C. PCT Int. Appl. WO, 93 12,468 (1993), *Chem. Abstr.* 120, 65911p (1994), assigned to Hoechst Celanese Corp.
203. Kim, S. J.; Kim, K. D.; Lee, H.; Lee, D. Y. U.S. US, 5,523,191 (1996), *Chem. Abstr.* 125, 154379s (1996), assigned to Korea Kumho Chemical Co., Ltd.
204. Ahn, D. K.; Kang, J. H.; Kim, S. J.; Park, B. S.; Park, C. E.; Park, C. G. *J. Photopolym. Sci. Technol.* **1992**, *5*, 67.
205. Fantin, G.; Gleria, M.; Medici, A.; Minto, F.; Fogagnolo, M. Italian Patent, 1,264,677 (1996), assigned to Consiglio Nazionale delle Ricerche.
206. Di Marco, P. G.; Giro, G.; Lora, S.; Gleria, M. *Mol. Cryst. Liq. Cryst.* **1985**, *118*, 439.
207. Beggiato, G.; Casalbore-Miceli, G.; Giro, G.; Lora, S. *Eur. Polym. J.* **1986**, *22*, 245.
208. Fattori, V.; Bürge, G.; Geri, A. *J. Imaging Technol.* **1986**, *12*, 334.
209. Allcock, H. R.; Austin, P. E.; Neenan, T. X.; Sisko, J. T.; Blonsky, P. M.; Shriver, D. F. *Macromolecules* **1986**, *19*, 1508.
210. Di Marco, P. G.; Giro, G.; Gleria, M.; Lora, S. *Thin Solid Films* **1986**, *135*, 157.
211. Di Marco, P. G.; Giro, G.; Gleria, M.; Lora, S. Italian Patent, 1,226,472 (1991), assigned to Consiglio Nazionale delle Ricerche.
212. Badesha, S. S.; Stolka, M.; Pai, D. M.; Abkowitz, M. A. U.S. US, 5,004,663 (1991), *Chem. Abstr.* 116, 95651p (1992), assigned to Xerox Co.

213. Yu, R. C. U.; Badesha, S. S. U.S. US, 5,008,169 (1991), *Chem. Abstr.* 116, 95652q (1992), assigned to Xerox Co.
214. Lora, S.; Minto, F.; Pezzin, G.; Giro, G.; Di Marco, P. G. Eur. Pat. Appl. EP, 200,854 A2 (1986), *Chem. Abstr.* 106, 177126d (1987), assigned to Consiglio Nazionale delle Ricerche.
215. Lora, S.; Minto, F.; Di Marco, P. G.; Giro, G.; Pezzin, G. Italian Patent, 1,183,333 (1987), assigned to Consiglio Nazionale delle Ricerche.
216. Lora, S.; Minto, F.; Di Marco, P. G.; Giro, G.; Pezzin, G. U.S. US, 4,657,993 (1987), *Chem. Abstr.* 106, 177126d (1987), assigned to Consiglio Nazionale delle Ricerche.
217. Zhan, C.; Wu, J.; Li, X.; Yang, X.; Qin, J. *Proc. SPIE-Int. Soc. Opt. Eng.* **1998**, *3554*, 252, *Chem. Abstr.* 130, 139742q (1999).
218. Neenan, T. X.; Allcock, H. R. *Biomaterials* **1982**, *3*, 78.
219. Facchin, G.; Minto, F.; Gleria, M.; Bertani, R.; Bortolus, P. *J. Inorg. Organomet. Polym.* **1991**, *1*, 389.
220. *Photochromism. Molecules and Systems*; Dürr, H.; Bouas-Laurent, H., Eds.; Elsevier: Amsterdam, 1990.
221. Allcock, H. R.; Manners, I.; Mang, M. N.; Parvez, M. *Inorg. Chem.* **1990**, *29*, 522.
222. Adrover, V.; Bertrand, G.; Majoral, J. P.; Rolland, H.; Potin, P. Fr. Demande FR, 2,682,386 (1993), *Chem. Abstr.* 120, 31595y (1994), assigned to Atochem.
223. Rolland, H.; Potin, P.; Majoral, J. P.; Bertrand, G. *Phosphorus, Sulfur and Silicon Relat. Elem.* **1993**, *76*, 471.
224. Rolland, H.; Potin, P.; Majoral, J. P.; Bertrand, G. *Inorg. Chem.* **1993**, *32*, 4679.
225. Rolsma, P. B.; Lee, J. N. *Opt. Lett.* **1990**, *15*, 712.
226. Hoyle, C. E.; Creed, D.; Subramanian, P.; Rufus, I. B.; Chatterton, P.; Bahadur, M.; Wisian-Neilson, P. *Macromolecules* **1994**, *27*, 4605.
227. Bortolus, P.; Dellonte, S.; Gleria, M.; S., L.; Minto, F., *Proceedings of the International Symposium on Macromolecules, IUPAC Macro Florence*, Florence, Italy, 7-12th September, 1980, Vol. 4, p 119.
228. Birks, J. B.; Dyson, D. J.; Munro, I. H. *Proc. Roy. Soc.* **1963**, *A275*, 575.
229. Flamigni, L.; Camaioni, N.; Bortolus, P.; Minto, F.; Gleria, M. *J. Phys. Chem.* **1991**, *95*, 971.
230. Itaya, T.; Sasaki, Y.; Tanigaki, T.; Matsumoto, A.; Inoue, K. *Chem. Lett.* **1996**, 305.
231. Inoue, K.; Sasaki, Y.; Itaya, T.; Tanigaki, T. *Eur. Polym. J.* **1997**, *33*, 841.
232. Allcock, H. R. *Acc. Chem. Res.* **1979**, *12*, 351.
233. Gleria, M. *Res. Adv. in Macromolecules* **2000**, *1*, 103.
234. Labarre, J. F. *Top. Curr. Sci.* **1982**, *102*, 1.
235. Labarre, J. F. *Top. Curr. Sci.* **1985**, *129*, 173.
236. Allcock, H. R.; Silverberg, E. N.; Dudley, G. K. *Macromolecules* **1994**, *27*, 1033.
237. Allcock, H. R.; Desorcie, J. L.; Riding, G. H. *Polyhedron* **1987**, *6*, 119.
238. Chandrasekhar, V.; Justin Thomas, K. R. In *Structure and Bonding. Structures and Biological Effects*; Clarke, M. J., Goodenough, J. B., Ibers, J. A., Jørgensen, C. K., Mingos, D. M. P., Neilands, J. B., Palmer, G. A., Reinen, D., Sadler, P. J., Weiss, R., Williams, R. J. P., Eds.; Springer-Verlag: Berlin, Germany, 1993, Vol. 81, p 41.
239. Chandrasekhar, V.; Justin Thomas, K. R. *Appl. Organometallic Chem.* **1993**, *7*, 1.
240. Mitjaville, J.; Caminade, A. M.; Majoral, J. P. *Tetrahedron Lett.* **1994**, *35*, 6865.

241. Hochart, F.; Mouveaux, C.; Levalois-Mitjaville, J.; De Jaeger, R. *Tetrahedron Lett.* **1998**, *39*, 6171.
242. Corda, L.; Delogu, G.; Maccioni, A.; Maccioni, E.; Podda, G. *Gazz. Chim. Ital.* **1995**, *125*, 491.
243. Landini, D.; Maia, A.; Podda, G. *Gazz. Chim. Ital.* **1995**, *125*, 583.
244. Pond, D. M.; Wang, R. S. H. U.S., 3,936,418 (1976), *Chem. Abstr.* 84, 151551r (1976), assigned to Eastman Kodak Co.
245. Pond, D. M.; Wang, R. S. H. Brit. UK Pat. Appl. GB, 1,524,996 (1978), *Chem. Abstr.* 84, 151551r (1976), assigned to Eastman Kodak Co.
246. Gleria, M.; Paolucci, G.; Minto, F.; Lora, S. *Chem. Ind. (Milan)* **1982**, *64*, 479.
247. Wiezer, H. Eur. Pat. Appl. EP, 64,752 (1982), *Chem. Abstr.* 98, 108408w (1983), assigned to Hoechst Aktiengesellschaft.
248. Wiezer, H. US Patent, 4,451,400 (1984), *Chem. Abstr.* 98, 108408w (1983), assigned to Hoechst Aktiengeselleschaft.
249. Gleria, M.; Minto, F.; Bortolus, P.; Lora, S. Italian Patent, 1,176,618 (1987), assigned to Consiglio Nazionale delle Ricerche.
250. Bortolus, P.; Busulini, L.; Lora, S.; Minto, F.; Pezzin, G. Italian Patent, 1,196,213 (1988), assigned to Consiglio Nazionale delle Ricerche.
251. Inoue, K.; Takahata, H.; Tanigaki, T. *J. Appl. Polym. Sci.* **1993**, *50*, 1857.
252. Inoue, K. Jpn. Kokai Tokkyo Koho JP, 06 135,977 [94 135,977] (1994), *Chem. Abstr.* 121, 179870a (1994), assigned to Nippon Shoe.
253. Gleria, M.; Minto, F.; Galeazzi, A.; Scoponi, M. Italian Patent, PD 98 000241 (1998), assigned to Consiglio Nazionale delle Ricerche.
254. Breslow, D. S. US Patent, 3,446,876 (1969), *Chem. Abstr.* 71, 30222g (1969), assigned to Hercules Inc.
255. Breslow, D. S. Brit. UK Pat. Appl. GB, 1,180,475 (1970), *Chem. Abstr.* 71, 30222g (1969), assigned to Hercules Inc.
256. Wang, R. H. S.; Irick, G. US Patent, 4,080,361 (1978), *Chem. Abstr.* 89, 111121x (1978), assigned to Eastman Kodak.
257. Goins, D. E.; Li, H. M. U.S. US, 5,105,001 (1992), *Chem. Abstr.* 117, 61662y (1992), assigned to Ethyl Corp.
258. Hartle, T. J.; McInstosh, M. B.; Allcock, H. R. *ACS Polym. Prep.* **1999**, *40(2)*, 908.
259. Yaguchi, A.; Mori, S.; Kitayama, M.; Onda, T.; Kurahashi, A.; Ando, H. *Thin Solid Films* **1992**, *216*, 123.
260. Mori, S.; Namiki, M. Jpn. Kokai Tokkyo Koho JP, 05 294,980 [93 294,980] (1993), *Chem. Abstr.* 121, 11867f (1994), assigned to Idemitsu Petrochemical Co. L.t.d.
261. Mori, S.; Ando, H. Jpn. Kokai Tokkyo Koho JP, 06 01,869 [94 01,869] (1994), *Chem. Abstr.* 121, 85900h (1994), assigned to Idemitsu Petrochemical Co. L.t.d.
262. Yokoshima, M.; Okubo, T.; Sasahara, K. Jpn. Kokai Tokkyo Koho JP, 04 225,038 [92 225,038] (1992), *Chem. Abstr.* 118, 82988v (1993), assigned to Nippon Kayaku Co., Ltd.

In: Applicative Aspects of Poly(organophosphazenes) ISBN 1-59454-025-X
Editors: M. Gleria and R. De Jaeger, pp. 85-99

Chapter 4

RADIATION CHEMISTRY OF POLYPHOSPHAZENES: FUNDAMENTALS AND APPLICATIONS

Silvano Lora and Mario Carenza
Istituto per la Sintesi Organica e la Fotoreattività, Consiglio Nazionale delle Ricerche, Viale dell'Università, 2 – 35020 Legnaro, Padova, Italy

ABSTRACT

Different reactions induced by high energy radiation, such as polymerization, crosslinking, chain scission and grafting, of polyphosphazenes are reviewed. Biomedical applications of modified polymers are summarized, together with properties and biocompatibility. Other recent applications are also briefly outlined.

INTRODUCTION

The most important high energy radiations used in radiation chemistry or for industrial processess are represented by gamma rays from Co-60 or Cs-137 and electrons produced in electron accelerators. Also X rays (Bremsstrahlung), produced when high energy electrons are decelerated by interaction with matter, *i.e.* on collision with heavy metals such as tungsten or uranium, are used.

Organic compounds exposed to electromagnetic radiation absorb energy and the Compton process, *i.e.* elastic collision between photon and electron, dominates in the region from 0.1 to 10 MeV. As a consequence, an ionized molecule is produced. In addition to the ionization process, also the excitation of the molecule, that requires lesser energy, usually occurs. In the case of fast electrons, the main mechanism of energy loss is the inelastic scattering by atoms or molecules. Except for high atomic number materials and high energy electrons, ionization is the primary means of electron energy loss. These ionized and excited molecules give eventually rise to new chemical species through the formation of free radicals or ions.

In the field of macromolecular science, radiation is used to synthesize high polymer by radiolysis of the corresponding monomer through both free radical and ionic mechanism.

High purity products can be easily obtained since the reaction does not require the presence of chemical initiators or other additives. Moreover, radiation allows the polymerization to be carry out in a wide temperature range.

When a macromolecular compound is irradiated, free radicals undergo different reactions which, depending on the molecular structure of the material and experimental conditions, cause modifications such as main chain scission and crosslinking with the consequence that the molecular weight decreases or increases, respectively. The net effect of these two reactions, often concurrently occurring, depends on the process which predominates under irradiation. Thus, a polymer will preferentially degrade if a tetrasubstituted carbon atom is present in the main chain while it will crosslink if each carbon atom carries at least one hydrogen atom. It follows that different properties, such as physical, mechanical, electrical and rheological properties, can be significatively affected.

Finally, another important modification that can be achieved by radiation is graft copolymerization. When a polymer selected for a given application is irradiated in the presence of a monomer, a material with new properties is obtained. For example, in the case of biomaterials, under proper irradiation conditions it is possible to preserve the original mechanical properties of the polymer and, at the same time, to considerably enhance the biocompatibility if a suitable monomer is used.

The science and technology of radiation of monomeric and polymeric systems is the object of many books or reviews published in the course of many years (1-9).

Polyphosphazenes, POPs, are a relatively new class of inorganic polymers that, due to their extreme versatility, are very interesting from both scientific and technological point of view. The reader interested in the last developments can refer to the other chapters of this book and very recent overviews (10,11).

Our purpose is to discuss the fundamentals of radiation chemistry of POP systems and to emphasize the most promising applications that this technique can offer.

POLYMERIZATION

The main route for synthesis of POPs is the thermal polymerization of the trimer hexachlorocyclotriphosphazene, HCCP, to give the very reactive polydichlorophosphazene followed by substitution of the chlorine atoms with nucleophilic groups.

In the literature a few works have been devoted to radiation-induced polymerization of HCCP. A thorough investigation on this subject was carried out by Liu and Stannett (12) who studied the polymerization induced by Co γ-rays of the trimer in the solid state, in the melt and in the solution in a superdried state after a careful purification.

The results of Figure 1 show a slight higher polymerization conversion with respect to those previously reported by Caglioti *et al.* (13) in the solid state using 50 kV X-rays.

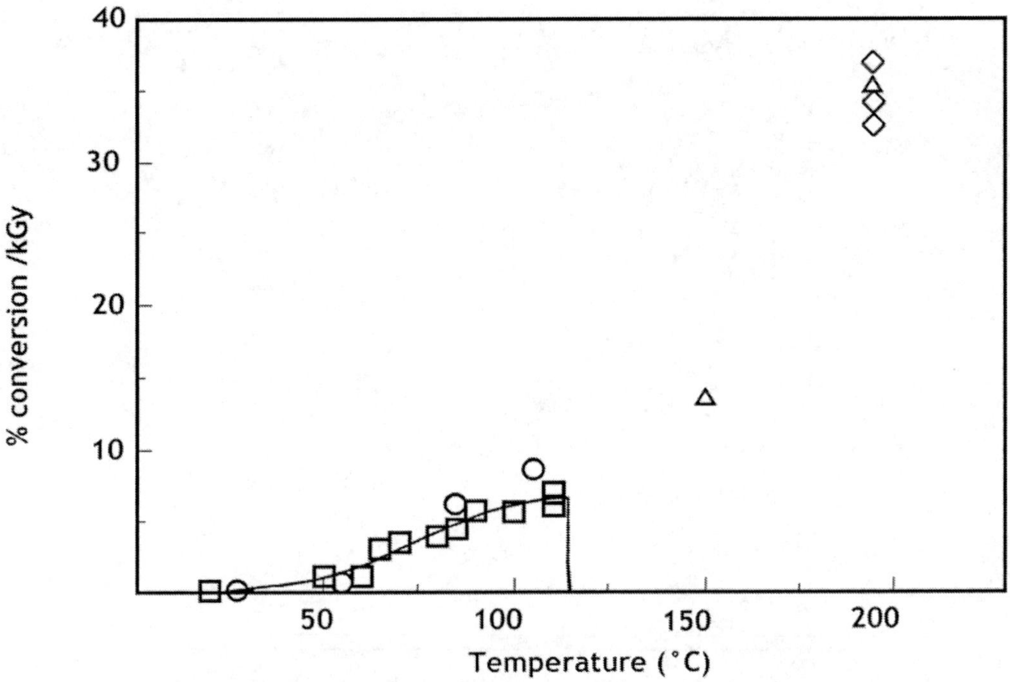

Fig. 1. Radiation-induced polymerization of HCCP: superdried trimer (○); in the melt (△); in 10% decalin solution (◇). For comparison, the results obtained by Caglioti *et al.* (13) are shown (□).

A substantial increment of the conversion was observed if the polymerization was carried out in the melt. However, the results were affected by an unacceptable lack of reproducibility, mainly at temperatures lower than 185°C, with low values of molecular weights.

On the contrary, radiation polymerization of HCCP in a concentrated solution of dry decalin at 195°C enabled a high polymer yield, as well as a good reproducibility, to be attained. Nevertheless, again low molecular weights were obtained, probably owing to chain transfer to monomer.

RADIOLYSIS OF POPS

The effect of γ irradiation of poly[bis(*p*-tolylamino)phosphazene] in the presence of oxygen was investigated (14). The degradation of the polymer dissolved in different solvents as a function of irradiation dose was followed by viscometry. It can be seen from Figure 2

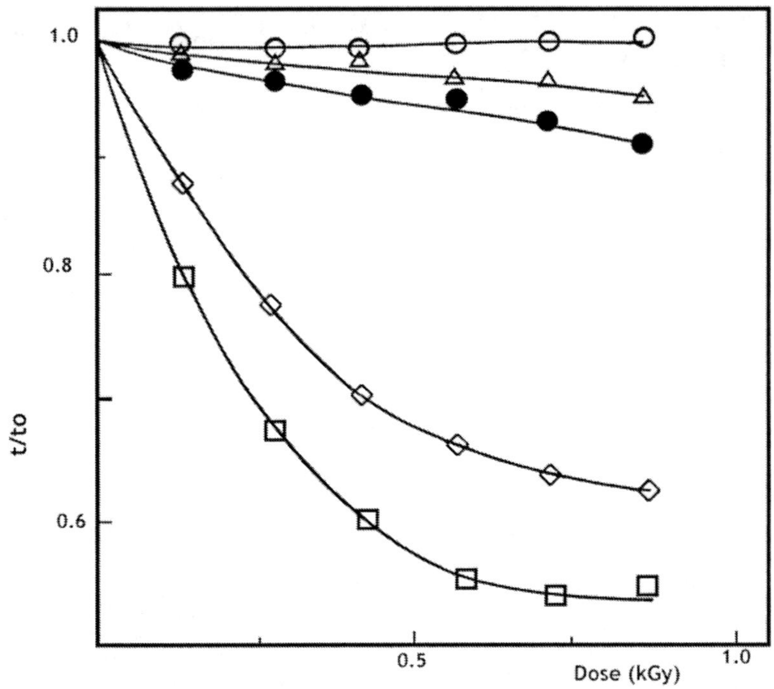

Fig. 2. Ratio of the efflux times after and before irradiation *vs.* dose plots for solutions of poly[bis(*p*-tolylamino)phosphazene] in different solvents: tetrahydrofuran (○); benzene (△); toluene (●); chloroform (◇); carbon tethachloride (□).

that viscosity decreases with all the solvents examined. It was assumed that a peroxide radical, formed by reaction between solvent and oxygen, can react with the polymer giving rise to main chain scission. On the contrary, no change in viscosity was detected in the case of tetrahydrofuran, THF. A possible mechanism of the stabilizing effect of THF consists in the reaction between a POP macroradical with the solvent followed by successive hydrogen abstraction from the THF radical by another POP macroradical. It should be noted that a similar mechanism was put forward for irradiated mixtures of poly(vinyl chloride) with THF which was found to be capable of preventing the polymer from discoloration (15). Infrared analysis gave further evidence for the oxidative character of the process because, after irradiation, a band at 1740 cm^{-1} attributed to carbonyl absorption was observed (14). After irradiation at 0.72 kGy, the molecular weight distribution, M_w/M_n, where M_w and M_n are the weight-average molecular weight and the number-average molecular weight before irradiation, respectively, changed from 16.5 to 4.6, suggesting a main chain scission reaction.

In a subsequent paper, the same workers carried out a study on radiation-induced degradation of dilute solutions of poly[bis(*p*-tolylamino)phosphazene] in methylene chloride using the pulse radiolysis technique (16). From molecular weight determinations of polymer samples irradiated in air saturated solutions, it was ascertained that crosslinking and degradation reactions took place at low and high doses, respectively. Pulse radiolysis technique demonstrated that the chlorine atoms formed by radiation can react with the polymer chain directly or after peroxidation. The aniline N-H and benzyl C-H bonds of the side groups were recognized as the labile sites for radical attack to POP polymer.

The effect of different substituents, such as methyl, ethyl and t-butyl groups, of poly(p-phenoxy phosphazenes) with $M_w/M_n > 10$ on the crosslinking and chain scission reactions was investigated under irradiation (17,18). Similarly to the results previously found by other authors (14), they observed a narrowing of M_w/M_n with increasing radiation dose.

A quantitative evaluation of the crosslinking and chain scission reactions concurrently occurring was obtained by using the well known Charlesby-Pinner equation (1):

$$s + \sqrt{s} = \frac{G(s)}{2G(x)} + \frac{4.82\ 10^6}{G(x)\ M_n\ D}$$

where s is the soluble fraction, $G(s)$ and $G(x)$ are the yields per 100 eV absorbed energy of chain scission and crosslinking, respectively, and D is radiation dose. $G(s)$ and $G(x)$ values showed approximatively the same value. Moreover, the latter is the higher the bulkier alkyl group is.

A thorough investigation of side group effect on some p-phenoxy-substituted POPs irradiated both in an inert atmosphere and in oxygen was carried out in our laboratory (19).

Poly[bis(phenoxy)phosphazene], PPP, poly[bis(4-methylphenoxy)phosphazene], PMP, poly[bis(4-t-butylphenoxy)phosphazene], PTP, and poly(bis(4-cumylphenoxy)phosphazene], PCP, in the form of films cast from a dichloromethane solution were irradiated at different doses and network formation was followed by UV determination of the sol fraction. As shown in Figure 3,

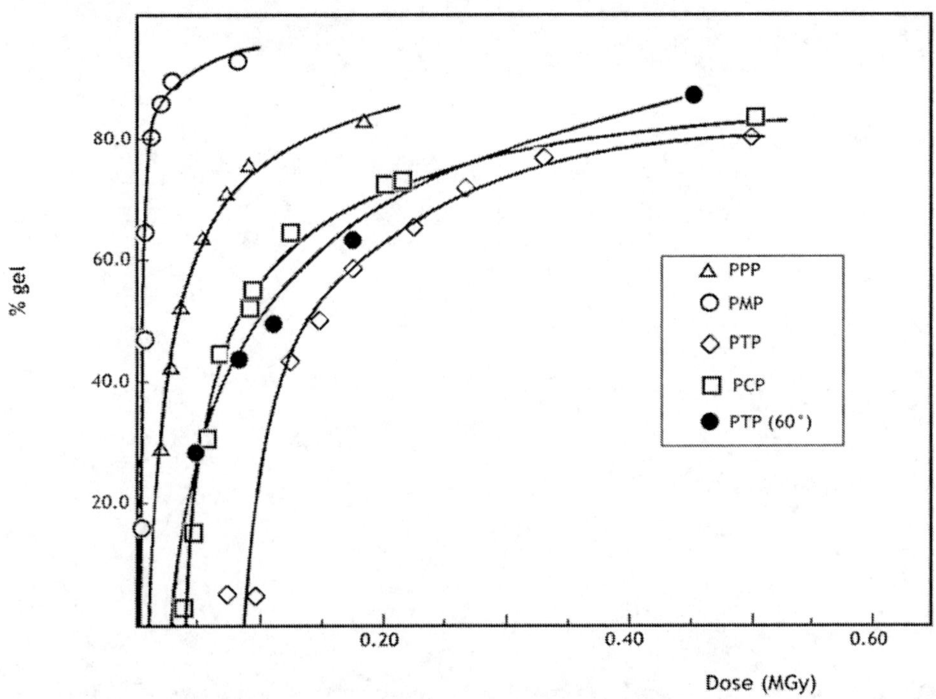

Fig 3. Gel percent *vs.* dose plots for the irradiation carried out under vacuum: at 20°C of PPP, PMP, PTP and PCP.

the gel fraction g, with g = 1-s, increases as a function of dose and the polymer samples exhibit a different susceptibility to form crosslinking, *i.e.* the rate of gel formation decreases in the order: PMP>PPP>PCP>PTP. At the same time, the gel dose values decrease, ranging from 19 kGy for PMP to 73 kGy for PTP. Unexpectedly, some temperature effect was observed because the gel fraction at 60°C for PTP was higher than that found at room temperature (20).

G(s) and *G(x)* values derived from the Charlesby-Pinner equation indicate that in all the examined polymers the chain scission and crosslinking reactions take place approximately in the same ratio. Moreover, it was ascertained a different susceptibility for the two reactions concurrently occurring.

The role played by the different free radicals produced in the radiolysis of the same POPs have also been investigated by ESR analysis (19,21). Formation of alkyl, benzyl and cyclohexadienyl type radicals indicates that most of the radicals is localized in the organic moiety of the polymer chains.

As an example, in Fig. 4

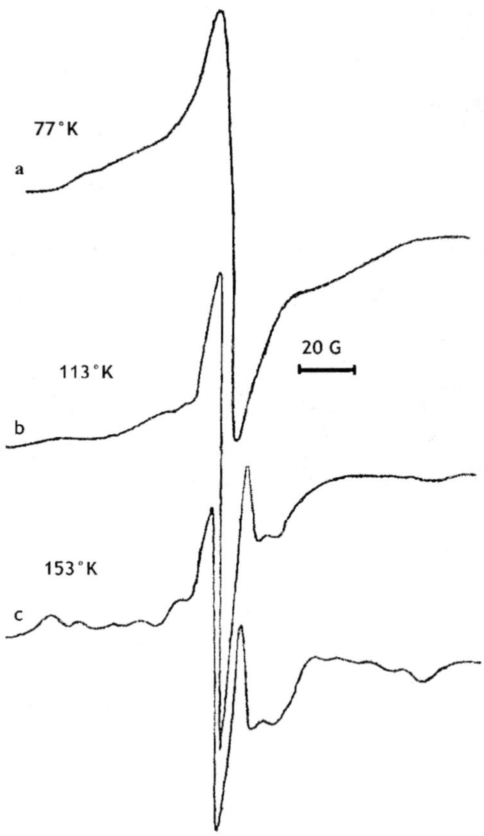

Fig. 4. ESR spectra of irradiated PPP: recorded at 77°K after the irradiation at the same temperature showing the singlet and the broad site patterns attributed to cyclohexadienyl radicals (a); low temperature spectra of cyclohexadienyl radicals from γ-irradiated benzene (b, c).

there are reported the ESR spectra of irradiated PPP at 77°K and recorded at the same temperature. A singlet at g = 2.003 (Fig 3(a)) is the main component of the spectrum that is attributable to phenyl and phenoxy radicals as well as radical ions. The signal is superimposed on a broad unresolved pattern extending over ±50G which, by comparison of low temperature ESR of irradiated solid benzene shown in Fig. 3(b, c), is assigned to cyclohexadienyl radicals (22).

In a recent work, Stewart *et al.* (23) studied the crosslinking induced by electron beam irradiation of fluoroalkoxy, methoxyethoxyethoxy and phenoxy-substituted POPs. They found that the average molecular weight between crosslinks, $(M_n)_c$, decreased with irradiation dose and, as a consequence, the crosslinking density increased. Moreover, thermal analysis of the polymer samples demonstrated that the glass transition temperature, T_g, did not change after irradiation, showing that the polymer backbone motion was not affected by crosslinking, in other words the amorphous nature was found to be maintained.

Applications of POPs in Biomedicine

Radiation-induced graft copolymerization is a well-established methodology for a wide variety of applications (24,25). It consists in the chemical modification of a polymer backbone as a result of the irradiation of a vinyl monomer of different nature.

Radiation grafting of hydrophilic monomers onto a variety of polymers enables a significant improvement in their biocompatibility to be attained with the maintenance of the original mechanical properties (5,26).

Many macromolecular compounds have been used as support for grafting of different monomers but those mostly used for biomedical applications are represented by silicone rubber and polyurethane.

In our laboratories some investigations have been carried out (27-30) on radiation grafting of some hydrophilic monomers, such as N-vinyl-2-pyrrolidone (NVP), acrylamide (AAm) N,N-dimethyl acrylamide (DMAA) and N,N-dimethylaminoethyl methacrylate (DMAEM), onto two semicrystalline and hydrophobic POPs, *i.e.* poly[bis(phenoxy)-phosphazene] (PPP) and poly[bis(trifluoroethoxy)phosphazene] (PTFP).

The direct grafting procedure consisting in the irradiation of POP films immersed in excess of monomer or its solution was adopted.

In order to obtain the most suitable irradiation conditions, swelling at equilibrium at room temperature of polymeric films immersed in the monomer-methanol or monomer-water mixtures were preliminarly carried out. It was found that all the monomers, except for DMAA, do not dissolve the polymers but swell them to a moderate extent. As expected, the extent of swelling increases as the amount of solvent in the monomer mixture decreases, with higher values for methanol solutions with respect to aqueous ones. As a consequence, under irradiation the polymerization takes place in two *loci*, *i.e.* a liquid phase and a swollen polymer phase where homopolymerization and graft copolymerization occur, respectively.

Radiation grafting at 20°C of NVP onto PPP and AAm onto PTFP were carried and the kinetic curves are reported in Figs. 5 and 6, respectively.

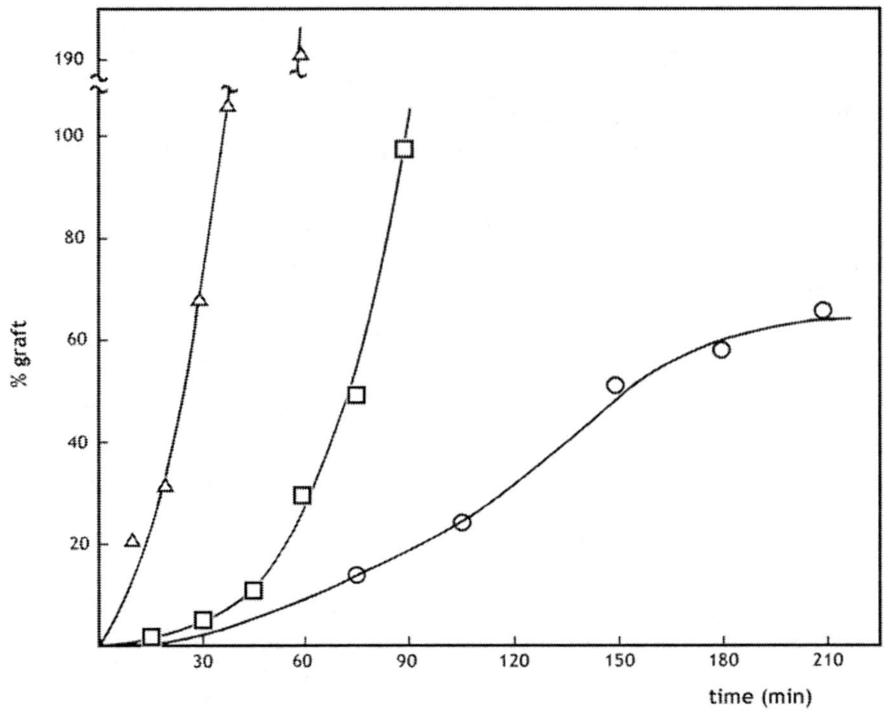

Fig. 5. Grafting percent *vs.* time plots at 0.50 Gy/s and 20°C for PPP films immersed in NVP/methanol mixtures in the following ratios (w/w): 100:0 (△); 80:20 (□); 50:50 (○).

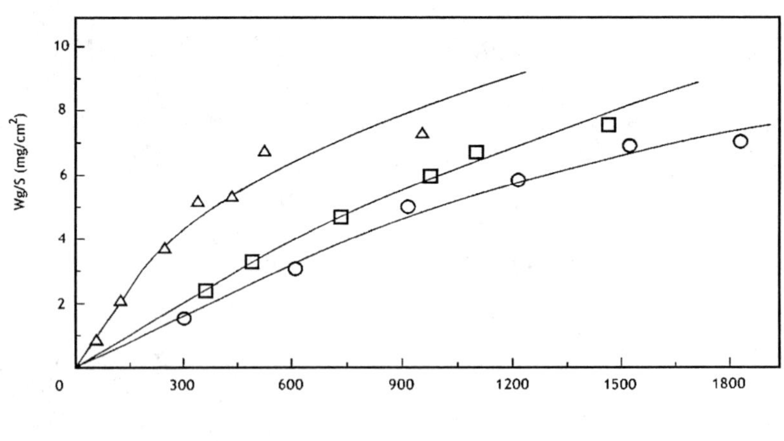

Fig. 6. Weight of graft per unit surface area *vs.* time plots for PTFP films irradiated in AAm-water mixtures in the ratio 30:70 (w/w) at: 0.116 Gy/s (△); 0.0304 Gy/s (□); 0.015 Gy/s (○) in 5.10^{-3} M aqueous solution of $CuCl_2$.

The grafting extent increases with both dose rate and monomer concentration. Moreover, as it can be seen in Fig. 5 for PPP-g-NVP, at the beginning an accelerating pattern is apparent due to the high swellability of PPP while a progressive deceleration owing to the monomer depletion takes place in the later stages, mainly at low monomer concentrations. This fact can be explained by a hindered diffusion of the monomer inside the film so that the grafting reaction is mainly confined to the film surface. This situation is still more favoured if the polymers are less swellable, as it is the case of monomer-water mixtures in which the surface grafting process will occur from the beginning (see Fig. 6).

To fulfil the requirements for a polymer modified by the radiation grafting for biomedical purposes, proper experimental conditions should be adopted in order to minimize the diffusional processess which play an important role. In other words, to confine the reaction to the film surface, parameters such as the extent of swelling before irradiation, monomer concentration and dose rate, which strongly affect the spatial distribution of hydrophilic grafted chains onto POP backbone, should be carefully controlled.

An evaluation of mechanical properties of the graft copolymers in the water-swollen state was made and the stress-strain curves concerning the PTFP-g-DMAEM with different percent of graft are shown in Fig. 7.

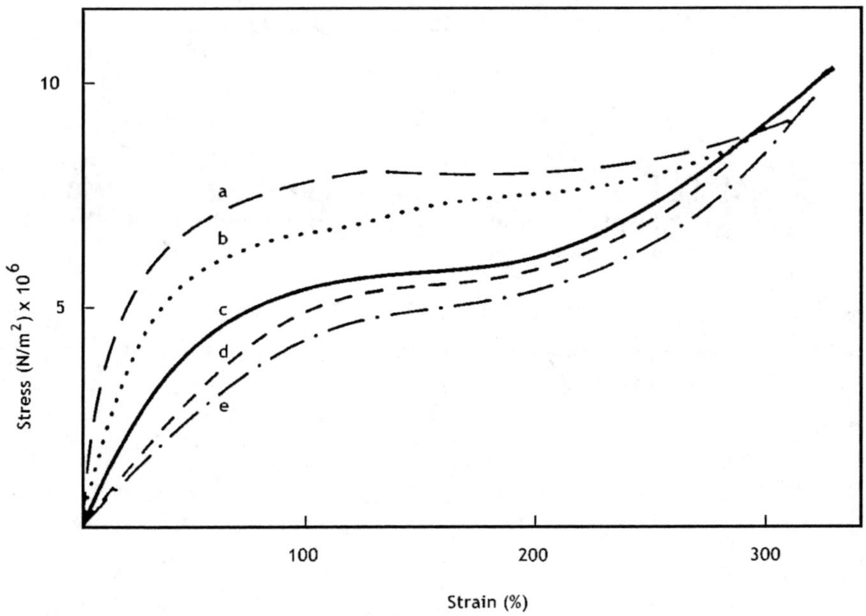

Fig. 7. Stress-strain curves for PTFP-g-DMAEM with grafting percentages of: 0% (a); 20% (b); 27% (c); 42% (d); 50% (e).

The curve referring to ungrafted PTFP reported for comparison is typical of a semicrystalline rubbery material, *i.e.* with rather high tensile strength, σ_b, and elongation at break, ε_b, values. On the contrary, the grafted samples show a continuous decrease of modulus as the percent of graft increase.

The values of modulus E relative to PTFP-g-DMAA samples both in dry and swollen state are reported in Fig. 8.

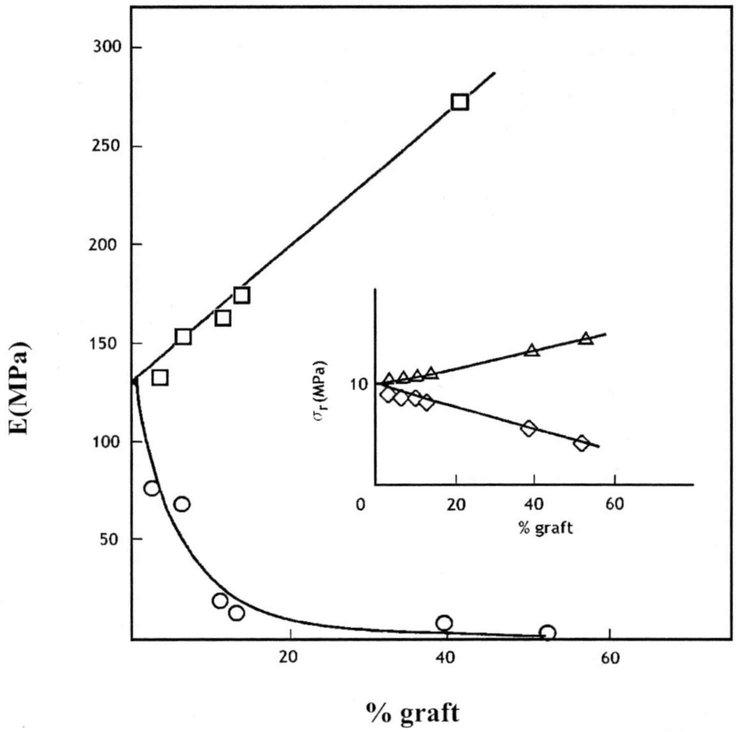

Fig. 8. Young modulus *vs.* grafting percentage for PTFP grafted with DMAA-methanol mixtures. The plots relative to tensile strength are shown in the inset. Dry state: (□) and (△); swollen state: (○) and (◇).

It should be remembered that the latter case is of relevance in the field of biomaterials because it is the state in which very often the polymeric materials occur. The figure shows that E values decrease as the grafting percentage increases, as it is expected for more and more plasticized samples. On the contrary, for the samples in the dry state a progressive increase of E is recorded. The pattern can be explained by the low T_g value of –66°C for PTFP and by high T_g of 89°C for the glassy homopolymer poly(DMAA). From the inset of Fig. 8 it can be seen that the behaviour of tensile strength both in the dry and in the swollen state is similar to that of modulus.

Biocompatibility and anticoagulating properties of the modified POP samples were also examined. *In vivo* tests were carried out with grafted PPP and PTFP films implanted intraperitoneally in rats for some weeks. While ungrafted polymer exhibited a very high extent of encapsulation, the grafted ones showed an excellent biocompatibility since no encystement was observed and their surface was free of fibroblasts.

Heparin, a well-known anticoagulant, was ionically bonded to PPP-g-DMAEM and PTFP-g-DMAEM films. It was found that heparinization induced a considerable anticoagulant effect to the polymer surfaces on the basis of the increase of clotting time of the blood from some healthy volunteers. Moreover, the photographic observations, not reported here, of the films after clotting time measurements enabled the absence of clots in the film surface to be ascertained.

Hydrogels are polymer networks which can absorb and retain significative amounts of water, generally >20% of their dried weight. In recent years, hydrogels have been used with increasing success as biomaterials in different fields such as, for example, in the manufacture of soft contact lenses, as non thrombogenic materials and as carriers for the controlled release of drugs (5-9).

Hydrogel membranes for biomedical applications based on mixed substituent polyphosphazenes containing both hydrophilic $NHCH_3$ and hydrophobic OCH_2CF_3 or OC_6H_5 groups in different proportions were synthesized and radiation crosslinked (31) A proper combination of the ratio of side groups and radiation dose enabled a suitable swellability of the membranes to be achieved for solute permeation.

In a subsequent paper (32), another method consisting in the radiation crosslinking of poly[*bis*(methoxyethoxyethoxy)phosphazene], MEEP, a water soluble polymer, was successfully reported to control water uptake.

pH-sensitive hydrogels obtained by irradiation of a water-soluble polyphosphazene bearing different proportions of sodium oxybenzoate and methoxyethoxyethoxy groups was the object of a interesting paper recently published (33). The degree of swelling in water of crosslinked poly[(76% sodium oxybenzoate) (24% methoxyethoxyethoxy)phosphazene] (A) and poly[(30% sodium oxybenzoate) (70% methoxyethoxyethoxy)phosphazene] (B) at different pH values are shown in Fig. 9.

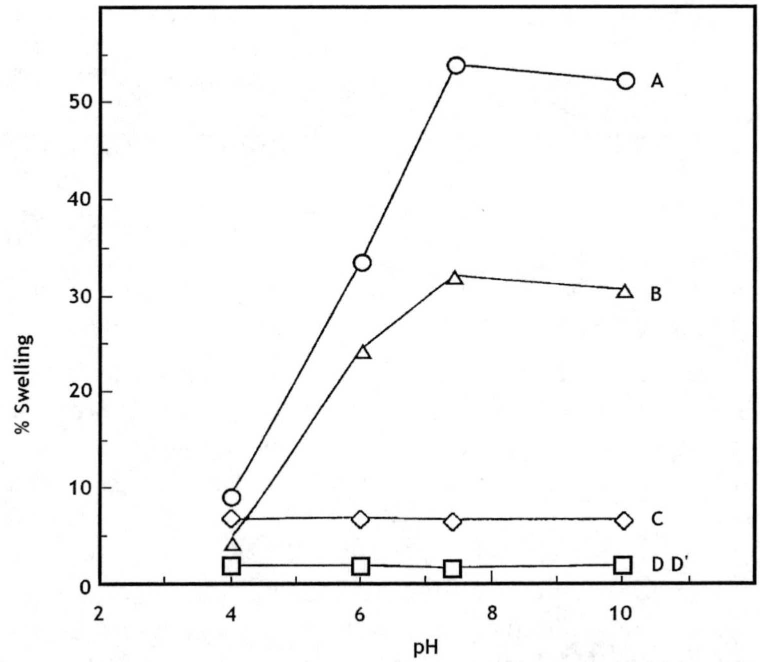

Fig. 9. Swelling percent *vs.* pH plots for hydrogels A, B, C, D and D′.

It can be seen that, due to their anionic character, these hydrogels the more swell the more basic pH is. Moreover, the swelling increases with increasing sodium oxybenzoate/methoxyethoxyethoxy ratio. The authors explained this behaviour with the occurrence of two factors: a lower extent of crosslinking owing to the reduced sensitivity to crosslinking of methoxyethoxyethoxy moiety bearing low concentrations of C-H bonds and a higher concentration of sodium oxybenzoate units that enhance the anionic character of the hydrogel. The swelling is independent of pH for crosslinked MEEP (C), poly[(76% propyl oxybenzoate) (24% methoxyethoxyethoxy)phosphazene] (D) and poly[(30 % propyl oxybenzoate) (70% methoxyethoxyethoxybenzoate] (D′) also reported for comparison.

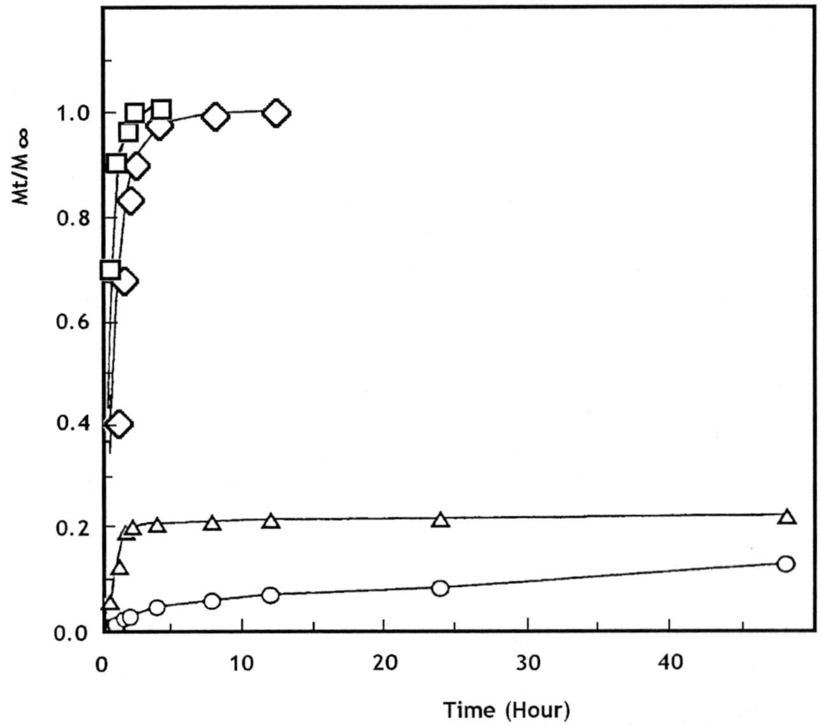

Fig. 10. Release of dye from hydrogels A (□) and B (◇) at pH = 7.4; A (○) and B (△) at pH = 2.

Figure 10 shows the release profile of the dye Briebich Scarlet, expressed as M_t/M_∞, where M_t and M_∞ are the amounts of dye released at time t and at infinite time, respectively. As expected from the results of Fig. 9, the extent of release was higher the more swellable the hydrogels were. In fact, at pH 7.4 the release was so rapid that within a few hours all dye was completely released.

Interpenetrating Polymer Networks, IPNs, of POPs were synthesized and characterized by Visscher *et al.* (34). They crosslinked by γ-radiation the hydrophilic MEEP to obtain an insoluble gel. Subsequently, the latter was swollen in a mixture of methyl methacrylate, MMA, or styrene, St, an initiator and a crosslinking agent to synthesize both the glassy, slightly adhesive PMMA/MEEP IPN (I) and the flexible, non adhesive PSt/MEEP IPN (II), respectively. Characterization of these IPNs by different techniques, such as IR, ^{1}H and ^{31}P NMR spectroscopy, and DSC were carried out. In particular, the thermograms of (I) showed T_g values corresponding to the component polymers, *i.e.* 112°C (PMMA) and –80°C (MEEP) while the T_g's found for (II) showed values considerably shifted, *i.e.* 57°C (PSt) and –70°C (MEEP).

OTHER APPLICATIONS

Another interesting application of irradiated POPs in microlithography has been reported by Stannett *et al.* (35,36). These authors investigated the behaviour of a multifunctional cyclic phosphazene monomer containing six hydroxyethyl methacrylate groups, 6-HEMA, to enhance radiation crosslinking of both elastomeric and glassy POPs.

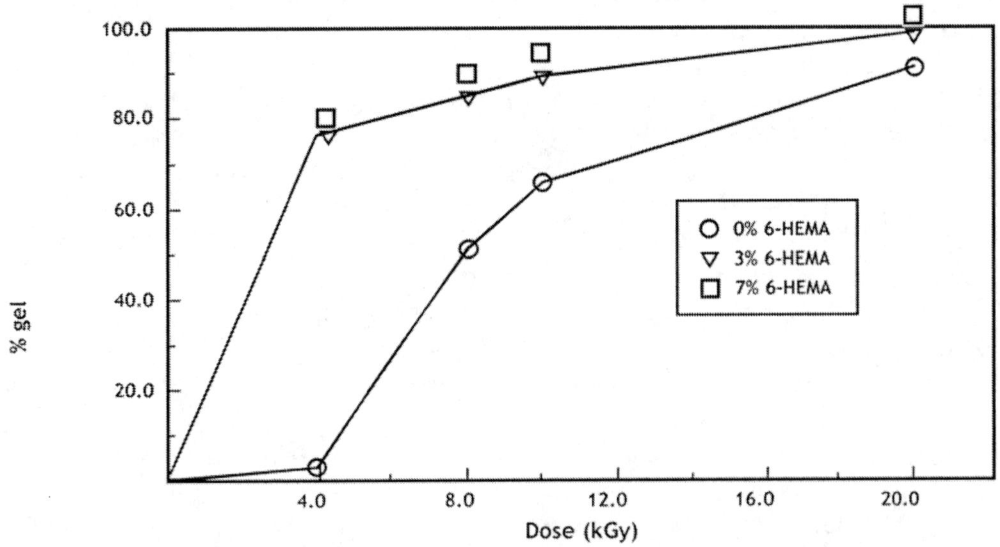

Fig. 11. Gel percent *vs.* dose plots of blends of p-ethyl-anilinophosphazene with 6-HEMA at different 6-HEMA percentages.

Figure 11 shows the percent gel *vs.* irradiation dose of films of elastomeric p-ethyl anilinophosphazene blended with 6-HEMA. It can be seen that the presence of the latter induces a very high increase of the crosslinking yield. Moreover, different amounts of 6-HEMA do not cause any significant variation. Similar results were obtained with the glassy allyl-amino-substituted POP, a suitable candidate as a negative resist for microlithographic applications. It was ascertained that SiO_2 wafers prepared with a mixture of this polymer with low amounts (1-10wt%) of 6-HEMA monomer induced an increase of at least a order of magnitude of radiation sensitivity with the consequence of an enhanced thermal stability and reactive ion etching.

REFERENCES

1. Charlesby, *Atomic Radiation and Polymers*, Pergamon Press, New York, 1960.
2. Chapiro, *Radiation Chemistry of Polymeric Systems*, Interscience, New York, 1962.
3. D.W. Clegg and A.A. Collyer, *Irradiation Effects on Polymers*, Elsevier, Londra, 1991.
4. V.S. Ivanov, *Radiation Chemistry of Polymers*, VSP, Utrecht, 1992.
5. M. Carenza, Recent achievements in the use of radiation polymerization and grafting for biomedical applications, *Radiat. Phys. Chem.* **39**, 485-493 (1992).

6. S. Machi, New trends of radiation processing applications, *Radiat. Phys. Chem.* **47**, 333-336 (1996).
7. Kaetsu, Biomedical materials, devices and drug delivery systems by radiation technique, *Radiat. Phys. Chem.* **47**, 419-424 (1996).
8. J.M. Rosiak, Gel/sol analysis of irradiated polymers, *Radiat. Phys. Chem.* **51**, 13-17 (1998).
9. V. Ya. Kabanov, Preparation of polymeric biomaterials with the aid of radiation-chemical technique, *Russ. Chem. Rev.* **67**, 783-816 (1998).
10. R. De Jaeger and M. Gleria, Poly(organophosphazene)s and related compounds: synthesis, properties and applications, *Progr. Polym. Sci.* **23**, 179-276 (1998).
11. H.R. Allcock, Inorganic-organic polymers as a route to biodegradable materials, *Macromol. Symp.* **144**, 33-46 (1999).
12. H.Q. Liu and V.T. Stannett, Radiation polymerization of hexachlorocyclotriphosphazene, *Macromolecules* **23**, 140-144 (1990).
13. V. Caglioti, D. Cordischi and A. Mele, Solid-state polymerization of phosphonitrilic chloride, *Nature (London)* **195**, 491-492, (1962).
14. G. Beggiato, P. Bordin, F. Minto and L. Busulini, Effects of ^{60}Co gamma rays on poly[bis(*p*-tolylamino)]phosphazene, *Eur. Polym. J.* **15**, 403-407 (1977).
15. G. Palma, M. Carenza and C. Pollini, Radiation crosslinking of poly(vinyl chloride) enhanced by a tetrafunctional coagent, *Eur. Polym. J.* **16**, 333-338 (1980).
16. G. Beggiato, S.S. Emmi, S. Lora, A. Martelli and F. Minto, Radiation induced oxidative degradation of poly[bis(*p*-tolylamino)]phosphazene in chlorinated hydrocarbons, *Eur. Polym. J.* **20**, 455-462 (1984).
17. S. Yanai, V.T. Stannett, D.R. Squire, G.L. Hagnauer and R.L. Singler, Radiochemical behaviour of some poly(aryloxy phosphazenes), *Radiat. Phys. Chem.* **23**, 489-490 (1984).
18. D. Babic, D.M. Souverain, V.T. Stannett, D.R. Squire, G.L. Hagnauer and R.E. Singler, The radiation crosslinking and scission of some polyphosphazenes, *Radiat. Phys. Chem.* **28**, 169-172 (1986).
19. S. Lora, F. Minto, M. Carenza, G. Palma and A. Faucitano, Radiation-induced modifications of some poly(aryloxyphosphazenes), *Radiat. Phys. Chem.* **31**, 629-638 (1988).
20. Charlesby, Crosslinking and degradation of polymers, *Radiat. Phys. Chem.* **18**, 59-66 (1981).
21. Faucitano, S. Lora, F. Minto, M. Carenza, G. Palma and A. Buttafava, ESR of intermediate radicals in the solid state radiolysis of polyorganophosphazenes, *Radiat. Phys. Chem.* **37**, 573-579 (1991).
22. K.U. Ingold, *J. Am. Chem. Soc.* **98**, 6803-6806 (1976).
23. F.F. Stewart, R.E. Singler, M.K. Harrup, E.S. Peterson and R.P. Lash, Electron beam crosslinking of fluoroalkoxy, methoxyethoxyethoxy, and substituted phenoxy polyphosphazenes: physical and chemical characterization and comparison to a thermally induced free radical process and ionic complexation, *J. Appl. Polym. Sci.* **76**, 55-66 (2000).
24. V.T. Stannett, Radiation grafting: State-of-the-art, *Radiat. Phys. Chem.* **35**, 82-87 (1990).

25. V.Ya. Kabanov, R.E. Aliev and V.N. Kudryavtsev, Present status and development trends of radiation-induced graft copolymerization, *Radiat. Phys. Chem.* **37**, 175-192 (1991).
26. A.S. Hoffman, Ionizing radiation and gas plasma (or glow) discharge treatments for preparation of novel polymeric biomaterials, *Adv. Polym. Sci.* **57**, 141-154 (1984).
27. S. Lora, M. Carenza, G. Palma and P. Caliceti, Radiation-induced graft copolymerization of hydrophilic monomers onto polyphosphazenes, *Radiat. Phys. Chem.* **35**, 117-121 (1990).
28. S. Lora, M. Carenza, G. Palma, G. Pezzin, P. Caliceti, P. Battaglia and A. Lora, Biocompatible polyphosphazenes by radiation-induced graft copolymerization and heparinization, *Biomaterials* **12**, 275-280 (1991).
29. G. Pezzin, G. Palma, S. Lora and M. Carenza, Films of poly[bis(trifluoroethoxy)phosphazene] radiation-grafted with hydrophilic monomers: thermal and mechanical characterization, *Chim. Ind. (Milan)* **73**, 185-189 (1991).
30. S. Lora, G. Palma, M. Carenza, P. Caliceti and G. Pezzin, Radiation grafting of hydrophilic monomers onto poly[bis(trifluoroethoxy)phosphazene], *Biomaterials* **15**, 937-943 (1994).
31. H.R. Allcock, M. Gebura, S. Kwon and T.X. Neenan, Amphiphilic polyphosphazenes as membrane materials: influence of side groups on radiation cross-linking, *Biomaterials* **9**, 500-508 (1988).
32. H.R. Allcock, S. Kwon, G.H. Riding, R.J. Fitzpatrick and J.L. Bennett, Hydrophilic polyphosphazenes as hydrogels: radiation cross-linking and hydrogel characteristics of poly[*bis*(methoxyethoxyethoxy)phosphazene], *Biomaterials* **9**, 509-513 (1998).
33. H.R. Allcock and A.M.A. Ambrosio, Synthesis and characterization of pH-sensitive poly(organophosphazene) hydrogels, *Biomaterials* 17, 2295-2302 (1996).
34. K.B. Visscher, I. Manners and H.R. Allcock, Synthesis and properties of polyphosphazene interpenetrating polymer networks, *Macromolecules* **23**, 4885-4886 (1990).
35. V. T. Stannett, G.L. Grune, R.T. Chem and H.R. Allcock, Radiation crosslinking of some polyphosphazenes and possible applications in microlithography, *ACS Polym. Prepr.* **35** (2), 876- 877 (1994).
36. G.L. Grune, R.W. Greer, R.T. Chern and V.T. Stannett, Sensitivity enhancement of polyorganophosphazenes to radiation with 2,2,4,4,6,6-hexakis(2-hydroxyethyl methacrylate)cyclotriphosphazene monomer and its application for negative resists, *J. Macromol. Sci.-Pure Appl. Chem.* **A31**, 1193-1206 (1994).

In: Applicative Aspects of Poly(organophosphazenes)
Editors: M. Gleria and R. De Jaeger, pp. 101-118
ISBN 1-59454-025-X

Chapter 5

PHOSPHAZENE HYBRIDS BY SOL-GEL

Giovanna Brusatin[1], Mario Gleria[2], Sandra Dirè[3], Giacomo Facchin[4] and Massimo Guglielmi[1]

1-Dipartimento di Ingegneria Meccanica-Settore Materiali, Università di Padova, Via Marzolo 9, 35131 Padova, Italy.

2-Istituto di Scienze e Tecnologie Molecolari del C.N.R., c/o Dipartimento C.I.M.A. dell'Università di Padova, Via Marzolo 1, 35131 Padova, Italy, and Istituto per la Sintesi Organica e la Fotoreattività, Consiglio Nazionale delle Ricerche, Viale dell'Università, 2 – 35020 Legnaro, Padova, Italy.

3-Dipartimento di Ingegneria dei Materiali, Università di Trento, Via Mesiano 77, 38050 Mesiano, Trento, Italy.

4-Centro di Chimica e Tecnologia dei Composti Metallorganici degli Elementi di Transizione del Consiglio Nazionale delle Ricerche, c/o Dipartimento di Processi Chimici dell'Ingegneria, Università di Padova, Via Marzolo 9, 35131 Padova, Italy.

SUMMARY

In this article the application of the sol-gel technique to phosphazene substrates to obtain a novel series of hybrid materials is reviewed. Different types of phosphazenes are suitable for this goal, i.e. hexachlorocyclophosphazene and polydichlorophosphazene, alkoxy-substituted phosphazene substrates, phosphazene polymers bearing oligoethyleneoxyde or trifluoroethoxide moieties, phosphazene derivatives containing alkoxy or aryloxy groups functionalized with free hydroxyl functions, and cyclophosphazenes containing trialkoxysilane functionalities. For some of these materials only weak forces are involved between phosphazenes and the inorganic network when hybrid materials are prepared. For others genuine chemical bonds are formed among them. In both cases relevant improvements for the applicability properties of phosphazene substrates are observed together with the onset of new features.

INTRODUCTION

New materials possessing interesting applicative characteristics are nowadays accessible by synthesis at the interfaces between ceramics, metals and polymers[1].

Very interesting perspectives, in fact, are provided by hybrid materials prepared through sol-gel technique[2] in which an organic part combined with an inorganic matrix. Properties of these new substrates inherently due to the organic component may be of interest for nonlinear optics[3-5], light emission[6,7], photochromism[8], elasticity, chemical reactivity, molecular recognition, etc., while the inorganic support may be selected depending on its refractive index, dimensional stability, mechanical properties, and thermal and flame resistance.

From the synthetic point of view two main classes of compounds are accessible by sol-gel technique. In the first one, no chemical bonds are formed between the organic and the inorganic partners of the hybrids, and the only interaction existing between these substrates being relatively weak are based on Van der Waals forces, hydrogen bonding or electrostatic attractions. In the second class, on the contrary, true covalent linkages are generated between the organic and the inorganic phases in the hybrids producing monophased materials, e.g. ORMOSIL (Organically Modified Silicates)[9], ORMOCERS (Organically Modified Ceramics)[10] or CERAMERS (Ceramics Polymers)[11].

In the case of phosphazene substrates the preparation of hybrid materials combined with silica, titania, zirconia or alumina networks is a relatively recent research field[12-14].

Four research groups are basically responsible for the advances obtained in this domain:

(a) the Eastman Kodak team in USA[15-23] mostly investigated the utilization of poly[bis(2,2',2"-trifluoroethoxy)phosphazene], PTFEP, containing a small percentage of hydrolyzed units,

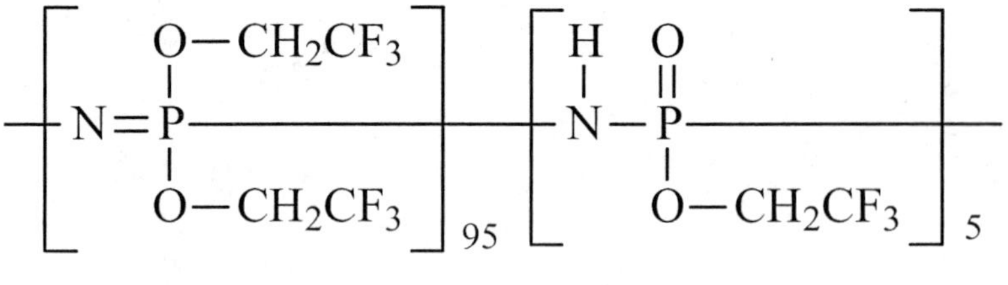

PTFEP

poly[bis(methoxyethoxyethoxy)phosphazene], MEEP, and octakis(methoxyethoxyethoxy)cyclotetra phosphazenes, MEEP Tetramer to prepare new hybrids in which no chemical bonds are formed between the phosphazene materials and the silica network. The good mechanical properties of these substrates are guaranteed by Van der Waals type interactions.

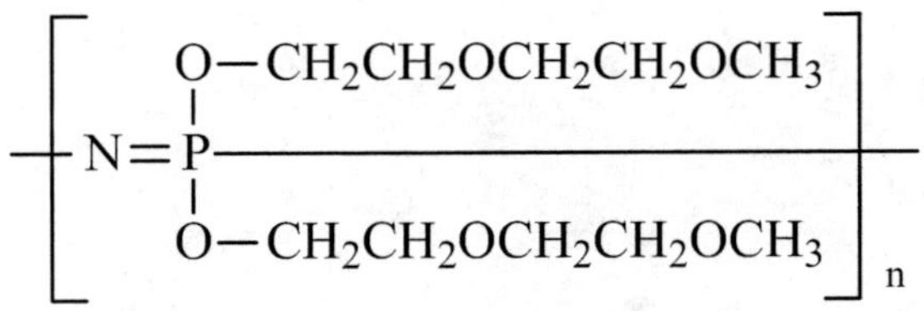

MEEP

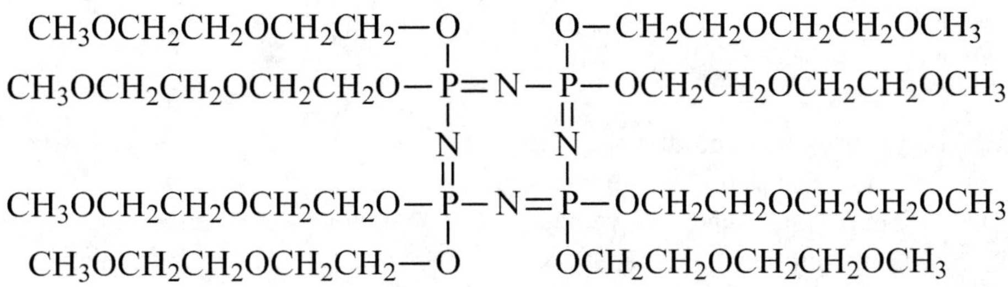

MEEP Tetramer

(b) our group in Italy[12-14,24-35] has exploited hydroxylated phosphazenes starting from hexakis(4-hydroxyphenoxy)cyclophosphazene, C6OH, poly[bis(4-hydroxyphenoxy)-phosphazene], POP-OH,

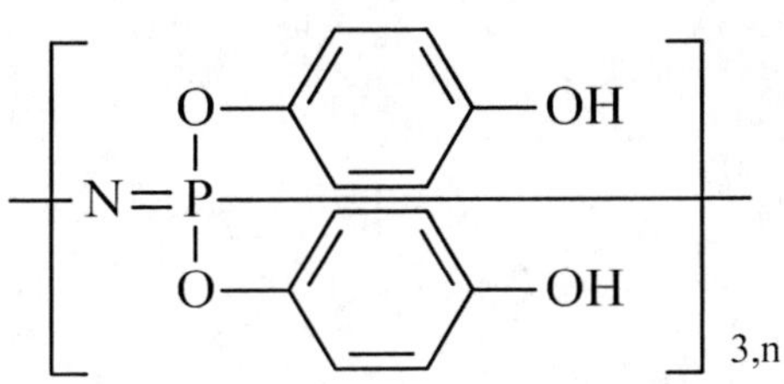

C6OH or POPOH

and other hydroxylated phosphazene copolymers such as

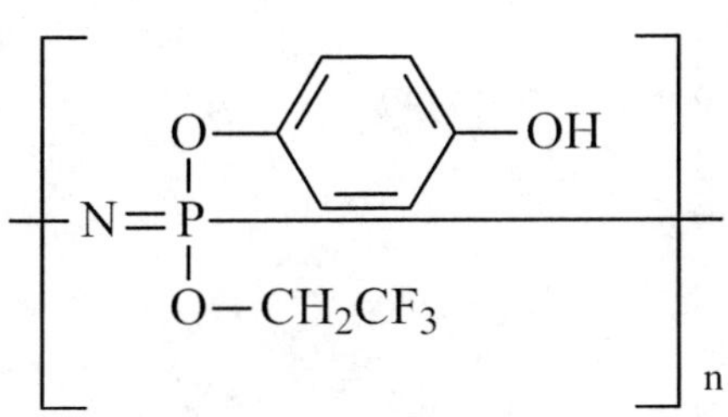

PhOH/TFE CoPOP

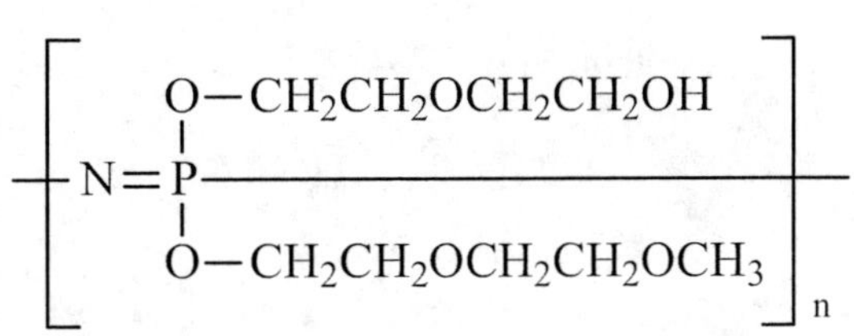

MEEPOH

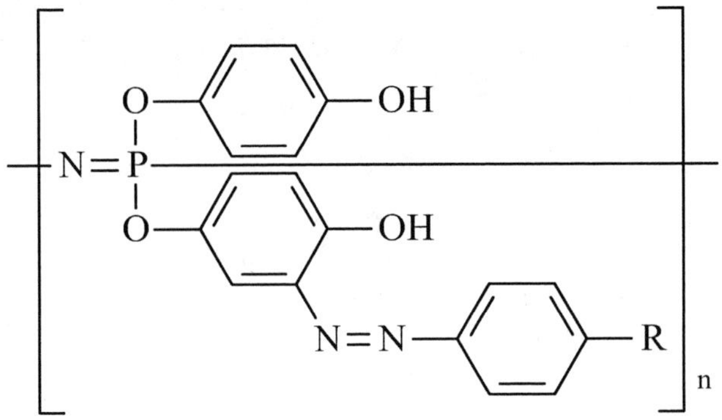

Azo CoPOP

in order to guarantee the formation of genuine chemical bonds between the phosphazene materials and the inorganic matrix.

(c) the group at the University of Montpellier (France)[36-39] in which hexachlorocyclophosphazene, polydichlorophosphazene, and cyclo- or polyphosphazenes substituted with alkoxy groups,

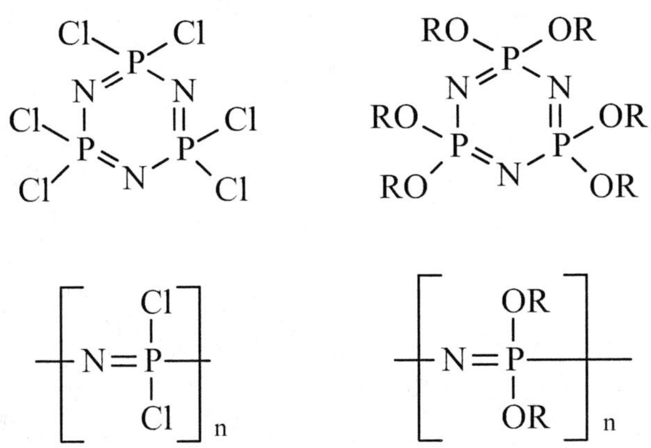

where R may be CH_3, iso-C_3C_7 or CH_2CF_3 were used for the preparation of hybrid materials using a non-hydrolytic condensation process.

(d) the group of Würzburg (Germany), which used cyclic phosphazenes derivatives containing free -$Si(OR)_3$ moieties[10] having the following structures

$N_3P_3[NHCH_2CH_2CH_2Si(OEt)_3]_6$ $N_5P_5[CH_2Si(OCH_3)_3]_{10}$

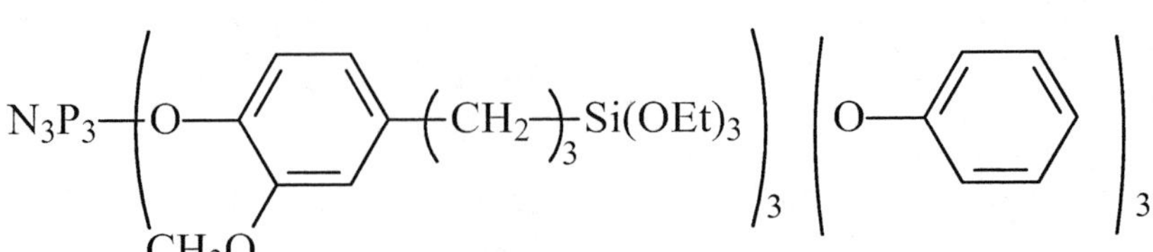

for the synthesis of sol-gel hybrid materials.

Few other strategies for obtaining new phosphazene/metal oxide hybrids through sol-gel have been explored[40,41] or have only been announced[42] without additional publication.

In this paper we would like to highlight the work carried out on phosphazene-based hybrids, with particular emphasis dedicated to possible practical utilization of these materials.

SHORT INTRODUCTION TO THE SOL-GEL TECHNIQUE

The sol-gel process can be defined as a gentle chemical route to prepare a variety of inorganic materials through a very broad range of different procedures[2,43]. According to the more general approach, the sol-gel process is based on the formation of inorganic networks starting from solutions of molecular precursors, the *sol*, followed by its gelation to form a continuous solid network in a continuous liquid phase, the *gel*. Among precursors, metal alkoxides are most popular as they can undergo inorganic polymerization by hydrolysis-condensation reactions. The hydrolysis reaction (A) through the addition of water replaces alkoxide groups with hydroxyl groups

$$-\overset{|}{\underset{|}{M}}-OR + H_2O \longrightarrow -\overset{|}{\underset{|}{M}}-OH + ROH \qquad (A)$$

suitable for condensation (B,C) leading to the formation of oxo-bridges and the elimination of water or alcohol by-products, respectively:

$$-\overset{|}{\underset{|}{M}}-OH + HO-\overset{|}{\underset{|}{M}}- \longrightarrow -\overset{|}{\underset{|}{M}}-O-\overset{|}{\underset{|}{M}}- + H_2O \qquad (B)$$

$$-\overset{|}{\underset{|}{M}}-OH + RO-\overset{|}{\underset{|}{M}}- \longrightarrow -\overset{|}{\underset{|}{M}}-O-\overset{|}{\underset{|}{M}}- + ROH \qquad (C)$$

Depending on the choice of precursors and synthetic conditions, the features of the gel network can be modulated in order to obtain different structures and functionalities in the final materials. This aspect has been extensively exploited not only to produce multi-component inorganic oxides but also to prepare the so-called organic/inorganic hybrid materials, where organic groups are dispersed within the inorganic network, on a scale from the sub-micronic range down to the nanometer level[43].

The sol-to-gel transition can be exploited with the aim of producing coatings, fibers, powders and monoliths and this accounts for the large potentiality of the sol-gel technique. The applications of sol-gel processes have found outstanding success in the field of functional coatings, ceramic fibers, advanced fine-grains ceramics and ordered mesoporous materials and are still in expansion[44].

Thanks to the wide potentialities of organic chemistry, the main development of sol-gel process is foreseen in the field of hybrid organic/inorganic materials, where organic and inorganic components can give a synergistic effect in tailoring new materials. The introduction of non-polymerizable organic groups as well as of reactive organic functionalities allows the production of new interesting materials particularly for the development of optical devices; the control of phase interaction in hybrids prepared from organic polymers and inorganic oxo-clusters leads to the production of materials with modulated mechanical properties[45,46]. The almost unlimited availability of organic and organometallic compounds and polymers gives the opportunity to design, by the sol-gel process, new and successful materials for the century to come.

SYNTHESIS OF CYCLO- AND POLY-PHOSPHAZENES

The cyclo (CPs) and poly (POPs) phosphazenes used in the preparation of hybrid materials through the sol-gel technique were synthesized according to standard procedures[47-49] based on the reaction of hexachlorocyclotriphosphazene, octachlorocyclotetraphosphazene or polydichlorophosphazene with the appropriate sodium salts of alkoxy or aryloxy derivatives, or with -NH_2-functionalized organic compounds. When necessary suitable copolymerization[50] and/or functionalization[51] reactions have been exploited.

In particular:

(a) Poly[bis(methoxyethoxyethoxy)phosphazene], MEEP, and octakis-(methoxyethoxyethoxy)cyclophosphazene (MEEP Tetramer) were prepared according to the procedure reported by Allcock [52,53] and Coltrain[15,16,18,19,54]. MEEP is an amorphous polymer[52], with a very low T_g value (-75°C)[53] which was selected for the preparation of hybrid materials using the sol-gel technique because of its remarkable importance as a water soluble material[55,56], biologically important macromolecule[57,58], ionic conductor when doped with Li^+ or Ag^+ salts[52,53,59], and phase transfer catalyst[60].

(b) Hydroxylated poly[bis(methoxyethoxyethoxy)phosphazene], MEEP-OH was prepared by De Jaeger[61,62] by reacting MEEP with trimethylsilyl iodide[63] to deplete CH_3 end groups of the oligoethyleneoxide moieties in this polymer, thus forming a novel functionalized polyphosphazene bearing free hydroxyl groups;

(c) poly[bis(2,2'2,"-trifluoroethoxy)phosphazene], PTFEP, was prepared according to the synthetic procedure established by Ferrar[64]. This polymer is one of the most important phosphazene synthesized because of its good film forming capability[65], micro-crystallinity[66], high thermal stability[67,68], good self-extinguishibility[69], remarkable flame retardant characteristics[70,71], noticeable mechanical properties[72,73], great hydrolytic stability[74] and very high inertness towards chemicals[65], biological agents[57,75], atomic oxygen[76], and light[77,78] or γ-rays[79] irradiation. Performances of this polymer as membrane were also evaluated[80].

(d) poly[bis(4-hydroxyphenoxy)phosphazene], POPOH, was prepared following the procedure established by Medici[81], by reaction of poly[bis(4-methoxyphenoxy)phosphazene][82] with BBr_3 under hydrolytic conditions. This functionalization leads to the removal of the methyl groups in the para position of the aromatic ring and the formation of free hydroxyl functions. The same procedure was adopted to prepare the corresponding phosphazene trimer hexakis(4-

hydroxyphenoxy)cyclophosphazene, C6OH[81], and PhOH/TFECoPOP[12-14], a phosphazene copolymer containing trifluoroethoxy groups and free hydroxyl functionalities;

(e) AzoCoPOP was prepared by coupling POPOH with diazotized aromatic amines to obtain a deeply colored phosphazene copolymer bearing azo dyes substituents on the phosphorus atoms of the phosphazene chain[26,83];

(f) in the recent work carried out by Leclercq[36-39], neat hexachlorocyclophosphazene and polydichlorophosphazene together with alkoxy-substituted phosphazene derivatives were exploited using a non-hydrolytic approach to the synthesis of phosphazene metal oxide hybrid derivatives;

(g) finally, the synthesis of cyclophosphazenes substituted with trialkoxysilane moieties have been reported by Rose[10] starting from $N_3P_3Cl_6$ and γ-aminopropyltriethoxysilane, cyclophosphazenes substituted with eugenol and hydrosilylation reaction with triethoxysilane $HSi(OC_2H_5)_3$, or, eventually from $N_5P_5(CH_3)_{10}$ by lithiation with lithium n-butyl of the methyl groups followed by reaction with chlorotrimethylsilane

Synthesis of Phosphazene/Inorganic Matrix Hybrid Materials

Starting from the above described cyclo- and poly-phosphazenes, research has been carried out to obtain new hybrid derivatives formed by POPs and inorganic matrices of SiO_2, TiO_2, ZrO_2 or Al_2O_3, respectively.

Thus Coltrain[17,21] first prepared MEEP/SiO_2 composites by dissolving MEEP (average molecular weight about $3x10^6$ g·mol^{-1}) in a 1:1 mixture of THF/methanol, adding tetraethoxysilane (TEOS) (50 % w/w respect to MEEP, resulting in composite film containing 23 % w/w of SiO_2) and HCl. After stirring 15 min and heating at 60 °C for 30 min the solution was poured in a teflon dish obtaining a 75 μm thick free standing film after 4 days. The resulting composite was a clear product, thus indicating a homogeneous dispersion of the polyphosphazene throughout the SiO_2 matrix. Thicker samples (up to 0.5 cm) were prepared by a longer process for Shore A hardness measurements. An analogous procedure has been adopted when MEEP/TiO_2, MEEP/ZrO_2 and MEEP/Al_2O_3 hybrids have been prepared[22,23].

Similarly, PTFEP/SiO_2 composites have been obtained[17,21] using PTFEP (molecular weight of $2x10^5$ gmol^{-1}), TEOS and alkaline hydrolysis of the system by heating at 60°C for 30 min, storage at room temperature for 14 days and eventual heating at 50°C in vacuum for 10 days.

POPOH/SiO_2 composites were prepared following two methods. In the first approach tetraethoxysilane (TEOS), the silica precursor exploited for these experiments, was dissolved in ethanol solutions, treated with water and HCl (TEOS/H_2O/HCl = 1/2/0.01), pre-hydrolyzed for 3, 6, 9, 12 h and eventually reacted with POPOH for an additional hour[24,25]. Following this procedure the hybrids obtained were often opaque materials frequently showing evident phase separation phenomena[28]. This method was successively improved by eliminating the pre-hydrolysis step, and simultaneously reacting POPOH, TEOS, water and HCl in ethanol for 6 h under inert atmosphere, and aging the resulting mixture for 24 h at room temperature[28]. Samples obtained in this case were transparent films or bulky materials showing no evident phase separation.

The second approach was also used for the preparation of MEEPOH/SiO_2[27,31,84], PhOH/TFECoPOP/SiO_2[13,14] and Azo CoPOP/SiO_2[26] hybrids.

When TiO_2 and ZrO_2 composites were prepared[25], acetylacetone was used to depress the chemical reactivity of the system.

New hybrid materials have been recently prepared by Leclercq[36-39] using chlorinated and alkoxy (ethoxy, isopropoxy and trifluoroethoxy) substituted cyclo- and poly-phosphazenes in combination with metal alkoxides, metal chlorides or metal chlorides in the presence of aliphatic ethers at 110°C in sealed ampoules. The three main types of reactions involved in this approach for cyclophosphazene derivatives are reported below:

$$N_3P_3Cl_6 + 6/n\ M(OR)_n \longrightarrow N_3P_3M_{6/n}O_6 + 6\ RCl$$

$$N_3P_3(OR)_6 + 6/n\ MCl_n \longrightarrow N_3P_3M_{6/n}O_6 + 6\ RCl$$

$$N_3P_3Cl_6 + 6/n\ MCl_n + 6\ R_2O \longrightarrow N_3P_3M_{6/n}O_6 + 12\ RCl$$

where M = Si or Ti

It is important to stress that in these processes metal alkoxide functionalities may be present in the phosphazene substrate, in the metal alkoxide or may be generated *in situ* by the presence of an ether compound.

Finally the acid-catalyzed hydrolysis of the trialkoxysilane-containing cyclophosphazenes in the presence of TEOS leads the preparation of bulky materials and coatings containing cyclophosphazene units inside the silica network[10].

Properties and Practical Applications of Phosphazene/Inorganic Matrix Hybrid Materials

Previous sections on the synthesis of cyclo- and polyphosphazenes and on the preparation of tri-dimensional inorganic matrices by sol-gel technique, evidenced that novel phosphazene-containing hybrid materials can be obtained in a rather simple way using mild experimental conditions, which show improved features with respect to the original substrates, and that are potentially useful for a large variety of practical applications.

Thermally Stable Materials

Investigations on the thermal characteristics of neat PTFEP[17,21], MEEP[17,21], POP-OH[12,21,28,34,35], and on the related SiO_2 hybrid materials were carried out using DSC, DTA and TGA analyses.

According to DSC investigations[17,21], neat MEEP is a polymer in the flow regime of viscoelasticity at ambient temperature, due to its very low T_g value. It also seems to possess a low degree of crystallinity as demonstrated by the existence of an endothermic peak at 20-

30°C in its thermogram. This material, furthermore, shows a thermal stability limit at about 100°C; above this temperature thermal decomposition of the macromolecule takes place.

Hybrid MEEP/SiO_2 substrates obtained by incorporating MEEP into silica networks by sol-gel technique show practically no changes both in the T_g and T_m values of this polymer, to indicate that the SiO_2 network does not influence the chain mobility of MEEP and that the crystallinity of the phosphazene material is not destroyed by the hybrid formation process. These facts, together with the spectroscopic evidence that the ^{31}P $\{^1H\}$NMR chemical shifts are not modified both in neat polymer and in the polymer incorporated into silica networks, suggests that only weak interactions take place between the phosphazene substrate and the inorganic matrix during hybrid preparation without onset of covalent chemical bonds. For the MEEP/SiO_2 hybrid, however, thermal decomposition by DSC starts to occur at about 200°C, indicating that the formation of hybrid materials introduced a strong increase in the thermal stability of MEEP and a remarkable enlargement of the service temperature range for the polymer.

Similar results are obtained for PTFEP, whose T_g is at -59°C, T(1) at 51°C and the thermal degradation starts at about 250°C[17,21].

As far as POPOH is concerned, the polymer is a colorless material, fairly reactive toward oxygen (reticulation of the macromolecule takes place after prolonged air exposure[28]), diazotized aromatic amines[26,83], formaldehyde[83], and organic acid chlorides or anhydrides[81]. POPOH, therefore, needs to be manipulated almost immediately after preparation. By DSC measurements its T_g was found to be +88°C[28], possibly by formation of hydrogen bonding.

The presence of free hydroxy groups in this polymer provided the opportunity of preparing new hybrid materials by sol-gel technique in which the phosphazene polymer is chemically linked to the inorganic network of silica[13,14,25,27-29,31-33,84].

This could be proved by considering:

(a) a)DSC experiments carried out on POPOH/SiO_2 hybrids[28], which evidenced the disappearance of the T_g transition of POPOH in these materials, and selective extraction tests[28] of POPOH by acetone in which a very small amount of polymer was selectively extracted from the hybrids, while an almost quantitative extraction of POPOH took place when a physical mixture of the two components was considered;

(b) spectroscopic (X-Ray, FTIR and NMR)[32,34,35] and thermal (DTA-TG)[32,34,35] data obtained by analyzing C6OH/SiO_2 hybrids of different compositions which unequivocally demonstrated the presence of ≡C-O-Si- bonds in the structure of these materials.

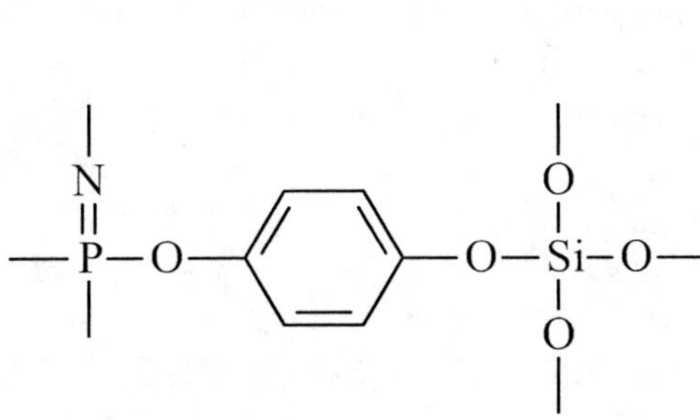

The thermal stability of POPOH/SiO_2 composites was investigated by IR spectroscopy by heating POPOH/SiO_2 monoliths and following the disappearance of the band at 1500 cm^{-1} (ν

C-C of the aromatic rings), and 960 cm^{-1} (ν P-O-Ph of the phosphazene system). These investigations evidenced that the bands are observable until 400C°, thus confirming that even in the case of POPOH the incorporation of the polymer in a hybrid structure results in the thermal stabilization of the macromolecule.

As far as the thermal stability of C6OH/SiO_2 hybrid materials is concerned, investigations have been carried out by TGA analysis and compared with the thermal behavior of the corresponding neat hydroxylated trimer[32,34,35]. From these researches it resulted that C6OH is completely degraded at temperatures at about 950°C, while the corresponding hybrid shows only a weight loss close to 20% w/w at comparable temperatures.

Similar results have been reported by Leclercq concerning the thermal stability of substituted cyclophosphazene/SiO_2 or TiO_2 hybrid derivatives[36,39].

An improved thermal stability of cyclophosphazene-containing SiO_2 composites with respect to that of the original substrates was observed also by Rose[10] for bulky materials prepared starting from cyclophosphazenes functionalized with -$Si(OR)_3$ residues. From these investigations it was concluded that the discriminating factor on the thermal stability of the final composite was the type of P-Substituent bond present in the cyclophosphazenes. In fact the thermal stability of the hybrids prepared increased from 220°C to 320°C and eventually to 430°C depending on the presence of P-NH, P-O or P-C bonds, respectively, in the phosphazene substrates. This is not surprising in view of the particular thermal stability of the phosphorus carbon bonds with respect to phosphorus-oxygen and especially phosphorus-nitrogen linkage[85].

Phosphazene/Silica Hybrids with Improved Mechanical Properties

In an analogous way the mechanical properties of the above described phosphazene/SiO_2 composites were investigated using DMTA and Stress-Strain techniques, together with surface hardness measurements.

It was found that MEEP/TEOS 50/50 composites cured at room temperatures were free-standing films with remarkable dimensional stability throughout the range of temperatures selected for measurement. The storage modulus of the hybrids started from about 10^{11} dynes/cm^2, decreased during glass transition of the polymer component, leveling at about 10^8 dynes/cm^2 and further increased with the increase in temperature up to 10^9 dynes/cm^2. The improvement in the mechanical properties of MEEP/SiO_2 composite was attributed to pseudo crosslinking phenomena between the SiO_2 network and the oligoethyleneoxide phosphazene substituent[17,21]. Flexible phosphazene coatings resisting to scratch and abrasion[15,16,18] or fibers[19,20] were also prepared starting from MEEP/SiO_2 composites to evidence the great flexibility of materials obtained starting from an unique MEEP chemical substrate.

Similar results were observed when MEEP/TiO_2, MEEP/ZrO_2 and MEEP/Al_2O_3 composites were exploited[22,23] as evidenced by Stress-Strain tests carried out on these materials. These tests showed the great influence on the final mechanical properties of the composite of the nature of the metal alkoxide exploited to prepare composites. The materials exhibited mostly a ductile behavior, especially for MEEP/SiO_2 and MEEP/ZrO_2 hybrids, sometimes (as in the cases of Al and Si oxides hybrids) with appearance of yield points, with tensile modulus ranging between 30-200 MPa, to indicate that the alkoxide has a significant effect in the reinforcement of MEEP polymer.

A Shore A Hardness of 82.5 on average was measured for MEEP/SiO_2 hybrid composites[17,21].

A significant improvement of surface hardness was observed also for coatings obtained by co-condensation processes of $N_3P_3[O(CH_2)_3Si(OC_2H_5)_3]_2$ and methacryloyl-propyl triethoxysilane (MEMO)

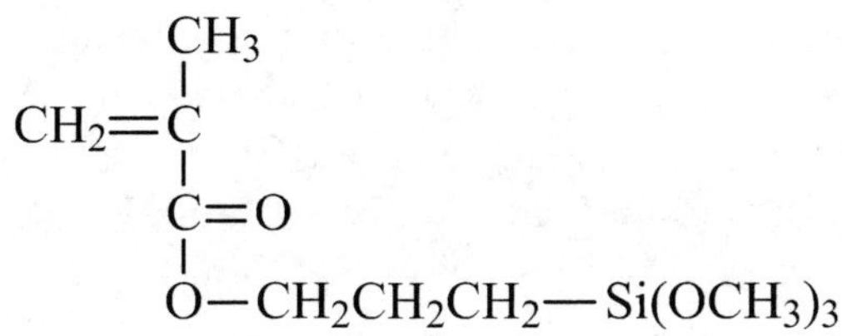

MEMO

and hybrid formation with hydrolyzed TEOS (256 N/mm^2) with respect to that formed by condensing MEMO alone with TEOS (208 N/mm^2) which was attributed to the presence of the cyclophosphazene residue in the composite[10].

Polyphosphazene Hybrids as Ionic Conductors

The ionic conductivity of sol-gel hybrid materials doped with Ag^+ or Li^+ salts was first investigated by Ferrar[22,23], who reported the addition of up to 17% (w/w) of low lattice energy salts to the MEEP/inorganic oxide composite. Thanks to the retainment of the low glass transition characteristics of MEEP polymer due to the lack of chemical linkages between MEEP and the silica matrix (*vide supra*), authors claimed that the composite should display similar charge carrying properties as MEEP solid electrolytes. The addition of potassium triflate to the MEEP/alkoxide solution does not result in remarkable differences in the formation process of the composite. The presence of this salt induces ionic conductivity phenomena with values in the range of 10^{-5} S/cm, similarly to what observed for MEEP homopolymer system[52,59]. Impedance measurements were carried out under ambient conditions and without an environmental control, to compare samples with and without salt addition. Although the general shape of the curves did not change, the conductivity of the composites containing salts were several orders of magnitude higher than those without salts. In the case of MEEP/TiO_2 composite a conductivity of $10^{-8.7}$ and $10^{-5.0}$ (ohm cm)$^{-1}$ was measured without and with salt, respectively[22,23].

These findings were recently confirmed by Kim[41] who investigated the ionic conductivity of MEEP/TEOS composites as a function of the lithium triflate concentration in the hybrid, TEOS percentage, and temperature.

It was found that the ionic conductivity of the above mentioned hybrids initially increases with the increase in the salt concentration up to 10-15% w/w of CF_3SO_3Li, to decrease successively by further addition of the salt. These facts were explained by considering that the initial increase in the number of charge carriers, which enhances conductivity until 15% w/w of Li^+ concentration (maximum conductivity), is followed by a progressive reduction in the segmental motions of the polymer host due to the formation of ion-pairs or larger agglomerates, which depress the ion transportation process (decreased conductivity).

The influence of the TEOS concentration on the ionic conductivity of MEEP/SiO_2 composites results in a decrease in this parameter by increasing TEOS during the synthesis. This fact was accounted for by assuming a lengthening of the conduction pathway due to phase separation between MEEP and TEOS.

The maximum ionic conductivity for MEEP/SiO_2 compositions was 4.6 x 10^{-5} S/cm, observed at 10% w/w TEOS and 15% w/w lithium triflate.

Finally, it was noted that higher conductivity phenomena are generated with increasing temperature. Most samples showed an Arrhenius type behavior, thus suggesting that ion diffusion is the main source of conductivity in these materials.

As reported above, the Glass Transition Temperature values of hybrid electrolytes were not significantly influenced by varying the amount of TEOS in the hybrid, thus confirming that only feeble interactions take place between MEEP polymer and the SiO_2 network and no true covalent linkages are produced between them.

The lack of chemical interactions between MEEP and the inorganic matrix was recently overcome by Brusatin[27,28], who used a MEEP derivative functionalized with free hydroxy groups, MEEPOH[61,62]. These functions, in fact, are expected to induce the formation of chemical bonds by reaction with condensed silica matrices, as previously demonstrated for C6OH[32,34,35].

The results of the conductivity measurements carried out on various samples of these materials as a function of lithium content, thickness and thermal treatment gave values in the order of 10^{-6} S/cm for lithium-containing samples and of about two order of magnitude lower for lithium-free samples. A diffusion coefficient $D=1.2x10^{-8}$ cm^2s^{-1} of the mobile charge carriers was also calculated.

An additional interesting application of MEEP/SiO_2 composites doped with inorganic salts was the possibility of obtaining antistatic coatings and fibers.

Antistatic coatings were prepared by Coltrain[15,16,18] by dissolving an etheric phosphazene material, typically MEEP or MEEP Tetramer, in ethanol while stirring, by adding a metal alkoxide precursor (e.g. TEOS) and hydrolyzing the material by addition of aqueous hydrochloric acid. CF_3SO_3K was eventually added to this mixture. The final solution was then spin coated at 2000 rpm for 30 sec and the coated sample cured at 250°C for 30 min. The final coating showed a surface resistivity of log 9 Ohm/sq at ambient conditions. Small differences in the experimental conditions were adopted when titaniun isopropoxide was exploited as a metal alkoxide precursor.

The preparation of conductive fibers was claimed by Landry[19,20,54] by blending MEEP with metal alkoxide (ZrO_2 and TiO_2) and salts (CF_3SO_3K) under sol-gel conditions. The synthetic approach used to obtain these materials was similar to that used for the preparation of antistatic coatings, but in this case the final solutions were formed into fibers by extrusion processes under appropriate experimental conditions.

Phosphazene Polymers as Colored Coatings

The synthesis of AzoCoPOP-containing hybrid materials was exploited using a polyphosphazene containing 15 % by mole of azo dyes attached to the inorganic -P=N- chain, the residual 85 % of possible functionalities in the polymer being free hydroxyl groups. In this way the reaction of an ethanol solution of AzoCoPOP, with a mixture of TEOS/H_2O/HCl = 1/4/0.01 on soda lime led to transparent colored films of hybrid materials showing considerably different UV visible spectroscopic features with respect to those of the original POPOH. In fact, the maximum in the UV absorption spectrum of POPOH, originally located at about 280 nm, was shifted towards the visible in a way which is respective of the possibility of electron delocalization along the azo dye/phosphazene structure, which in turn depends on the type of diazotized aromatic amine exploited for the coupling reaction with POPOH. In this way a relevant shift of the absorption spectrum of the pristine compound to form deeply colored materials could be observed.

POPOH/SiO_2 Hybrid Materials under High Energy Irradiation

Irradiation experiments using C or Au ions were performed on poly[bis(4-hydroxyphenoxy)phosphazene]/silicon oxides hybrid by J.C. Pivin[29,30]. The hybrid were obtained by dissolving POPOH in a solution of TEOS and ethanol, with different molar ratios between phosphazene and silica. After hydrolysis, films were deposited on silicon substrates. The analysis of the composition of films after irradiation showed that all parts of the polymeric chains are progressively destroyed, with increasing fluences of implanted ions. Carbonaceous clusters were formed by the decomposition of the phenyl rings which have an optical absorption spectrum typical of semiconductors. At higher irradiation fluences -P=N- and P-O- bonds were broken and, as a consequence, isolated silica clusters were formed. The demixing of SiO_4 groups from the phosphazene was detected by means of optical spectroscopy for a critical amount of energy deposited to each atom of the order of 100 eV, when most of polymer bonds was broken. In this way the structure was hardened by the carbonaceous component. Submicroscopic indentation tests showed that the films obtained were five times harder than the original samples.

Phosphazene Hybrids for Biomedical Applications

Phosphazene polymers for biomedical applications have been explored by Park[40] using a phosphazene macromolecule containing free triethoxysilane groups having the following chemical structure

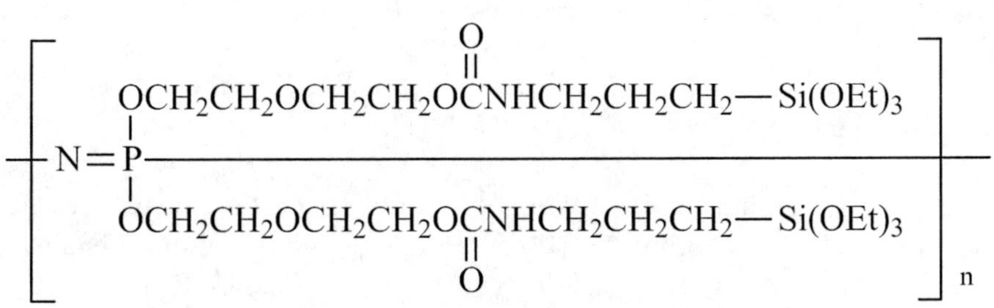

to form a phosphazene/silicate network susceptible of reacting with hexamethylene diisocyanate through the urethane functionalities of the hybrid surface, eventually linking heparin residues on the surface of the phosphazene composite. The surface was fully characterized by ATR-IR spectroscopy and the final concentration of heparin groups determined by standard techniques. The biological activity of the immobilized heparin was evaluated to be 4 % that of the free heparin.

Phosphazene Hybrid Materials as Gas Separation Membranes

POPOH/SiO_2 hybrid substrates have been exploited for the preparation of membranes for the selective separation of gas mixtures[33]. The general goal for this study was to obtain hybrid phosphazene-based membranes capable of performing separation processes under experimental conditions that are hardly afforded by purely polymeric membranes, i.e. high temperatures (above 100°C) and in the presence of aggressive chemicals. It has been thought, in fact, that the presence of phosphazene macromolecules in the silica hybrids should be able to improve the toughness of the ceramic network, while the stiffness of the phosphazene phase should be largely improved by the presence of the silica matrix. As a consequence, the overall mechanical and thermal properties of the composite are expected to be greatly enhanced with respect to the features of the single hybrid constituents[29].

The membrane preparation process has been carried out in two steps. In the first step POPOH was dissolved in ethanol together with a TEOS solution hydrolyzed at pH = 2 (HCl). Then the reaction mixture was poured into tubular α-Al_2O_3 supports (pore size 4 nm) for membrane deposition. The morphology of the deposited membrane layer was successively analyzed in dependence on the aging time and contact time in the support, with the aim of determining the best experimental conditions to prepare phosphazene-based hybrid membranes. In the second step the composite support used for POPOH membrane preparation was changed to TiO_2 (pore size 10 nm) and the new series of membranes obtained were tested for the selective separation of He/N_2 and He/CO_2 mixtures.

It was found that the POPOH/TEOS hybrid material forms a layer on the α-Al_2O_3 support and that it infiltrates the pores of the TiO_2 material. The membranes show an improved selectivity for the He/N_2 mixture, probably due to the binding action of the phosphazene polymer that limits the formation of cracks during the drying step, with an ideal selectivity very close to that of the Knudsen limit (2.65). On the contrary, in the case of He/CO_2 mixtures, the separation factor is far from this limit (3.65) due to an increased contribution of the viscous flow to the overall transport process.

Conclusion

In this chapter we highlighted the work carried out in the field of phosphazene hybrid materials as obtained by exploiting the sol-gel technique, with particular emphasis on the synthetic approaches used for the preparation of these substrates and on the new possible practical applications of the materials obtained. It is significant that a consistent improvement of both thermal stability and mechanical properties was obtained for the large majority of materials prepared in this way, regardless of the type of phosphazene substrate and metal alkoxide used for the materials tested and the experimental conditions adopted to synthesize them. It is also of importance that new applications (e.g. scratch-resistant coatings) have become accessible for phosphazenes and that tested applications (e.g. membranes, biomedical utilization and ionic conductors) have already found remarkable extension of their potential.

The impression is that this topic is still in its infancy and that its potential is very high for the production of new substrates for innovative applications in electro-optical, biomedical, and catalytic fields.

REFERENCES

1. Allcock, H. R. In *Inorganic and Organometallic Polymers*; Zeldin, M., Wynne, K. J., Allcock, H. R., Eds.; ACS Symp. Series: Washington DC, USA, 1988; Chapt. 19, Vol. 360, p 250.
2. Brinker, C. J.; Sherer, G. W. *Sol-Gel Science*; Academic Press: Boston, USA, 1990.
3. Mackenzie, J. D. *J. Sol-Gel Sci. Technol.* **1993**, *1*, 7.
4. Oviatt, H. W.; Shea, K. J.; Kalluri, S.; Shi, Y.; Steier, W. H.; Dalton, R. L. *Chem. Mater.* **1995**, *7*, 493.
5. Lebeau, B.; Brasselet, S.; Zyss, J.; Sanchez, C. *Chem. Mater.* **1997**, *9*, 1012.
6. Shi, Y.; Seliskar, J. *Chem. Mater.* **1997**, *9*, 821.
7. Corriu, R. J. P.; Hasemann, P.; Lanneau, G. *J. Chem. Soc. Chem. Commun.* **1996**, 1845.
8. Schaudel, B.; Guermeur, C.; Sanchez, C.; Nakatani, K.; Delaire, J. A. *J. Mater. Chem.* **1997**, *7*, 61.
9. Schmidt, H. *J. Non-Cryst. Solids* **1985**, *26*, 300.
10. Schneider, A.; Kairies, S.; Rose, K. *Monatsch. Chem.* **1999**, *130*, 89.
11. Wilkes, G. L.; Orler, B.; Huang, H. *ACS Polym. Prep.* **1985**, *73*, 681.
12. Guglielmi, M.; Brusatin, G.; Facchin, G.; Gleria, M.; De Jaeger, R.; Musiani, M. *Proceedings of the I[st] National Workshop on Cyclo- and Poly- Phosphazenes*, Padua, Italy, February 15-16, 1996, Comm. C 14.
13. Guglielmi, M.; Brusatin, G.; Facchin, G.; Gleria, M.; De Jaeger, R.; Musiani, M. *J. Inorg. Organomet. Polym.* **1996**, *6*, 221.
14. Guglielmi, M.; Brusatin, G.; Facchin, G.; Gleria, M. *Appl. Organometallic Chem.* **1999**, *13*, 339.
15. Coltrain, B. K.; Ferrar, W. T.; Landry, C. J. T. PCT Int. Appl. WO, 90 11,323 (1990), *Chem. Abstr.* 114, 165903g (1991), assigned to Eastman Kodak Co.
16. Coltrain, B. K.; Ferrar, W. T.; Landry, C. J. T. Eur. Pat. Appl. EP, 465,555 (1990), *Chem. Abstr.* 114, 165903g (1991), assigned to Eastman Kodak Co.
17. Coltrain, B. K.; Ferrar, W. T.; Landry, C. J. T.; Molaire, T. R. *ACS Polym. Prep.* **1991**, *32(3)*, 477.
18. Coltrain, B. K.; Ferrar, W. T.; Landry, C. J. U.S. US, 5,010,128 (1991), *Chem. Abstr.* 114, 165903g (1991), assigned to Eastman Kodak Co.
19. Landry, C. J. T.; Ferrar, W. T.; Coltrain, B. K. PCT Int. Appl. WO, 91 09,080 (1991), *Chem. Abstr.* 115, 234535y (1991), assigned to Eastman Kodak Co.
20. Landry, C. J. T.; Ferrar, W. T.; Coltrain, B. K. U.S. US, 5,104,602 (1992), *Chem. Abstr.* 115, 234535y (1991), assigned to Eastman Kodak Co.

21. Coltrain, B. K.; Ferrar, W. T.; Landry, C. J. T.; Molaire, T. R.; Zumbulyadis, N. *Chem. Mater.* **1992**, *4*, 358.
22. Coltrain, B. K.; Ferrar, W. T.; Landry, C. J. T.; Molaire, T. R.; Schildkraut, D. E.; Smith, V. K. *ACS Polym. Prep.* **1993**, *34(1)*, 266.
23. Ferrar, W. T.; Coltrain, B. K.; Landry, C. J. T.; Long, V. K.; Molaire, T. R.; Schildkraut, D. E. In *Inorganic and Organometallic Polymers II. Advanced Materials and Intermediates*; Wisian-Neilson, P., Allcock, H. R., Wynne, K. J., Eds.; ACS Symposium Series: Washington, DC, USA, 1994; Chapt. 20, Vol. 572, p 258.
24. Facchin, G.; Fantin, G.; Gleria, M.; Guglielmi, M.; Spizzo, F. *ACS Polym. Prep.* **1993**, *34(1)*, 322.
25. Guglielmi, M.; Colombo, P.; Brusatin, G.; Facchin, G.; Gleria, M. *J. Sol-Gel Sci. Technol.* **1994**, *2*, 109.
26. Facchin, G.; Gleria, M.; Minto, F.; Bertani, R.; Guglielmi, M.; Brusatin, G. *Macromol. Rapid Commun.* **1995**, *16*, 211.
27. Brusatin, G.; Guglielmi, M.; De Jaeger, R.; Facchin, G.; Gleria, M.; Musiani, M. In *Syntheses and Methodologies in Inorganic Chemistry. New Compounds and Materials*; Daolio, S., Tondello, E., Vigato, P. A., Eds.: Bressanone, Italy, 1996, Vol. 6, p 315.
28. Guglielmi, M.; Brusatin, G.; Facchin, G.; Gleria, M. *J. Mater. Res.* **1996**, *11*, 2029.
29. Pivin, J. C.; Brusatin, G.; Zalczer, G. *Thin Solid Films* **1996**, *287*, 65.
30. Pivin, J. C.; Brusatin, G.; Guglielmi, M.; Facchin, G.; Gleria, M. *Nucl. Instr. and Meth. B* **1996**, *112*, 294.
31. Brusatin, G.; Guglielmi, M.; De Jaeger, R.; Facchin, G.; Gleria, M.; Musiani, M. *J. Mater. Sci.* **1997**, *32*, 4415.
32. Facchin, G.; Guarino, L.; Gleria, M.; Dirè, S.; Ceccato, R. *Phosphorus Res. Bull.* **1999**, *10*, 724.
33. Golemme, G.; Facchin, G.; Gleria, M.; Guarino, L.; Drioli, E. *Phosphorus Res. Bull.* **1999**, *10*, 736.
34. Dirè, S.; Facchin, G.; Ceccato, R.; Guarino, L.; Sassi, A.; Gleria, M. *J. Inorg. Organomet. Polym., 12*, 59 **(2002).**
35. Dirè, S.; Facchin, G.; Ceccato, R.; Guarino, L.; Sassi, A.; Gleria, M. In *Syntheses and Methodologies in Inorganic Chemistry. New Compounds and Materials*; Armelao, L., Barreca, D., Daolio, S., Tondello, E., Vigato, P. A., Eds.: Bressanone (Bz), Italy, 2000, Vol. 9, p 145.
36. Crouzet, L., *Thesis*, Université de Montpellier 2, France, July 10, (1998).
37. Crouzet, L.; Leclercq, D.; Mutin, P. H.; Corriu, R. J. P.; Vioux, A. *Mater. Res. Soc. Symp. Proc.* **1998**, *519*, 51, *Chem. Abstr.* 130, 70068 (1999).
38. Crouzet, L.; Leclercq, D. *Phosphorus Res. Bull.* **1999**, *10*, 714.
39. Crouzet, L.; Leclercq, D. *J. Mater. Chem.* **2000**, *10*, 1195.
40. Park, S.; Kim, J. S.; Chang, Y.; Lee, S. C.; Kim, C. *J. Inorg. Organomet. Polym.* **1998**, *8*, 205.
41. Kim, C.; Kim, J. S.; Lee, M. H. *Synthetic Metals* **1998**, *98*, 153.
42. Allcock, H. R.; Kuharcik, S. E. *J. Inorg. Organomet. Polym.* **1995**, *5*, 307.
43. Pierre, A. C. *Introduction to Sol-Gel Processing*; Kluwer Academic Publishers: Norwell, Massachusetts, USA, 1998, Chapt. 1,6, and references therein.
44. *Proceedings of XI International Workshop on Glasses, Ceramics, Hybrids and Nanocomposites from Gels*, Abano Terme (Padova), September 16-21, 2001.

45. Sanchez, C.; Lebeau, B.; Ribot, F.; In, M. *J. Sol-Gel Sci. Technol.* **2000**, *19*, 31.
46. Schmidt, H.; Jonschker, G.; Goedicke, S.; Mennig, M. *J. Sol-Gel Sci. Technol.* **2000**, *19*, 39.
47. Allcock, H. R.; Kugel, R. L. *J. Am. Chem. Soc.* **1965**, *87*, 4216.
48. Allcock, H. R.; Kugel, R. L.; Valan, K. J. *Inorg. Chem.* **1966**, *5*, 1709.
49. Allcock, H. R.; Kugel, R. L. *Inorg. Chem.* **1966**, *5*, 1716.
50. Rose, S. H. *J. Polym. Sci., Part B* **1968**, *6*, 837.
51. De Jaeger, R.; Gleria, M. *Prog. Polym. Sci.* **1998**, *23*, 179.
52. Blonsky, P. M.; Shriver, D. F.; Austin, P. E.; Allcock, H. R. *J. Am. Chem. Soc.* **1984**, *106*, 6854.
53. Allcock, H. R.; Austin, P. E.; Neenan, T. X.; Sisko, J. T.; Blonsky, P. M.; Shriver, D. F. *Macromolecules* **1986**, *19*, 1508.
54. Landry, C. J. T.; Ferrar, W. T.; Coltrain, B. K. U.S. US, 5,190,819 (1993), assigned to Eastman Kodak Co.
55. Allcock, H. R.; Pucher, S. R.; Turner, M. L.; Fitzpatrick, R. J. *Macromolecules* **1992**, *25*, 5573.
56. Allcock, H. R.; Pucher, S. R.; Fitzpatrick, R. J.; Rashid, K. *Biomaterials* **1992**, *13*, 857.
57. Allcock, H. R. In *Biodegradable Polymers as Drug Delivery Systems*; Chasin, M., Langer, R., Eds.; Marcel Dekker: New York, 1990; Chapt. 5, p 163.
58. Scopelianos, A. G., In *Biomed. Polym.*; Shalaby, S. W., Ed.; Hanser: Munich, 1994; p 153.
59. Blonsky, P. M.; Shriver, D. F.; Austin, P. E.; Allcock, H. R. *Solid State Ionics* **1986**, *18/19*, 258.
60. Bonsignore, L.; Corda, L.; Maccioni, E.; Podda, G.; Gleria, M. *Gazz. Chim. Ital.* **1991**, *121*, 341.
61. Delprato, C.; De Jaeger, R.; Houalla, D. *ACS Polym. Prep.* **1993**, *34(1)*, 320.
62. Delprato, C.; De Jaeger, R.; Houalla, D.; Potin, P. *Macromolecules* **1995**, *28*, 2500.
63. Bhatt, M. V.; Kulkarni, S. U. *Synthesis* **1983**, 249.
64. Ferrar, W. T.; Marshall, A. S.; Whitefield, J. *Macromolecules* **1987**, *20*, 317.
65. Allcock, H. R. *Phosphorus-Nitrogen Compounds. Cyclic, Linear, and High Polymeric Systems*; Academic Press: New York, USA, 1972.
66. Allcock, H. R. *Phosphorus-Nitrogen Compounds. Cyclic, Linear, and High Polymeric Systems*; Academic Press: New York, 1972, p 358.
67. Kyker, G. S.; Antkowiak, T. A. *Rubber Chem. Technol.* **1974**, *47*, 32.
68. Allcock, H. R.; Cook, W. C. *Macromolecules* **1974**, *7*, 284.
69. Lawson, D. F.; Cheng, T. C. *Fire Research* **1977/78**, *1*, 223.
70. DeEdwardo, A. H.; Zitomer, F.; Stuetz, D.; Singler, R. E.; Macaione, D. *Org. Coat. Prep.* **1976**, *36*, 737.
71. DeEdwardo, A. H.; Zitomer, F.; Stackman, R. W.; Kramer, C. E. US Patent, 4,042,561 (1977), *Chem. Abstr.* 87, 118861f (1977), assigned to Celanese Corp.
72. Penton, H. R. In *Inorganic and Organometallic Polymers*; Zeldin, M., Wynne, K. J., Allcock, H. R., Eds.; ACS Symposium Series: Washington, DC, USA, 1988; Chapt. 21, Vol. 360, p 277.
73. Pezzin, G.; Palma, G.; Lora, S.; Carenza, M. *Chem. Ind. (Milan)* **1991**, *73*, 185.
74. Mark, J. E.; Allcock, H. R.; West, R. *Inorganic Polymers*; Prentice Hall: Englewood Cliffs, New Jersey, USA, 1992, Chapt. 3, p 61.

75. Wade, C. W. R.; Gourlay, S.; Rice, R.; Hegyeli, A.; Singler, R. E.; White, J. In *Organometallic Polymers*; Carraher, C. E., Sheats, J. E., Pittman, C. U., Eds.; Academic Press: New York, USA, 1978, p 289.
76. Fewell, L. L. *J. Appl. Polym. Sci.* **1990**, *41*, 391.
77. O'Brien, J. P.; Ferrar, W. T.; Allcock, H. R. *Macromolecules* **1979**, *12*, 108.
78. Gleria, M.; Bortolus, P.; Minto, F. *This book.*
79. Allcock, H. R.; Gebura, M.; Kwon, S.; Neenan, T. X. *Biomaterials* **1988**, *9*, 500.
80. Golemme, G.; Drioli, E. *J. Inorg. Organomet. Polym.* **1996**, *6*, 341.
81. Medici, A.; Fantin, G.; Pedrini, P.; Gleria, M.; Minto, F. *Macromolecules* **1992**, *25*, 2569.
82. Singler, R. E.; Hagnauer, G. L.; Schneider, N. S.; LaLiberte, B. R.; Sacher, R. E.; Matton, R. W. *J. Polym. Sci., Polym. Chem. Ed.* **1974**, *12*, 433.
83. Facchin, G.; Guarino, L.; Modesti, M.; Minto, F.; Gleria, M. *J. Inorg. Organomet. Polym.* **1999**, *9*, 133.
84. Brusatin, G.; Guglielmi, M.; De Jarger, R.; Facchin, G.; Gleria, M.; Musiani, M. *Proceedings of the 6th Meeting on Syntheses and Methodologies in Inorganic Chemistry. New Compounds and Materials*, Bressanone (Bz), Italy, December 18-21, 1995, p 110, Comm. P 23.
85. Neilson, R. H.; Wisian-Neilson, P. *Chem. Rev.* **1988**, *88*, 541.

In: Applicative Aspects of Poly(organophosphazenes)
Editors: M. Gleria and R. De Jaeger, pp. 119-137
ISBN 1-59454-025-X

Chapter 6

THE USE OF PHOSPHAZENES AS FLAME RETARDANTS

Christopher W. Allen and Daniel Hernandez-Rubio
Department of Chemistry, University of Vermont,
Burlington, Vermont 05405–0125, U.S.A.

I. INTRODUCTION

A. Problems Associated with Commercial Flame Retardants

The world of contemporary manufactured items has a broad array of components derived from polymeric materials. These range from hard, molded objects (thermosets) and flexible items (thermoplastics) to films, fibers and foams. The vast majority of these polymers are based on organic compounds and consequently are thermodynamically unstable in an oxidative environment. The most common consequence of this instability is fire. The flammability of materials presents a general threat to both integrity of the product and human health. Therefore, the search for fire retardants and inherently inflammable polymers represents a major technological challenge. The ideal solution is the utilization of a polymer which in itself is resistant to fire (the reasons for which will be explored below). Unfortunately, these materials are not common or are not economically viable for mass production (specialty polymers). The alternative is incorporation of an additive with flame retardant properties to the mixture of agents which transform the raw polymer to a manufacturable material (compounding). The current market for all varieties of flame retardants has been estimated to be in the range of $2.2 billion which translates to approximately 2.2 billion pounds per year.[1] Traditionally, halogenated materials have played a major role in flame retardant additives. However the release of toxic materials, particularly dioxins, from incineration of halogenated compounds, has lead to the removal of chlorinated materials from the market.[2] The obvious move to the corresponding brominated compounds occurred.[1,2] Recent concerns about the endocrine disrupter activity and other health related issues of selected brominated flame retardants[3,4] as well as the potential for release of polybrominated dibenzo-p-dioxins and dibenzofurans under pyrolytic conditions[1] has resulted in the reconsideration of the viability of these materials. The EU has recently placed restrictions on selected brominated flame retardants.[3] The regulatory activity associated with materials which have, or are perceived to have, a negative impact on the environment will result in considerable pressures to find a "green chemistry" solution to these problems. Phosphorus compounds, and compounds containing phosphorus and nitrogen are two major target areas for development of non-halogenated flame retardants. Rhodia Corporation offers several of these in the Proban and Antiblaze series of products. Given this success, phosphazenes represent a logical target for next generation of green flame retardant additives and materials.[5]

B. Fire and Fire Retardancy

A brief discussion of fire and the combustion process in polymeric materials[2,6] is a necessary prelude to a consideration of flame retardants and their modes of action. The application of a source of thermal energy, usually originating external to the polymeric material, will initiate thermal decomposition events. These processes may lead to crosslinking and char formation, volatilization or a combination of both. The char formation at the surface often produces a foamed matrix in a process referred to as intumescence. The combination of the volatile organic matter with oxygen leads to oxidation, ultimately to carbon dioxide and water, of the organic material. This is a very exothermic process, e.g. H° for the reaction of methane with oxygen to yield carbon dioxide and water is -890.3 kJ/mole. The excess heat released in the combustion process leads to additional thermal decomposition of the polymer thus accelerating burning of the material. This simple model allows for a rational approach to the design of fire retardant systems. There are two points in the flammability cycle where it can be interrupted and consequently quenched. If, in the initial thermal decomposition step of the polymer, the crosslinking and decomposition events are extensive enough to form an impenetrable layer on the surface, volatilization of additional organic matter which can act as fuel for the burning process will be inhibited. The formation of a protective surface layer is similar to the well known passivation of active metals such as aluminum in an oxidative environment. Alternatively, if the volatile material inhibits rather than promotes the oxidative process, the flame will also be quenched. The chemical processes in flames are very complex, but free radicals are believed to be an important component. Consequently, chemical entities which will terminate the radical propagation reaction will, in turn, be fire suppressants. Halogens are typical examples of agents which are believed to act in this fashion.

C. Evaluation of Flame Retardants

The next question which arises is: how does one evaluate the effectiveness of flame retardant systems? The "zero level" approach is to simply expose the sample to a flame and see if it burns (simple flame test). Certainly low tech but ultimately related to real world conditions. Also, it is the most widely used test. Standardized conditions have been established for this test.[7] One can move from the yes/no result of the simple flame test, to a quantitative scale by use of the limiting oxygen index (LOI) methology. In this case, burning is conducted in a controlled environment wherein the percent of oxygen can be controlled.

The minimum amount of oxygen required to sustain burning is the limiting oxygen index. Thus, the more oxygen required (higher LOI), the stronger the flame retardancy effect. There is a standardized ASTM methodology and slight variations in oxygen index determination.[8] In this discussion, all data will be reported as LOI independent of measurement method used. Additional measurement technologies involve cone calorimetry[9,10], in which the enthalpy change of the burning process is evaluated and smoke tests[9] another chemical issue in real world fire behavior evaluation. The results of these latter techniques are far less commonly reported and are often part of studies which come at a latter stage of analysis than that in which the polymer chemist is involved. The other evaluation methodology which is commonly involved in evaluation of potential flame retardancy is thermogravimetric analysis (TGA). This measurement will provide data on the amount of char formation and hence, in principle, the ability of formation of a protective surface layer in the thermal decomposition process[10]. However, it has been noted that there is no direct correlation between flammability and TGA results for many polymeric systems.

D. Fundamentals of Commercial Nitrogen-Phosphorus Based Flame Retardants

As mentioned in Section I.A., materials containing both nitrogen and phosphorus have been commercialized as flame retardant additives. The archetype compound in this series of products is ammonium polyphosphate (Rhodia antiblaze Cl, Antiblaze FSD and Antiblaze LR3). The combination of nitrogen and phosphorus in the same compound may lead to a synergistic effect wherein the combined effect of two components is greater than a simple additive combination of their individual flame retardant effects.[12-14] Knowledge of the mechanism of action of ammonium polyphosphate may prove useful in understanding the behavior of the more structurally complex phosphazene based flame retardants. Generally, it is believed that nitrogen-phosphorus additives form phosphoric or polyphosphonic acid which in turn plays a catalytic role in char formation.[15-17] A study of the effect of ammonium phosphate on polyolefin flammability showed that the addition of a nitrogen containing material derived from melamine significantly enhanced the flame retardant effect.[16] This behavior was related to release of volatile nitrogen compounds into the gas phase during the decomposition event. These volatile materials inhibit radical processes. The role of char formation was shown to be crucial in the action of ammonium phosphate on polyurethane flammability[18]. It was shown that the additive accelerates surface decomposition and increases the amount of the high temperature residue. Of particular interest was the detection of trapped radical species in the char which also may contribute to flame inhibition[18]. These studies show the complexity of flame retardant effects even in simple additive systems.

E. Classes of Phosphazenes Exhibiting Flame Retardancy

Phosphazenes have been considered separately as additives or inherently flame retardant materials. In the additive class, both small molecule (cyclophosphazenes) and polymeric species (polyphosphazenes) have been studied. Inherently flame retardant phosphazenes include cyclophosphazene fluids, carbon chain polymers with cyclophosphazenes as substituents, polymers in which the cyclophosphazene is part of the backbone (cyclolinear and cyclomatrix materials) and, of course, the polyphosphazenes. Each of these will be considered in turn in the following sections. Earlier reviews of the flame retardant properties of various phosphazene derivatives have appeared[5,19] many of which are in less accessible sources[20-28].

II. PHOSPHAZENES AS ADDITIVES

A. Cyclophosphazenes

The commercial application of the propoxy derivatives $[NP(OC_3H_7)_2]_{3,4}$ plus linear oligomers, as a flame retardant additive for rayon was an early application of phosphazene derivatives[29]. This application suggests that the entire range of cyclophosphazenes should be considered in this field of application. The nucleophilic substitution reactions of $N_3P_3Cl_6$ and its higher homologs lead to literately thousands of derivatives which have the potential to function as flame retardants. This chemistry is adequately discussed in this monograph and in the core reference sources in the area.[30-32] An extensive but not comprehensive survey of cyclophosphazene derivatives which have been shown to exhibit flame retardancy and the polymers in which this effect has been manifested may be found in Table 1.

A cursory survey of the entries in Table 1 leads to a few general observations. Earlier studies employed mixtures of oligomeric cyclo- and linear phosphazenes while later investigations focused on specific species, usually a cyclotriphosphazene derivative. From a conceptual perspective examination of a defined molecular unit is the most logical approach in that the number of variable contributing to the flame retardant effect are minimized. However, from an applied perspective utilization of the phosphazene oligomer mixture, and thus avoiding the often tedious and complex procedures[31] involved in isolation of the pure cyclotriphosphazenes is preferable. Unfortunately, a controlled study investigating the relative flame retardant effects of individual cyclophosphazene oligomers and linear derivatives has not appeared. If it could be established that the effect is independent of structural unit, then the economic advantage is dealing with unpurified mixtures is apparent. It also should be noted that several of the phosphazene moities in Table 1 contain halogens, either as direct substituents i.e. chlorophosphazenes or on the organophosphazene substituent e.g. haloalkoxy or aryloxy units. As noted in Section I.A., halogen based flame retardants are either banned on being phased out, so introduction of a new halogen containing flame retardant additive into the marketplace is unlikely. Finally, one can subdivide the phosphazenes in Table 1 into two groups, those with inert substituents with respect to the host polymer e.g. phenoxy and those which have the potential for electrostatic or covalent interactions with the host polymer e.g. amidophosphazenes. It has been noted[64] that an advantage of the nonfunctional phosphazenes is the absence of degradation of key polymer physical properties such as modulus, moisture uptake, fracture toughness, etc. by the incorporation of the phosphazene additive. There are also potential benefits of a reactive phosphazene substituent. A flame retardant in fibers used for clothing and the like has the potential of being removed by the cleaning process thus increasing the flammability of the material over time. A distinct advantage of amido phosphazenes such as $[NP(NH_2)_2]_{3,4}$ is that stability to washing, and interestingly, shrink proofing, is observed for cotton and rayon with these additives.[41,43,49] Presumably, electrostatic interactions such as hydrogen bonding between the basic amido functionality and polar components of the host polymer promote additive binding. However stability to washing and dry cleaning has also been observed for aryloxy[72] and alkoxylcyclophosphazene[29] additives. Phosphazenes with more reactive substituents which are directly incorporated into the backbone structure via chemical reactions will be considered separately along with other cyclolinear polymers (Section III. B).

Table 1. Flame Retardant Cyclophosphazene Additives

Cyclophosphazene Substituents	Polymer Matrix	Reference
Cl	Lignin, Lignosulfonates	33, 34
$N_3P_3Cl_6$ + $Cu(OAc)_2$	Nylon	35
$(NPCl_2)_{3,4}$ + orthophosphoric acid esters	General flame retardants	36
Oligomeric chlorophosphazenes	Chlorosulfonated polyethylene (flame proof hose covers)	37
Chloro oligomers and polymer	Cellulose	38
Chloro oligomers and polymer	Thermoplastics	39
Oligomeric chlorophosphazenes + NH_3; $N_3P_3Cl_6$ + NH_3	Cotton Cotton	40 41,42
NH_2	Wood	43
NH_2	Cellulose fibers	44
NH_2	Aromatic polyamide fibers	45
$(NH_2)_3PNNP(NH_2)_3$	Cotton, polyester	46
NH_2, OMe; oligomers and linear	Cotton	47
NH_2, OPh	Polyurethane	48
$N_3P_3(OPh)_4(NH_2)_2$	Polyethylene terephthalate	49
NH_2, $NHCH_2OH$	Cotton	50
$(NPCl_2)_{3,4}$ + alkyl, arylamines	Polyurethanes	51
Oligomers + N-methylol derivatives of ureas	General additive	52
$NHCH_2OH$, OR	Polyurethane	53
C_6H_5NH, $C_6H_4(NH)_2$, $C_6H_4(O)NH$	Poly(butylene terephthalate)	54
Chloro oligomers and polymer + OH^-	Cotton	55
2,3-dichloropropoxy, n-butylamino, ethylene diamino	vicose fiber	56
Alkoxy, haloalkoxy	Rayon	29
2,3-dibromopropoxy, alkyloxy	Cotton	57
$OCH_2(CF_2)_mY(Y{=}H,F)$	Fire retardant plasticizer	58
OR, OR'(haloalkyl)	Hydrophobic fibers	59
$N_3P_3(OCH_2CCl_3)_6$	Polyester	60
$(NPCl_2)_{3,4}$ + mono and polyalkylene glycols	General additive	61
Chloro cyclics and oligomers	Polyurethane foams	62

$N_3P_3(OPh)_6$	Poly(butylene terephthalate)	54
$N_3P_3(OPh)_6$	Expoxy resins	63
$N_3P_3(OPh)_6$	Polyarylamides	64
$N_3P_3(OPh)_6$	Poly(dicyclopentadiene)	65
OC_6H_4Br	Poly(ethlene terephthalate) Rayon, poly(dicyclopentadiene), polyesters	66,67 68 65 69
OC_6H_4Cl	Phenolic resins	70
$OC_6H_4NH_2$	Polystyrene blends	71
$N_3P_3Cl_6 + BrC_6H_4OH$ + bisphenol A	Dacron	72
Chlorooligomers + polyhydroxyaromatics	Fire resistant materials	73, 74
o-$O_2C_6H_4$	Cyanoacrylates	75
o-$O_2C_6H_4$	Poly(butylene terephthalate)	54
$N_3P_3(OC_6H_4OP(O)(OR)_2)_6$	Polystyrene	11
$N_3P_3(OMe)_5OPh$	Polyester	76
$N_3P_3Ph_6$	Polyester	77

A variety of LOI measurements have been reported for cyclophosphazene fire retardant additives. Typical examples are discussed below. A series of alkoxy and fluoroalkoxy ($Rf{=}CH_2(CF_2)_nH$, n=2,4,6) chlorocyclotriphosphazenes have have been examined with LOI values ranging from 24.0 for $N_3P_3(OMe)_6$ to 38 for $N_3P_3(OCH_2C_2F_4H)_3Cl_3$.[78] The relative order of effectiveness of substituents in increasing the LOI value is: NP(ORf)Cl>$NPCl_2$>$NP(ORf)_2$>NP(ORf)(OR)>NP(OPh)Cl>$NP(OR)_2$ The LOI of polyester resin blends goes from around 18 to 21.6 and 24.0 by addition of trichlorethanol[60] and p-bromophenol[69] derivatives of $N_3P_3Cl_6$ respectively. The $N_3P_3[OC_6H_4P(O)(OR)_2]_6$ (R=Et,Ph) derivatives are effective flame retardants for poly(styrene) with the R=Ph derivative being the better additive. LOI values at 10 and 20% loading (24 and 26) are better than a standard additive, triphenylphosphate. There is a leveling off of performance at this level so 30% loading is minimally less effective.[11] The addition of $N_3P_3(OR)_6$ ($R{=}Ph,CH_2CF_3$) to poly(styrene) at a moderate loading level produces impressive LOI values of 45 and 35 for the phenyl and trifluoroethoxy derivatives respectively. Similar effects, however, were not observed for poly(methylmethacrylate).[79] A series of aromatic, $N_3P_3(EPh)_6$ (E=NH,O) and spiroaromatic N_3P_3(ortho L,L'C_6H_4)$_3$ (L,L'=O,O; NH, NH;O,NH) derivatives were evaluated as fire retardant additives for poly(butylene terephthate) at loading levels of 10 and 20%[54]. An increase of LOI values from 22 (untreated) to 29 was obtained with $N_3P_3[o\text{-}(O,NH)C_6H_4]_3$ being the most effective agent. However, dripping of flame for vertically positioned samples (UL94 protocol) reduced the overall ratings of the materials. Oxygen index data for these cyclophosphazenes with co-additives was also collected. Oxidizing agents such as metal oxides were ineffective but high charring polymers such as poly(phenylene oxide) and novolac resins were effective[54].

A general mechanism to explain the flame retardant effect of cyclophosphazene derivatives has not appeared. Rather, it appears that different processes are important depending both on the phosphazene substituent and the polymer matrix. Protective surface layer formation has been indicated in several studies. In the combustion of wood treated with chloro, methoxy or phenoxy phosphazenes, the residues show evidence for POC bond formation i.e. reaction of the phosphazene with the matrix. The gases given off in the pyrolysis of these materials show a decrease in the amounts of combustible CH_4 and CO. This general process, i.e. interaction of the phosphazene with the matrix to produce a new type of char has been noted in other systems. For example, the combination of $[NP(OPh)_2]_3$ with polymer additives such as polyphenylene oxide and novolac produces a high char level and hence a protective barrier[54]. Similarly, $[NP(OC_6H_4OP(O)(OR)_2)_2]_3$ combines with polystyrene to produce an increase in char yield [11]. A detailed study of mixed substituent tris(alkoxyl), aminophenoxy cyclotriphosphazene pyrolytic decomposition process has shown the production of a phosphorus oxynitride product bonded with carbonized polyaromatics which could model the protective layer found in the thermal decomposition of selected cyclophosphazene derivatives [81]. An interesting variant on the phosphazene additive-polymer matrix reaction mechanism is one in which the resulting crosslinking suppresses production of flammable volatiles from the polymer itself. In turn, volatiles from the degradation of the phosphazene act to quench the flame in a vapor phase mechanism. This process might be expected to be enhanced in highly functional materials such as cellulose and related materials.[33,34] Direct volatization of the phosphazene additive provides a gas phase material which will inhibit the development of the flame. Examples include $N_3P_3(OPh)_6$ in poly(butyleneterephthalate)[54] and polystyrene.[79] If the polymer matrix depolymerizes to produce oxidizable volatiles at a temperature lower than that required to volatilize the cyclophosphazene additive then the flame retardant effect will not be operative. This balance of properties has been observed when poly(methylmethacrylate) (PMMA) was used as the polymer matrix.[79]

B. Polymeric Additives

Polymeric phosphazene additives are another potential source of inducing fire retardancy. Linear polyphosphazenes with inert substituents can, in principle, be blended with flammable organic polymers. Potential phosphazene targets which could be used in this approach are discussed in Section III. E. Polymeric phosphazene additives which undergo reactions with the polymer matrix are considered in Section III.E. Two ceramic materials with high thermal stability derived from, or related to, cyclophosphazenes, phospham (PN_2H) and phosphorus oxynitride (PON), have been studied and employed as flame retardant additives. The efficiency of phospham in nylon 4,6 is similar to nylon with red phosphorus as an additive [14,82] (red phosphorus is a strong flame retardant and has been used in nylon[81]). Phospham is not particularly effective in poly(butylene terephthalate). The LOI for 20% loading of the additive is increased by only approximately 1% with a low degree of char formation.[54] Phosphorus oxynitride is a somewhat better char former for poly(butylene terephtalate) but the LOI at 20% loading is actually lower than the untreated polymer. At 30% loading a modest improvement in the LOI is noted.[54] The low effectiveness was attributed to the acceleration of production of combustible aliphatic moieties.[83] Phosphorus oxynitride alone or with coadditives has been shown to be an effective fire retardant for polyamide 6. The results of IR and TGA experiments suggest that PON is good char producer in polyamide 6.[83]

III. Phosphazenes as Inherently Flame Retardant Materials

A. Small Molecule Species

In addition to the use of small molecule, i.e. cyclic and short linear, as opposed to polymeric phosphazenes as additives, the inflammable nature of these materials has been recognized as a valuable property in bulk materials. Non-ionic linear phosphazene bases, e.g. $((Me_2N)_3P{=}N)_3P{=}NCMe_3$, have attracted significant attention as non-nucleophilic bases in organic E2 elimination reactions.[84] The non-flammable nature of these materials is a useful value added property in their use in synthesis.[85] Crystallization of cyclophosphazenes can be inhibited by a combination of mixed substituents, the use of alkoxy and aryloxy substituents rather than substituents derived from amines which tend to support hydrogen bonding, or the use of fluorinated organic fragments on the substituent. The resulting materials are oils which have the potential for a variety of applications in the area of synthetic lubricants and higher performance functional fluids. This work has been summarized in a valuable review.[86] One property of interest for the cyclophosphazene fluids is the spontaneous inignition termperature as measured by a standard ASTM (D-2155-66) test. For materials with combinations of aryloxy (OAr; Ar=C_6H_5, C_6H_4-m-CF_3, C_6H_4-m-Cl) and fluoroalkoxy (OR_f; $R_f{=}CH_2(CF_2)_nH$(n=2,4), $CH_2(CF_2)_nF$(n=1,3)) substituents, these values range from 538 to 677 °C [86] which clearly demonstrates the difficulty in initiating burning of these fluids. A specific application in this area is non-toxic fire resistant hydraulic fluids which were of specific interest to the U.S. Navy. Trisubstituted aryloxy/fluoroalkoxy cyclotriphosphazenes exhibited the optiminal property profile (including fire resistance) in this series.[87] Fluoralkoxy phosphazene fluids have been shown to be effective fire suppressants which are safe from environmental and toxicological perspectives.[88] All of these studies indicate the high applications potential of the cyclophosphazene derivatives but cost considerations continue to present the most significant barrier to commercial explotiation.[86]

B. Polymers with Cyclophosphazenes as Substitutents

A logical extension of the additive approach would be to develop systems in which the cyclophosphazenes flame retardant is covalently bound to the polymer chain. In principle, this approach would prevent leaching of the additive from the polymer over time and promote char formation. On the other hand, this type of system maybe less effective in cases where the additive works via a vapor phase mechanism. Systems of this type were first designed and prepared in our laboratories[89] and have been since reviewed[90,91] as well as being discussed elsewhere in the monograph. The fundamental approach employed in these studies involves preparation of a cyclophosphazene with an olefin containing substituent followed by olefin addition polymerization or copolymerization leading to an organic polymer with a cyclophosphazene substituent. A recent alternative approach has also been reported in which an azidocyclophosphazene is allowed to react with a polymer containing a diphenylphosphine substituent. The resulting Staudinger reaction gives rise to a phosphoranimine linkage between the cyclophosphazene and the polymer.[79] Examples of homopolymers containing cyclophosphazene substituents which have exhibited flame retardant properties in simple flame tests include the following: $[CH_2CH(RON_3P_3X_5)]_n$ (X=Cl; F; $R=C_6H_4C_6H_4$)[92], $[(CH_2CH(ON_xP_xCl_{2x-1})]_n$ (x=3,4)[93], $[(CH_2CMe(C(O)ORON_3P_3Cl_5)]_n$ ($R=(CH_2)_2$[94], $(CH_2)_{3,4}$)[95], $[CH_2CH(C_6H_4CH_2P(Me)N(NPCl_2)_2)]_n$.[96] These materials all exhibit high char yield in TGA studies. Some species give off flammable volatiles but self-extinguish upon removal of the flame.[96] This suggests a protective barrier mechanism however in other cases rapid evolution of HCl is observed[93] which could also contribute to the flame retardancy.

A more interesting example of the potential applications of polymers with cyclophosphazene substituents can be found in copolymers where there is a limited amount of the expensive cyclophosphazene component. Selected examples of fire retardant materials obtained from the copolymerization of olefin containing cyclophosphazenes and standard organic comonomers may be found in Table 2. Incorporation of only a few mole percent of $N_3P_3F_5CMe{=}CH_2$ in styrene copolymerization yields copolymers which do not burn or give off smoke in air when exposed to flame.[89] Often increased thermal stability (over the organic homopolymer) is observed[99] and high char yields may be[96] but are not always[97,98] obtained. The fact that flame retardancy is observed in cases when thermal decompositon by random sission is observed indicates that the vapor phase flame inhibition process can occur. In the hydroxyethylmethacrylate derivative/MMA copolymers only modest effects are noted.[94] This may be related to high level of combustibility of the MMA homopolymer.[79] A quantitative and systematic studies of the copolymers derived from examination of the copolymers derived from MMA or styrene with diphenyl-p-styrylphosphine followed by the Staudinger reaction with azidocyclophosphazenes $N_3P_3(OR)_5N_3$(R=Ph, CH_2CF_3), to introduce the phosphazene into the copolymer has appeared.[79] In the styrene system with 10% loading the LOI increases from 22 (in the absence of the phosphazene) to 28 ($R=CH_2CF_3$) and 33 (R=Ph). In the MMA system, LOI increases from 17 to 22 (R=Ph) and 27 ($R=CH_2CF_3$). In the styrene system, the observation of a high char yield suggests a barrier mechanism whereas in MMA a vapor phase mechanism has been proposed.[79] Reactions of a cyclophosphazene containing the 3,5-dihydroxyphenyloxy substitutents with diisocyanates produces heat resistant polyurethanes which, presumably, are also fire resistant.[101]

Table 2. Flame Retardant Phosphazene Containing Copolymers[a]

$N_3P_3X_4YR$			Comonomer	Reference
R	X	Y		
$C(Me)=CH_2$	F	F	Styrene	89
$C(OEt)=CH_2$	F	F	Styrene	97
$C_6H_4CMe=CH_2$	F	F	Styrene	98
$CH_2C_6H_4CH=CH_2$	Cl	Me	Styrene, MMA[b]	96
$C(OC(O)Me)=CH_2$	Cl	i-Pr	Styrene	96
$OC_6H_4C_6H_4CH=CH_2$	Cl	Cl	Styrene	99
$OC_6H_4C_6H_4CH=CH_2$	Cl	Cl	MMA[b] Acrylates	100
$O(CH_2)_2OC(O)CMe=CH_2$	Cl	Cl	MMA[b]	94

[a] derived from olefin addition copolymerization
[b] methylmethacrylate

C. Cyclolinear Phosphazenes

Another variation on the theme of covalent binding of the cyclophosphazene flame retardant to the polymer chain is the formation of cyclolinear polymers. In this approach a cyclophosphazene with two substituents with reactive moieties is allowed to react with a difunctional reagent leading, via a step polymerization process, to a polymer with the cyclophosphazene as part of the backbone. Oligomeric cyclolinear materials consisting of mixed chloro/aryloxy units show LOI values in the range of 23 to 29 which are raised to 32 to 36 when the bridging groups between the phosphazenes contain halogens (F, Br).[78] Well defined, characterized materials are available from the reactions of 2,4-bis-(p-aminophenoxy)tetraphenoxyclyclotriphosphazene with diacid chlorides ClC(O)RC(O)Cl (R=$(CH_2)_8$, C_6H_4, $C_6H_4CH_2C_6H_4$, $C_6H_4SO_2C_6H_4$). The resulting polyamides have high decomposition temperatures and char yields as well as being self extinguishing in simple flame tests.[102] Similarilly, diphenylcyclophosphazenes with bis-(p-aminophenoxy) or bis(p-phenoxyacid chlorides) gives cyclolinear polyamides and polyesters. Copolymers with a limited amount of phosphazene incorporation have also been prepared.[103] Cyclophosphazene derivatives containing two p-hydroxyphenoxy units undergo reactions with a diisocyante, $OCN(CH_2)_6 NCO$, to give cyclolinear phosphazene polyurethanes with high char yields and decomposition temperatures. These properties have been proposed to be related to the flame retardant effect of the phosphazenes.[101,104] A three component polymerization of the phosphazene, and organic diol and the diisocyante allows for the introduction of variable amounts of the phosphazene additive.[101] A new methology for producing cyclolinear phosphazene polymers *via* a metal catalyzed olefin metathesis reactions of cyclophosphazenes containing two alkoxy units with olefins at the terminal position has been described as a route to new fire resistant polymers.[105] The successful preparation of cyclomatrix cyclophosphazenes with polyimide linkage which have flame retardant activity (Section III.D) has lead to similar cyclolinear materials. Thus the geminal diamine 2,2-bis(p-aminophenoxy)4,4,6,6-bis[spiro(dioxy-1,1-biphenylyl)cyclotriphosphazene undergoes a two step polymerization with a benzophenone-tetracarboxylic anhydride followed by diphthalic anhydride to give polyamic acids. Thin films of these materials undergo thermal cyclodehydration to yield the cyclolinear polyimide which exhibits high char yields at 800°C.[106]

D. Cyclomatrix Phosphazenes

If one has three or more polymerizable substituents on the cyclophosphazene ring, chain growth will be in three directions leading to crosslinked cyclomatrix materials. These intractable materials present challenges to characterization at the molecular level and hence details of many of the products prepared in early studies are vague. More recently, renewed interest has been shown in cyclomatrix phosphazene polymers, particularly as flame resistant materials. Of particular interest has been polyimides and related materials reported by Kumar. For example, aminophenoxycyclophosphazenes react with maleic anhydride to give maleamic acids which upon heating give maleimides. Thermal curing leads to polymaleimide formation. Starting from this phosphazene maleimide, a highly crosslinked resin is obtained. This material has an 82% char level at 800° and graphite cloth laminates of the material exhibit an LOI of 100.[107] A wide variety of resins with similar properties have been reported from poly(bismaleimides)[107-110], polyimides,[111] poly(ether imides)[112], epoxy[113] and ethynl[114] resins. A hard, optically transparent film exhibiting burn resistance has been reported from the thermal or photochemical curing of $N_3P_3(OCH_2CH_2OC(O)CMe{=}CH_2)_6$.[115] Flame retardant epoxy resins are obtained from cyclophosphazenes having both $OC_6H_4OP(O)(OEt)_2$ and OC_6H_4OH substituents by allowing epichlorohydrin to react with the free phenolic reside followed by diamine curing agents. Compared to a standard epoxy polymer, the phosphazene material exhibited higher char yields and LOI values. The LOI values ranged from 32 to 36.[116] Epoxyclophosphazenes have also been considered as fillers, coatings etc. which exhibit high fire and heat resistance.[117] Cyclophosphazenes with phenolic residues are converted to cyanates by reaction with CNBr. Thermal curing leads to cyclotrimerization of the cyanate thus providing cyclomatrix materials containing both cyclophosphazenes and cyclotriazines. The LOI values range from 32 to 39. The flame retardancy improved with increased crosslink density and char yield of the polymer.[118] Composites derived from cyclophosphazenes have been developed as thermosetting polymers[119] with potential fire retardant systems. The addition of the hexaziridino derivative, $N_3P_3(NC_2H_4)_6$, to aqueous polyurethane solutions leads to cross-linked flame retardant materials. Cone calorimetry studies show a significant delay in ignition time. The materials are more thermally stable and less flammable than non-phosphazene containing polymers. However, larger degrees of smoke formation were noted.[120] It should be noted that the cytotoxic nature of aziridinocyclophosphazene means that caution should used in handling these materials.

E. Polyphosphazenes

The low flammability of a wide range of polyphosphazenes is one of the most important properties driving the attempts at commercialization of these materials. Synthesis and characterization of polyphosphazenes is covered in other sections of this monograph. The mechanism of the thermal decomposition of polyphosphazenes while ultimately relevant to flammability is a separate area and interested readers can explore this in several key references.[121-124] The elastomeric polyphosphazenes can be processed as films or foams with applications as flame retardant insulating foams, cable coatings, floor covering, acoustic sheeting and products for military needs. For a limited period of time, commercial products derived from aryloxy polyphosphazenes were available as flame retardant foams from Ethyl Corporation (EYPEL-A) and Elf Atochem (ORGANOFLEX). Numerous earlier reviews are available which have, in part, a focus on fire retardant applications.[5,125-129] A wide range of aryloxypolyphosphazenes and formulated mixtures thereof have been investigated as potential flame resistant polymers.[11,130-145] While the bis(phenoxy) derivative is the prototype material, polymers containing mixed substituents $[NP(OC_6H_4R)_{2-x}(OC_6H_4R')_x]_n$ where R and R' can be (in addition to H), alkyl, alkoxy and halogens have been the focus for development. Vulcanization/crosslinking can be attained by the use of peroxides, sulfur or radiation[129] on the materials indicated above[131,133-135,144] or in materials which have a crosslinking site, e.g. $OC_6H_4(OMe)CH_2CH{=}CH_2$,[141] as an additional substituent. The mixed substituent derivatives can be prepared by the traditional method of sequential addition of the corresponding aryloxides to polydichlorophosphazene or by polymerization of $N_3P_3Cl_5(OPh)$ followed by aryloxide addition to give more well defined structures[139,142-145] with better properties.[144]

Properties relevant to utilization of these materials in flame retardant applications have been reported. These include LOI values, NBS smoke density tests, flame spread indecies and evolved gas toxicity. Several of these measurements are on a formulated products rather than pure polymer. LOI values typically range from 23 to 65.[131-135,138] The alkoxy substituted aryloxy polymers have lower LOI values than the alkylaryloxy materials[138] and as expected, haloaryloxy species exhibit higher LOI values (38-65) than the non-halogenated materias.[135] Detailed LOI studies at a variety of heat fluxes show that the polyphosphazenes require more oxygen than is available in ambient air to burn. Fillers influence (usually increase) LOI values and foams burn more readily than films.[146] However, closed cell foams have been prepared with a LOI value of 44.[129,145] In general low smoke evolution is observed in these polymers.[129] Less smoke is released in alkoxyarylaryloxy than from alkylaryloxy derivatives which is opposite of the trend in the LOI values.[138] The resulting off gas toxicities are generally low [129,146] but depend strongly on the nature of the phosphazene substituent.[146] The halogen free polyphosphazenes are favored over the halogen containing polymers in this regard.[146] Smoke emission can also be mediated with the addition of certain additives such as organotitanate treated mineral fillers[140] or solid carboxylic acids.[137] Polyphosphazenes in which the known flame retardant phosphate and phosphonate moieties are incorporated on to the aryloxy unit have also been reported.[11] The synthetic methodology involves reactions of halogenated organophosphates with hydroxyl or lithio functionalized aryloxy units. The resulting polymers exhibit minimal weight loss until 490°C and produce a 40% char yield. Aryloxyclophosphazenes have been used as plasticizers for polyaryloxyphosphazenes.[147]

Significantly less interest has been shown in alkoxy or fluoralkoxyphosphazenes as materials where the primary focus is on low flammability. However, in many of the applications which exist for fluoroalkoxy derivatives e.g. seals, gaskets and fuel lines[129] certainly flame resistance is an additional valuable property. The mixed trifluoroethoxy-octafluoropentoxy polyphosphazene is reported to be non-flammable.[148] Fluoroalkoxypolyphosphazenes have been used as fire resistant coatings for polyester fabrics. The coated materials have better flame resistance and lower flame propagation than analogous materials coated with chlorinated organic polymers.[149]

An alternative strategy which can be employed in the utilization of the flame retardant properties of polyphosphazenes is to blend them with standard, commercial polymers. This is, of course, ultimately using the polyphosphazene as an additive although other fundamental physicochemical properties of polymers may be improved by using polymer blends. Blends which are physical mixtures as well as cases where there is covalent bonding between the phosphazene and the parent polymer will be considered. The 1:1 blend of p-sec-butylphenoxy phosphazene polymers and phenylvinyl dimethylsilicone gives fire retardant low smoke materials.[150] The same polyphosphazene was blended with a nonelastomeric polyaryloxyphosphazene.[151] The acrylonitrile butadiene copolymer can be blended with a p-sec-butylphenoxy, 4-methoxyphenoxy polyphosphazene to give a flame retardant (LOI 35.5) foam which exhibits little or no smoke emission on burning.[152] A careful study of poly[bis(p-chlorophenoxy)phosphazene] blends with polystyrene shows full compatability up to 50% polyphosphazene with weak interactions between the two polymers. The phosphazene raises the LOI values from 17.0 for polystrene to a range of 24.6 to 34.6. The value for the polyphosphazene itself is 41.2. The LOI values and char yields increase linearly and in nearly a proportionate fashion with phosphazene content.[153] The bis(propoxy)polyphosphazene has been shown to be an effective flame retardant when blended with methyl methacrylate. The thermal decomposition of the same polyphosphazene blended with expoxy resins has been examined by evaluation of the evolved gas and residual char. A one stage thermal degradation occurs at a temperature lower than that of the pure resin but with a higher char yield. The phosphazene accelerates release of small molecules into the gas phase. The phosphorus moieties remain in the char while other (presumably nitrogen containing) phosphazene components enter the gas phase.[155] Thus, it is possible that both vapor phase and passivation layer mechanisms may be operative in this case. The problem of compatibility of the blend components can be obviated by covalently bonding the two polymers in the blend. Polystyrene and other commercial organic polymers have been grafted onto polydichlorophosphazene. The reactions of the dichlorophosphazene component with the trifluoroethoxide ion yields thermally stable polymers with high LOI values.[156] The alkyl groups in poly[bis(p-isopropylphenoxy)phosphazene] have been partially converted to a hydroperoxide and used to initiate (and hence grafting of) polystyrene. The resulting polymer has increased thermal stability and presumably flame resistance.[157] Graft phosphazene copolymers are considered in detail in other sections of this monograph. The reaction of poly[bis(carboxylatophenoxy)phosphazene] with an isocyanate terminated prepolymer leads to polyurethane/polyphosphazene blends. The LOI value of the polyurethane is 20.0. Above 20% phosphazene loading, the LOI increases to 21.5 and higher and the blend was self-extinguishing. Horizontal flame tests were also conducted. TGA studies indicate the formation of a protective char layer on thermolysis.[158] Recent reports of new approaches to block and graft copolymers of polystyrene and polyphosphazenes[159] also raise the potential for materials with improved fire resistance.

IV. Conclusions

Clearly cyclo- and polyphosphazenes have the potential to be superior, green flame retardant additives and inherently flame retardant materials. As is always the case in applications of phosphazene chemistry, cost considerations present the major barrier to commercialization. In the polyphosphazene area, new more cost effective, synthetic routes to polydichlorophazene from the Allcock group[159] and from our laboratories[160] may provide an incentive to further development. Blends are also a very promising approach. In the small molecule additive area, studies to establish structure-activity relationships between substituents and flame retardancy would be of value. It would also be of value to explore the question of any differences in flame retardancy of a pure cyclophosphazene and mixture of cyclic and linear oligomers. If similar behavior is observed, then the financial advantage of dealing with the mixture is apparent. Finally any studies directed towards true commercialization will require non-halogenated, green materials (Section I.A.).

References

1. A.H. Tullo *Chem. Eng. News,* **2000**,*78* (49), 21.
2. M. Kesner and W. deVos *J. Chem. Educ.,* **2001**, 78, 41.
3. *Chem. Eng. News,* **2001**, *79* (38), 37.
4. Schubert *Science News,* **2001**, *160* (10/13/01), 238.

5. C.W. Allen *J. Fire Sci.,* **1993**, *11*, 320.
6. A.J. Garvin, E.-S. Yoo, R. Harris, E.B. Coughlin *Polym. Prepr.,* **2001**, 42(2), 374.
7. The UL94 protocol which utilizes vertical samples with a thickness of 1.6mm.
8. The ASTM D2863 procedure.; J.-M, Catala and J. Brossas, *Proc. 18th Int. Conf. Org. Coat. Sci. Technol.*, **1992**, 87; C.P. Reghunadhan Mair, G. Clouet and Y. Guilbert *Polym. Degrad. Stab.,* **1989**, *26*, 305.
9. M.A. Delichatsios *J. Fire Sci.,* **1993**, *11*, 287 .
10. T. Ohlemiller, T. Cleary, J. Brown, J. Shields *J. Fire Sci.,* **1993**, *11*, 308.
11. H. R. Allcock and J.P. Taylor *Polym. Eng. Sci.,* **2000**, *40*, 1177.
12. Camino and L. Costa *Reviews in Inorganic Chemistry,* **1986**, *8*, 69.
13. M. Banks, J.R. Ebdon and M. Johnson *Polymer,* **1993**, *34*, 4547.
14. E.D. Weil, S.V. Levchik, M. Ravey and W.M. Zhu *Phosphorus, Sulfur Silicon,* **1999**, *146*, 17.
15. N. Grassie and G. Scott *Polymer Degradation and Stabilization,* **1985**, Cambridge University Press, Cambridge.
16. V.V. Bogdanova *Polymer Science, Ser. B.,* **2001**, *43*, 105.
17. E.D. Weil in *Flame Retardancy of Polymeric Materials,* W.C. Kuryla and A.V. Papa, eds., **1976**, 123, Marcel Dekker, N.Y.
18. S. Duquesne, M. LeBras, S. Bourbigot, R. Delobel, G. Camino, B. Eling, C. Lindsay, T. Roels and H. Vezin *J. Appl. Polym. Sci.,* **2001**, *82*, 3262.
19. S.J. O'Brien *Textile Chemist and Colourist,* **1970**, *2*, 201.
20. J. Cermak, R. Kopecny, J. Kroupa, J. Kunicky and A. Novetny, *Nehorlavost Plast. Hmrot, Dreva Text.*, **1974**, *85*; *Chem. Abstr.,***1975**, *83*, 80036t.
21. M. Kajiwara, *Sen-i Kako*, **1975**, *27*, 195, 319, 381, 507, 631, 693; **1976** *28*, 21, 138, 194, 250, 312, 380, 504, 675; **1977** *29*, 9, 131, 201, 319, 495, 543, 596, 650; **1978** *30* 61, 115, 285, *Chem. Abs.*, **1975** *83*, 98955, 207416, 207420; **1976** *84*, 32397, 91421, 137022, 181456; **1976** *85*, 33757, 109869, 109237, 124917, 178231; **1977** *86*, 55984, 140727; **1977** *87*, 23989, 152756, 137113, 168643; **1978** *88*, 51821, 106035, 75211, 122254; **1978** *89*, 25849, 25701, 60218, 147986.
22. J. Kuncicky, *Horenie Org. Mater.,* **1975**, *121; Chem. Abstr.,* **1976**, *85*, 47469e.
23. P.A. Tatem and F.W. Williams, *Fire Saf. Combust. Mater. Int. Symp.*, **1975**, 341, *Chem. Abstr.*, **1976**, *85*, 143951b.
24. C. Gherorghiu and M. Chirca, *Rev. Chim. (Bucharest),* **1978**, *29*, 210; *Chem. Abstr.,* **1978**, *89*, 75671r.
25. H. Struszczky, *Polimery (Warsaw)*, **1978**, *23*, 41; *Chem. Abstr.,* **1978**, *89*, 25479p.
26. J.A. Beckman, *Rev. Gen. Caoutch, Plast. Chim.,* **1978**, *29*, 210; *Chem. Abstr.,* **1978**, *89*, 75671r.
27. T. Tanaka, *Kagaku Kojo,* **1977**, *21*, 89; *Chem. Abstr.,* **1978**, *89,* 147412g.
28. D. Popescu and A. Zaharia, *Ind. Usoara: Text. Tricotaje, Confectii Text.,* **1978**, *29*, 226; *Chem. Abstr.,* **1979**, *90*, 73148x .
29. L.E.A. Godfrey and J.W. Schappel *Ind. Eng. Chem., Prod. Res. Develop.,* **1970**, *9*, 426.
30. H.R. Allcock *Phosphorus-Nitrogen Compounds* **1972**, Academic Press, New York, N.Y.
31. C.W. Allen in *The Chemistry of Inorganic Homo- and Heterocycles*, I. Haiduc and D.B. Sowerby, eds. **1987**, Vol. 2, Chapter 20, 501, Academic Press, New York, N.Y.
32. C.W. Allen, *Chem. Rev.,* **1991**, *91*, 119.

33. H. Struszczyk, PL **1990**, 149,680; *Chem. Abstr.*, **1991**, *114*, 209445r ; H. Struszczyk *Fire and Mat.*, **1982**, *6*, 7**;** H. Stuszczyk *J. Macromol. Sci. Chem.*, **1986**, *A23*, 973.
34. G.G. Allan, E.J. Gilmartin and H. Struszczyk, *Macromol. Sci. Chem.*, **1981**, *A15*, 599; G.G. Allan and H. Struczczyk, *J. Thermal. Anal.*, **1981**, *21*, 113.
35. Imperial Chemical Industries, Ltd., Belg. Pat. **1964**, 638, 574 ; *Chem. Abstr.*, **1965**, *62*, 44714
36. R.F.W. Rätz, U.S. Pat. **1964**, 3,131,207; *Chem. Abstr., 61, 40076.*.
37. G. Singleton and J.A. Ross, *Brit. Pat.* **1963**, 919,377; *Chem. Abstr.*, **1963**, *58*, 11555b
38. X. Bilger, Fr. Pat. **1963**, 1,327,119; *Chem. Abstr.*, **1963**, *59*, 6052d.
39. E.G. Hoffman, Brit. Pat. **1961**, 876,035; *Chem. Abstr.*, **1962**, *56*, 8229.
40. X. Bilger, Fr. Pat. **1958**, 1,157,097; *Chem. Abstr.*, **1960**, *54*, 20336e.
41. X. Bilger, Fr. Pat. **1964**, 1,375,029; *Chem. Abstr.*, **1965**, *62*, 5382d.
42. S. Kubota,. JP **1989**, 01/040,673; *Chem. Abstr.*, **1989**, 111, 41320h.
43. T. Sasakura, and Y. Hayashi, JP **1989**, 01/209,101, *Chem. Abstr.*, **1990**, 112, 100997e.
44. R. Sato, T. Hara, T. Iizuka, S. Oe and H. Iwai, JP **1988**, 63/145,480, *Chem Abstr.*, **1988**, 109, 232719p; T. Sasakura and O. Yamazoe, JP **1988**, 63/105,179, *Chem. Abstr.*, **1988**, 109, 172096w; T. Sasakura, JP **1988**, 63/264,973, *Chem. Abstr.*, **1989**, 110, 59504k; T. Sasakura, O. Yamazoe, Y. Otsuka, T. Ikeda, M. Daimon and S. Tamura, JP **1988**, 63/120,173, *Chem. Abstr.*, **1988**, 109, 192106c; T. Sasakura and S. Ito, JP **1986**, 61/266,668, *Chem. Abstr.*, **1987**, 107, 24772g; T. Nishikawa, M. Samejima and M. Yoshida, JP **1986**, 61/296,177, *Chem. Abstr.*, **1987**, *107*, 79446j; T. Sasakura, M. Daimon, S. Tamura and O. Yamazoe, JP **1987**, 62/078,271, *Chem. Abstr.*, **1987**, 107, 200337e; Y. Hayashi, T. Sasakura, O. Yamazoe, M. Daimon, S. Tamura and S. Ito, JP **1987**, 62/191,572, *Chem. Abstr.*, **1988**, 108, 7469s.
45. A.Y. Garner, U.S. Pat. **1975**, 3,890,092; Chem. Abstr., **1975**, 83, 133324u.
46. A.Y. Garner, U.S. Pat. **1976**, 3,955,029; *Chem. Abstr.*, **1976**, *85*, 64743g.
47. C. Senez, Fr. Pat. **1964**, 1,357,831; *Chem. Abstr.*, **1964**, *61,* 12139d.
48. Y. Kurachi, T. Okuyama and T. Oohasi, *J. Mater. Sci.*, **1989**, *24*, 276.
49. M. Fujimoto, T. Sugino, M. Ikeyama, JP **1998** 10/268,871; *Chem. Abstr.*, **1998**, *129*, 317559z.
50. S. Kubota, JP **1989**, 01/040,674, *Chem. Abstr.*, **1989**, *111*, 79861y; S. Ito JP **1987**, 61/167,082, *Chem. Abstr.*, **1987**, *107*, 24768k.
51. Neth. Pat. Appl. **1965**, 6,400,735; *Chem. Abstr.*, **1965**, *62*, 1567d.
52. R.G. Rice and B. Grushkin, Fr. Pat. **1965**, 1,393,692; *Chem. Abstr.*, **1966**, *64*, 8328c.
53. N. Sakamoto, K. Okamoto and N. Kjiwara, JP **1989**, 01/190,718; *Chem Abstr.*, **1990**, *112*, 120231d.
54. G.F. Levchik, Y.V. Grigoriev, A.I. Balabanovich, S.V. Levchik and M. Klatt, *Polym. Int.*, **2000**, *49*, 1095.
55. M.A. Lyzentseva, R.K, Valetdinov and E.V. Kuznetsov, *Tr. Kazansk. Khim. Technol, Inst.*, **1962**, 170; *Chem. Abstr.*, **1964**, *61*, 9626b.
56. B. Lazkiewicz and H. Struszczyk, *J. Thermal. Anal.*, **1981**, *19*, 425.
57. J.B. Bullock, C.M. Welch and J.D. Guthrie, *Textile Res. J.* **1964**, *34*, 691; *Chem. Abstr.*, **1964**, *61*, 9626a.
58. R.F.W. Rätz and C.J. Grundman, U.S. Pat. **1959**, 2,876,248; *Chem. Abstr.*, **1959**, *53*, 13055g.

59. S. Ito, S. Tamura, M. Diamon, T. Sasakura and O. Yamazoe, JP **1986**, 61/266,669, *Chem. Abstr.*, **1987**, *107*, 24771f.
60. R.M. Murch, U.S. Pat. **1977**, 4,002,596; *Chem. Abstr.*, **1987**, *86*, 73826j.
61. Neth. Appl. **1965**, 6,400,735; *Chem. Abstr.*, **1965**, *62*, 1567d.
62. Neth. Appl. **1965**, 6402,984; *Chem. Abstr. **1965**, 62*, 59728.
63. S. Hagiwara, H. Saito and H. Sue, JP **1998**, 10/259,292; *Chem. Abstr.*, **1998**, *129*, 303445x.
64. W.D. Coggio, W.J. Schultz, D.C. Ngo, R.D. Waid and V.M. Juvin-Pedretti, U.S. Pat **1997**, 5,652,285; *Chem. Abstr.*, **1996**, *125*, 88142f.
65. P.A. Silver, A. Lukacs III, U.S. Pat. **1986**, 4,607,077; *Chem. Abstr.*, **1986**, *105*, 227945x.
66. B. Laszkiewicz, and R. Kotek, *Hung. J. Ind. Chem.*, **1989**, 17, 221; *Chem. Abstr.*, **1990**, *112*, 8647t.
67. P.L. Meredith, U.S. Pat. **1977,** 4,029,634; *Chem. Abstr.*, **1977**, *87*, 54138r.
68. L.V. Kryuchenkova, M.A. Tyuganova and Ya. A. Ivanova, *Plast. Massy.*, **1988**, 14; *Chem. Abstr.*, **1988**, *109*, 150611h.
69. D.R. Brackenridge, U.S. Pat. **1978**, 4,079,035; *Chem. Abstr.*, **1978**, *89*, 44646g.
70. N. Sakamoto, K. Okamoto and N. Kajiwara, JP **1989**, 01/062,333; *Chem. Abstr.*, **1989**, *111*, 175722z.
71. M.D. Benzoari, U. S. Pat. **1990**, 4,864,047; *Chem. Abstr.*, **1990**, *113*, 24242h.
72. R.J. Guschl, U.S. Pat. **1978**, 4,117,041; *Chem. Abstr.*, **1979**, *90*, 7517r.
73. C.A. Redfarn, U.S. Pat. **1958**, 2,866,773; *Chem. Abstr.*, **1959**, *53*, 9726e.
74. C.A. Redfarn, M. Appley, J.R. Alexander, Brit. Pat. **1963**, 940648; *Chem. Abstr.*, **1964**, *60*, 3912.
75. K. Nagasawa and C. Nakada, JP **1986**, 61/40,386**;** *Chem. Abstr.*, **1986**, 105, 98812a.
76. K. Ikehata, K. Midori, S.H. Uchisa, H. Nakacko and Y. Tada, J.P. **1998**, 10/298,188; *Chem. Abstr.*, **1999**, *130*, 39748c.
77. R.J. Clutter, U.S. Pat. **1975**, 3,865,783; *Chem. Abstr.*, **1975**, *82*, 157773v.
78. S. G. Fedorov, G.S. Gol`din, V.A. Rusakov, E.V. Kotova, S.F. Kravtsova, G.S. Nikitina, A.L. Lomonosov. N.L. Sinko and N.E. Tikhonenkova, *J. Appl. Chem.*, (Engl. Transl.) **1986**, *61*, 327.
79. H.R. Allcock, T.J. Hartle, J.P. Taylor and N.J. Sunderland, *Macromolecules*, **2001**, *34*, 3896.
80. E.N. Pokrovskaya, T.G. Bel`tsova and V.I. Shidorov, *Khim. Drev.*, **1989**, 101; *Chem. Abstr.*, **1989**, *111*, 41578y.
81. S.V. Levchik, G. Camino, M.P. Luda, L. Costa, A. Lindsay and D. Stevenson, *J. Appl. Poly. Sci.*, **1998**, *67*, 461.
82. E.D. Weil and N.G. Patel *Fire and Materials,* **1994**, *18*, 1.
83. S.V. Levchik, G.F. Levchik, A.I. Balabanovich, E.D. Weil and M. Klatt *Angew. Makromol. Chem.*, **1999**, *264*, 48.
84. R. Schweisinger and H. Schlemper, *Angew. Chem., Int. Ed. Engl.* **1987**, *26*, 1167.
85. *Aldrichemica Acta,* **2001**, *34*, 26.
86. R.E. Singler and F.J. Gomba in *Synthetic Lubricants and High-Performance Functional Fluids*, L.R. Rudnick and R.L. Shubkin, eds. **1999**, Chapter 13, 297, Marcel Dekker, Inc. New York, NY.
87. R.E. Singler, A.J. Deome, D.A. Dunn and M.J. Bieberich, *Ind. Eng. Chem. Prod. Res. Dev.*, **1986**, *25(1)*, 46.

88. R.E. Tapscott, S.R. Skaggs and D. Dierdorf, *ACS Symposium Series,* **1995**, *611,* 151.
89. J.G. DuPont and C.W. Allen, *Macromolecules*, **1979**, *12*, 169.
90. C.W. Allen, *Trends. Polym. Sci*, **1994**, *2*, 342.
91. C.W. Allen, *MGCN, Main Group Chemistry News,* **1998**, *6*, 6.
92. K. Inoue, M. Nakano, M. Takagi and T. Tanigaki, *Macromolecules,* **1989**, *22*, 1530.
93. D.E. Brown, K. Ramachandran, K.R. Carter and C.W. Allen, *Macromolecules*, **2001,** *34*, 2870.
94. D.E. Brown and C.W. Allen, *J. Inorg. Organomet. Polym.*, **1991**, *1*, 189.
95. C.W. Allen, R.F. Hayes, C.N. Myer, A.S. Freund and M.E. Kearney, *Phosphorus, Sulfur Silicon, Relat. Elem.,* **1996**, *109-110*, 79.
96. G. Bosscher, R.H. Wieringa, A.P. Jekel and J.C. van de Grampel, *J. Inorg. Organomet. Polym.*, **1996**, *6*, 255.
97. C.W. Allen and R.P. Bright, *Macromolecules*, **1986**, *19*, 571.
98. C.W. Allen, J.C. Shaw and D.E. Brown, *Macromolecules,* **1988**, *21*, 2653.
99. K. Inoue, H. Nakamura, S. Ariyoshi, M. Takagi and T. Tanigaki, *Macromolecules*, **1989**, *22*, 4466.
100. I. Immanuel Selvaraj and V. Chandrasekhar, *Polymer*, **1997,** *38*, 3617.
101. I. Dez, N. Henry and R.De Jaeger, *Polym. Degrad. Stab.*, **1999**, *64*, 433.
102. Y.W. Chen-Yang and Y.H. Chuang, *Phosphorus, Sulfur Silicon Relat. Elem.,* **1993**, *76*, 261.
103. D. Kumar and C.W. Allen, Unpublished work.
104. I. Dez and R. De Jaeger, *Phosphorus, Sulfur Silicon, Relat. Elem.,* **1997**, *130*, 1.
105. H.R. Allcock, E.C. Kellam and M.A. Hoffmann, *Macromolecules*, **2001**, *34*, 5140.
106. D. Kumar and A.D. Gupta, *Macromolecules,* **1995**, *28*, 6323.
107. D. Kumar, G.M., Fohlen and J.A. Parker, *Macromolecules*, **1983**, *16*, 1250.
108. D. Kumar, G.M. Fohlen and J.A. Parker, *J. Polym. Sci. Polym. Chem. Ed.,* **1983**, *21*, 3155.
109. D. Kumar, G.M. Fohlen and J.A. Parker, *J. Polym. Sci., Polym. Chem. Ed., **1984***, 22, *927.*
110. D. Kumar, G.M. Fohlen and J.A. Parker, *J. Polym. Sci., Polym. Chem. Ed.*, **1984**. *22*, 1141
111. D. Kumar, *J. Polym. Sci., Polym. Chem. Ed.*, **1984**, *22*, 3439.
112. D. Kumar, M. Khullar and A.D. Gupta, *Polymer,* **1993**, *34*, 3025.
113. D. Kumar, G.M. Fohlen and J.A. Parker, *J. Polym. Sci., Polym. Chem. Ed.*, **1986**, *24*, 2415.
114. D. Kumar, A.D. Gupta and M. Khullar, *J. Polym. Sci, Part A: Polym. Chem.,* **1993,** *31*, 707.
115. A. Yaguchi, S. Mori, M. Kitayama, T. Onda, A. Kurahashi and H. Ando, *Thin Solid Films*, **1992**, *216*, 123.
116. Y.W. Chen-Yang, H.F. Lee and C.Y. Yuan, *J. Polym. Sci., Part A: Polym. Chem.,* **2000**, *38*, 972.
117. M. Uchida, M. Noji and R. Ogiwara, JP **1999**, 11/43,495; *Chem. Abstr.,* **1999**, *130*, 197401p.
118. D. Matthew, C.P. Reghunadhan Nair and K.N. Ninan, *Polym. Int.,* **2000**, *49*, 48.
119. M.L. Stone, *Proc. Mater. Res. Soc. Symp.,* **1993**, *305*, 85.
120. W.K. Huang, J.T. Yeh, K.J. Chen and K.N. Chen, *J. Appl. Polym. Sci.,* **2001**, *79*, 662.

121. H.R. Allcock, G.Y Moore and W.J. Cook, *Macromolecules,* **1974**, *7*, 571.
122. S.V. Peddada and J.H. Magill, *Macromolecules*, **1983**, *16*, 1258.
123. H.R. Allcock, G.S. McDonnell, G.H. Riding and I. Manners, *Chem. Mater.,* **1990**, *2*, 425.
124. T.N. Bowmer, R.C. Haddon, S. Chichester-Hicks, M.A. Gomez, C. Marco and J.C. Fatou, *Macromolecules,* **1991**, *24*, 4827.
125. D.P. Tate, *J. Polym. Sci., Symp.,* **1974**, *48*, 33.
126. E.J. Quinn and R.L. Dieck, *Fire Retard Proc. Int. Sym. Flammability Fire Retard., 1977* **1979**, 198.
127. R.E. Singler, G. Hagnauer and R.W. Sicka, *ACS Symp. Ser.,* **1982**, *193*, 229.
128. H.R. Allcock, *Makromol. Chem. Macromol. Symp.,* **1986**, *6*, 101.
129. H.R. Penton, *ACS Symp. Ser.,* **1988**, *360*, 277.
130. E.J. Quinn and R.L. Dieck, *J. Fire Flammability,* **1976**, *7*, 5.
131. K.A. Reynard and J.C. Vicic, *Gov. Rep. Announce Index (U.S.),* **1976**, *76*, 194; *Chem. Abstr.,* **1977**, *86*, 56819s.
132. E.J. Quinn and R.L. Dieck, *J. Fire Flammability,* **1976**, *7*, 358.
133. R.L. Dieck and E.J. Quinn, U.S. Pat. **1977**, 4,053,456; *Chem. Abstr.,* **1977**, *87*, 202571u.
134. R.L. Dieck and E.J. Quinn, Ger. Offen. **1977**, 2,708,319; *Chem. Abstr.,* **1977**, *87*, 169034w.
135. J.E. Thompson and K.A. Reynard, *J. Appl. Polym. Sci.,* **1977**, *21*, 2575.
136. E.J. Quinn and R.L. Dieck, *J. Cell. Plast.,* **1977**, *13*, 96.
137. R.L. Dieck and E.J. Quinn, U.S. Pat. **1977**, 4,026,838; *Chem. Abstr.,* **1977**, *87*, 24275r.
138. R.L. Dieck and E.J. Quinn, *J. Fire Flammability,* **1977**, *8*, 412.
139. R.L. Dieck and E.J. Quinn, Ger. Offen. **1977**, 2,712,542; *Chem. Abstr.,* **1978**, *88,* 7890r.
140. R.L. Dieck and E.J. Quinn, U.S. Pat. **1978**, 4,083,820; *Chem. Abstr.,* **1978**, *89*, 111396r.
141. R.L. Dieck, E.J. Quinn and A.B. Magnusson, Ger. Offen. **1978**, 2,758,747; *Chem. Abstr.,* **1978**, *89*, 181071e.
142. R.L. Dieck, E.J. Quinn, Ger. Offen. **1979**, 2,838,905; *Chem. Abstr.,* **1979**, *90*, 187610x.
143. R. L. Dieck and E.J. Quinn, U.S. Pat. **1978**, 4,108,805; *Chem. Abstr.,* **1979**, *90*, 72830b.
144. R.L. Dieck and E.J. Quinn, Can Pat. **1982**, 1,123,451; *Chem. Abstr.,* **1982**, *97*, 93041e.
145. W.B. Mueller, *J. Cell. Plast.,* **1986**, *22*, 53.
146. P.J. Lieu, J.H. Magill and Y.C. Alarie, *J. Fire Flammability*, **1980**, *11*, 167.
147. R.L. Dieck and E.J. Quinn, Ger. Offen. **1978**, 2,758,748; *Chem. Abstr.,* **1978**, *89*, 180852y.
148. G.S. Kyker and T.A. Antkowiak, *Rubber Chem. Technol.,* **1974**, *47*, 32.
149. A. Wilson, *J. Coated Fabr.,* **1978**, *7*, 233.
150. R.L. Dieck and E.J. Quinn, U.S. Pat **1977**, 4,026,839; *Chem. Abstr.,* **1977**, *87*, 24536b.
151. R.L. Dieck and E.J. Quinn, U.S. Pat. **1977**, 4,055,520; *Chem. Abstr.,* **1978**, *88*, 7903x.
152. R.L. Dieck and E.J. Quinn, U.S. Pat. **1977**, 4,061,606; *Chem. Abstr.,* **1978**, *88*, 51619h.
153. Y.W. Chen-Yang, H.F. Lee and T.T. Wu, *ACS Sym. Ser.,* **1994**, *572*, 295.
154. B.L. Deng, W.Y. Chiu and K.F. Lin, *J. Appl. Polym. Sci.,* **1997**, *66*, 1855.
155. B.L. Deng, W.Y. Chiu, K.F. Lin and M.R.S. Fuh, *J. Appl. Polym. Sci.,* **2001**, *81*, 1161.
156. J. Brossas, G. Clouet and M. Tissot, *Fr. Demand* **1983**, 2,510,128, *Chem. Abstr.,* **1983**, *99*, 71610d.
157. M. Gleria, A. Bolognesi, W. Porzio, M. Catellani, S. Destri and G. Audisio, *Macromolecules,* **1987**, *20*, 469.

158. C. S. Reed, J. P. Taylor, K.S. Guigley, M. M. Coleman and H.R. Allcock, *Polym. Sci. Eng.,* **2000**, *40*, 465.
159. R. Prange, S.D. Reeves and H.R. Allcock, *Macromolecules,* **2000**, *33*, 5763.
160. C.W. Allen and A.S. Hneihen, *Phosphorus, Sulfur Silicon Relat. Elem.* **1999**, *144-46*, 213; C.W. Allen, A.S. Hneihen and E.S. Peterson, U.S. Pat. **2001**, 6,30,9,619, *Chem. Abstr.,* **2000**, *132*, 93835.

In: Applicative Aspects of Poly(organophosphazenes)
Editors: M. Gleria and R. De Jaeger, pp. 139-168
ISBN 1-59454-025-X

Chapter 7

POLYPHOSPHAZENE BLENDS

Luciano Canova[1], Luisa Fiocca[1], Giorgio Giannotta[1], Riccardo Po'[1] and Mario Gleria[2*]

1-Istituto G. Donegani, Polimeri Europa S.p.A.,Via Fauser 4, 28100 Novara, Italy.
2-Istituto di Scienze e Tecnologie Molecolari del C.N.R., c/o Dipartimento CIMA, Università di Padova, Via Marzolo 1, 35131 Padova, Italy. Istituto per la Sintesi Organica e la Fotoreattività del C.N.R., Viale dell'Università 2, 35020 Legnaro, Padova, Italy

SUMMARY

In this review the state-of-the-art on research in the field of polyphosphazene-based blends is presented. Both scientific and patent literature is considered. The results achieved by the different research groups in studying blends with organic or inorganic polymers or other polyphosphazenes, reactive blending processes and the use of low molecular weight analogues of polyphosphazenes is discussed; particular attention is paid to those systems containing commercial polymers where a polyphosphazene is added to improve the functional performances. The blend preparation, the techniques used to asses blend properties and compatibility, the main features of the blends and some proposed applications are illustrated.

INTRODUCTION

Polymer blends are formed by mixing two (or more) different macromolecules[1]. Their practical importance derives from the fact that they represent a simple and elegant way to meet the scientific and technologic demands[2] of modern civilization for sophisticated materials.

The preparation of blends is a relatively easy process because of the large availability of well known polymeric materials, for which technological developments, industrial plants and commercialization are already well established. In comparison the development of alternative

* To whom correspondence should be addressed

processes to synthesize new materials, such as novel classes of macromolecules, is a much more difficult task, being costly in term of time, labor and financial investments[3,4].

Once blended, two macromolecules may form totally immiscible, two-phase materials (where one polymer preferentially interacts with macromolecules of the same nature), partially miscible products, or single-phase substrates, if interactions between different macromolecules dominate[5]. In this last case, the free energy of mixing

$$\Delta G_m = \Delta H_m - T\Delta S_m$$

becomes negative because of the decrease of the mixing enthalpy ΔH_m of the system due to the onset of dispersion forces, dipoles, hydrogen bonding, charge-transfer phenomena or electrostatic attractive forces between the two components of the blend[6]. In some cases, moreover, reactive blending processes[7] are used to form covalent bonds between the blend components, thus avoiding phase segregation in the new materials and permanently establishing their morphology. In other cases, the utilization of additional substrates which act as polymer compatibilizers[7], is necessary to favor the formation of blends between macromolecules of different structure.

Poly(organophosphazenes), POPs

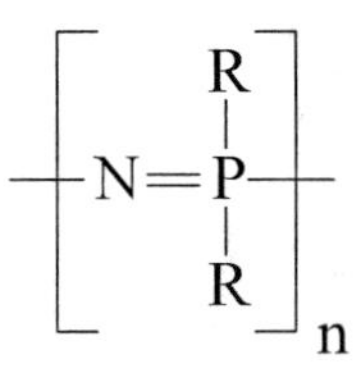

Figure 1

are very important inorganic-organic polymeric materials[8] which have been intensively studied since the mid 1960's (their discovery is generally fixed around 1965 by H.R.Allcock[9-11]) due to their great versatility[12], and to their outstanding properties as flame retardant[13] and self-extinguishing materials[14], bioactive polymers[15-17], photosensitive macromolecules[18-20], thermally stable substrates[21], etc. For these products, blending processes are a very important research area that has been investigated since the beginning of the history of phosphazenes[22-32] and continued to be developed until quite recently[33-41].

The reason of such intense and continuous effort is twofold:

(1) investigation into blending processes involving phosphazenes as a general way to prepare new materials[5];
(2) evaluation of the possibility of transferring the useful properties of POPs to commercial organic macromolecules[42], even using phosphazenes in a rather limited quantity. In this way cost problems related to POPs[43] will be hopefully overcome and their industrial development and commercialization facilitated.

In this chapter we will critically highlight the work that has been carried out in the field of POPs blends with other phosphazene macromolecules[5,22,27,30,32,44-46], and with a variety of commercial organic[5,6,23,25,28,29,33,35-41,44,47-93] or inorganic[24,26,31] polymers; analyze the properties of the obtained materials together with their practical exploitation; and describe the utilization of low molecular weight cyclic phosphazene analogues, the cyclophosphazenes, as possible additives to promote the modification[94-97], chain extension[88,98,99] and compatibilization[35,100] of different macromolecules.

Polyphosphazene Blends with Organic Polymers

Blends of POPs with commercial organic polymers have attracted the interest of several research groups[5,6,23,25,28,29,33,36-39,41,44,47-81,83-86,89-92,101,102].

There are a number of important reasons for this, including the possibility of shedding light on the mechanism of polymer/polymer compatibilization processes[5,33,36,37,44,58,64,67], preparing new materials that show improved flame retardant or heat resistant features[23,25,28,29,41,53,56,59,61,62,75-77,85-87,91], sustaining the dimensional stability of phosphazene based electric conductors[6,39,48-50,54,55,57,60,66,89], improving ionic conductivity of poly(ethylene oxide) (PEO)[65,68-74,80,81,83,84,90,92,102], or use of the phosphazene/organic polymer blends as membranes for separation processes[40] or for biological purposes[38,78,79].

Research in Phosphazene Compatibilization Processes

Attempts to explore the compatibilization processes of phosphazene/organic polymers blends led in the past to the preparation of binary mixtures involving several different thermoplastic organic macromolecules[5,44]. The organic polymers used were chosen to cover a wide range of properties in term of T_g, hydrophilicity, and hydrogen bonding formation: poly(vinyl chloride), polystyrene (PS), poly(methyl methacrylate) (PMMA), poly(4-vinylpyridine) (PVP), poly(vinyl alcohol) (PVA), poly(acrylic acid), PEO. In general, it was found that molecular interactions between the blend components (hydrogen bonding, van der Waals interactions, dipole-dipole interactions) favor miscibility[1]. When the interaction is strong, the miscibility is retained even after several years of ageing[5]; otherwise, slow separation occurs. For instance, poly[bis(phenoxy)phosphazene] (PBPP)/PS blends were found to be semimiscible in the range of concentrations of 30-90% of PS. Two T_g values were detected, the first one at -6°C, corresponding to the T_g of pure PBPP, the second one between 50°C and 90°C, due to a mixed phase; the latter T_g disappears after 5 years, indicating a complete phase separation. More recently, Gleria[36,37] studied the PBPP/PS system in detail, trying to promote the miscibility among these components. All of the different blends were prepared by both casting from chloroform solutions on glass slides and melt blending, and were analyzed by DSC, SEM and XPS. It was found that the two polymers were immiscible, and PBPP tends to be segregated on the upper surface of the film (which is in contact with the air)[36]. With the aim of inducing compatibility among PBPP and PS, poly(2,6-dimethylphenylene ether) (PPE) was added to the blends. PPE itself exhibits a limited compatibility with PBPP, possibly due to the interaction between the phenoxy groups of the two polymers. This is indicated by XPS data, that gives evidence of a variation of the phosphorus content at the surface, and also by the lowering of the T_g of PPE; the PBPP phase, however, maintains an unaltered T_g.

In ternary blends, the values of DSC parameters did not show any significant variation. XPS data showed that the film surfaces are composed of pure PBPP with low PPE contents; at higher PPE concentration, there is also PPE at the surface, which most likely contains dissolved PS. Melt mixed blends showed a different behavior, indicating a strong dependence of the blend properties on the preparation technique: PBPP was absent from the lateral surface of the pellets, but tended to emerge during compression molding.

The binary blend of poly[bis(4-phenoxy-phenoxy)phosphazene] (PBPPP) and PS exhibited two T_g values. The first one (19°C) corresponded to PBPPP, the second one was slightly below 100°C and decreased with composition, but the decrease was too small to give any indication of compatibility[37]. On the other hand, ternary blends PBPPP/PS/PPE have a limited compatibility, as shown by the values of T_g and by the presence of PS at the surface along with the other blend components in noticeable amounts. The reason for this greater compatibility was believed to arise from the greater similarity between the structure of PPE and the PBPPP side chains as compared to PBPP

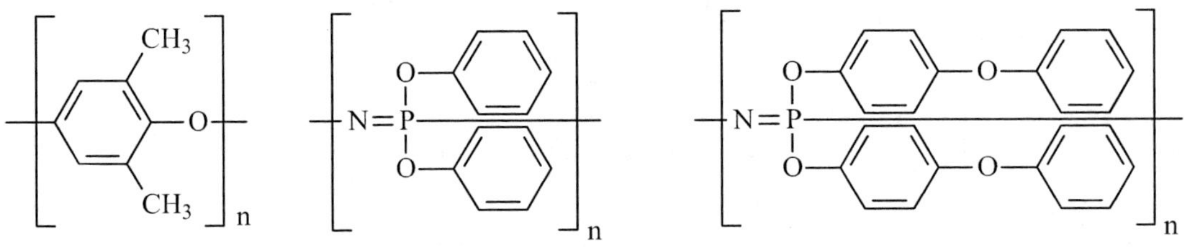

Figure 2

Under certain conditions, Chen-Yang[64,67] found that PS was miscible with poly[bis(p-chlorophenoxy)phosphazene] (PBCPP). Blends were prepared by solution casting or coagulation methods (THF) and analyzed by DSC, TGA, SEM, FTIR and LOI measurements. A single T_g was found in the first heating of DSC scans for blends containing more than 50% of PS (61°C-74°C), in agreement with the Couchman equation[103], while SEM micrographs showed that homogeneous dispersions of polyphosphazene in PS were attained[67]. For compositions with more than 60% of polyphosphazene component, the presence of two T_g values, intermediate to those of the pure polymers, indicated the presence of two phases. During the second scan, T_g values close to those of the parent polymers were observed, suggesting that the compatible phase was thermodynamically unstable and separated on heating. The P-O-C stretching band and C-H out of plane bending bands of the aromatic rings (both of PS and POP) exhibited characteristic shifts in the compatible blends, confirming the interaction between the polymers, whereas no interactions existed in the phase separated blends. Finally, an improved thermal stability and an enhanced flame resistance of the resulting material was demonstrated due to the presence of polyphosphazene in the blend: while flame resistance increased linearly with polyphosphazene content, thermal stability was improved due to the compatibility but did not depend linearly on composition.

Interactions between the polymer components played a peculiar role in the mechanism of compatibilization of poly[bis(5,7-dioxadibenzo[a,c]cycloheptenylidene)phosphazene] with poly(tetrahydrofuran) (PTHF)[33]. Blends were obtained either by coprecipitation or solvent evaporation, but were also unwillingly formed *in situ* during the reaction of poly(dichlorophosphazene) (PDCP) with 2,2'-dihydroxybiphenyl in THF solvent. Indeed, acidic groups in PDCP were responsible for cationic polymerization of THF[104].

Separation of the blends into the pure components by solvent extractions, was not possible due to the strength of the interactions between the different polymer chains. DSC, IR and X-ray data also confirmed that PTHF was amorphous in the blends, as a consequence of such interactions. The glass transition temperature of poly[bis(5,7-dioxadibenzo[a,c]cycloheptenylidene)phosphazene] (140°C) underwent a pronounced decrease in the blends, due to the compatibility effect. The decrease (ΔT = -60°C) was consistent with a content of 10% PTHF, as calculated through the Fox equation[105]. Compatibilization was further demonstrated by the stabilization of the mesophase of polyphosphazene (T = 145°C-190°C). Blends prepared *in situ* were significantly more compatible than those obtained by solvent evaporation (CH_2Cl_2), as revealed by the thermal behavior that evidenced the existence in the latter blends of some crystalline PTHF regions.

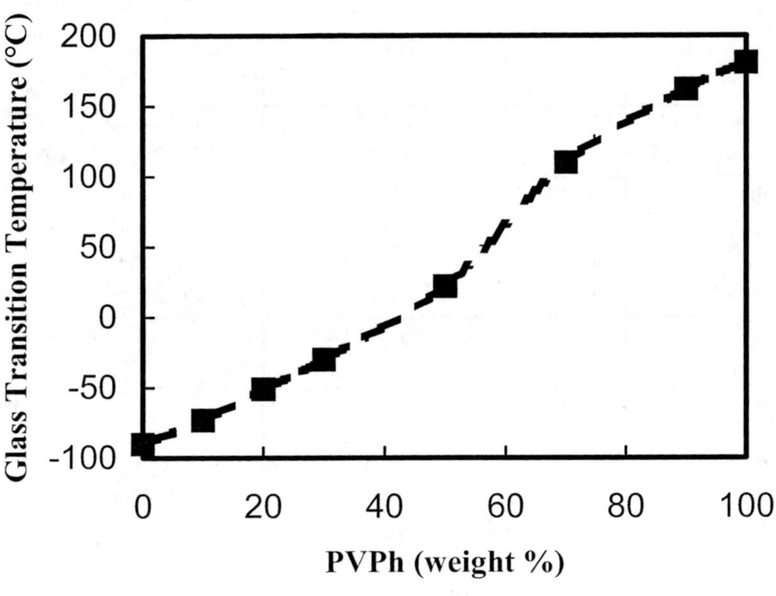

Figure 3

As observed by Allcock[5,44], hydrogen bonding is very important in establishing strong interactions in polyphosphazene-organic polymer systems, especially those containing ether groups. In fact, Landry[6,57,60,66] studied the miscibility of poly(vinylphenol) (PVPh) with four different polyphosphazenes that differ with respect to the presence of variable length etheric groups in the side chains (see Figure 3). DSC and FTIR were used to assess the physical properties. PBPP and poly[bis(2,2',2"-trifluoroethoxy)phosphazene] (PTFEP) were immiscible with PVPh, whereas blends with poly[bis(methoxyethoxyethoxy)phosphazene] (MEEP) and poly[bis(methoxyethoxy-ethoxyethoxy)phosphazene] (MEEEP) exhibited a single T_g and gave clear films cast from methyl ethyl ketone.

FTIR spectra and studies on heat of mixing on the model compounds showed that the latter polymer pairs interact strongly. The dependence of T_g on composition exhibited a strong negative deviation from the additivity rule at high polyphosphazene contents. This was explained on the basis of the increase of the mobility of PVPh chains caused by the decrease of the degree of self-association. When THF was used as a solvent in the preparation of the blends, a great tendency to phase separation was observed, due to the establishment of polymer-solvent (instead of polymer-polymer) interactions. MEEP was not miscible with PS and PMMA. However, modification of the two polymers by copolymerization with protic monomers (vinylphenol and acrylic or methacrylic acid, respectively), led to a substantial increase of the miscibility that depended on the comonomer content, thus confirming the paramount importance of the interactions between the etheric side groups of polyphosphazene and proton donating moieties. Blends of MEEP and a styrene/4-styrenesulfonic acid copolymer gave transparent films but no T_g was detected in the DSC curves[6].

Possible applications suggested for these formulates include the manufacture for gas or ionic transport membranes and electrolytes upon the addition of appropriate doping salts (*vide infra*).

Flame Retardant Phosphazene Blends

Most conventional polymers posses useful structural and mechanical properties but their use is limited by their low thermo-oxidative stability and high flammability[2]. So there is a large interest in the development of flame retardant polymeric compositions for a wide range of applications[21].

It is generally accepted that the incorporation of phosphorous, nitrogen and halogens into polymers enhances flame retardant properties[106]. By measuring Limiting Oxygen Index (LOI), DeEdwardo[23,25] reported interesting results for POPs bearing fluoro-anilino groups having the following general structure

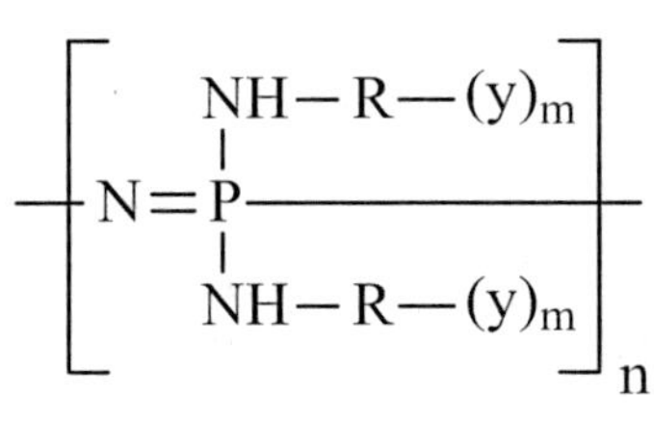

where: y=F, Cl, or Br $R=C_6H_4$

Figure 4

blended with poly(ethylene terephthalate) (PET), cellulose acetate, cellulose triacetate, polyolefins and PS. The data obtained show that POPs are effective flame retardants when incorporated into the above mentioned polymeric substrates. PET containing 10% of poly[bis(4-fluoroaniline)phosphazene] showed a LOI of 22.9, as compared with 18.8 for the unstabilized polymer.

Denq studied the blends of poly(dipropoxyphosphazene) (PDPP) with PMMA[75,77] PS[76], and epoxy prepolymer[87] using different experimental techniques.

Investigations into the thermal degradation behaviour of PMMA and PMMA/PDPP blends[75,77] showed that the main difference between these two samples was the formation of methanol and propylene in PMMA blends accompanied by the stabilization of the unzipping reaction of PMMA in the presence of PDPP. An anhydride structure was also observed in the residual yield during the thermal degradation of PMMA-polyphosphazene blend[75]. Since PDPP has a very low T_g (-100°C), the glass transition temperature of PMMA blends decreases with increasing PDPP content. However, the T_g of these blends began to stabilize after 20 phr of PDPP. Micrographs confirmed that PDPP was uniformly dispersed in PMMA matrix when its content was below 20 phr. A comparison between experimental and theoretical T_g calculated according to the Fox equation[105], indicated that there was a discrepancy between these T_g values due to the fact that PMMA blends were partially miscible at a the PDPP content lower than 20 phr, and incompatible at higher PDPP levels. Since FTIR studies of these blends did not evidence a change in intensity or in the position of the -CO- (1730 cm^{-1}) band, it was concluded that there were no interactions between PMMA and PDPP[77].

Similar studies were reported for PDPP/PS blends[76], showing that neat PDPP possessed three degradation steps up to 550°C (with propylene as the major gas evolved), polystyrene produces styrene, α-methyl-styrene and benzene, while blends PDPP/PS exhibited a two step reaction, with formation of char, the amount of which increased with the increase in PDPP content.

Finally, the influence of different amounts of PDPP on the curing kinetics and physical properties of an epoxy prepolymer, diglycidyl ether of bisphenol A, cured with diethylenetriamine (DETA) was investigated. The kinetic analysis of curing revealed that the effect of PDPP as a catalyst was higher than its effect as a diluent when the temperature of curing was below 70°C, with a PDPP content lower than 10 phr. SEM characterization of the fracture surface of the epoxy blends, under liquid nitrogen, showed that there was a correspondence between cavitated holes PDPP particles, and their number increased with increasing PDPP content. As FTIR spectra did not revealed any change in band intensity or position due to addition of PDPP, it was suggested that only weak interactions existed in the epoxy blends. DMTA measurements showed that T_g decreased with increasing PDPP content but did not obey to the Fox equation. Thus the SEM, FTIR and DMTA evidence can be used to explain the partial miscibility of epoxy blend system. The mechanical characterization showed that the modulus decreased with the increasing in PDPP content, and the elongation at break increased[87].

Chiu[86] described a two-step polymerization process to synthesize polyurethane (PU) containing different amounts of PDPP flame retardant. 4,4'-Methylenebis(phenylisocyanate) was first reacted with poly(tetramethylene glycol) 1000 to form an isocyanate-terminated prepolymer. The prepolymer was then reacted with a 1,4-butanediol/PDPP homogeneous mixture at 70°C for 100 minutes, followed by 1 hour at 100°C.

It was also confirmed that PDPP contains small amounts of isomeric compounds the quantity of which increases under thermal treatment. The corresponding phosphazene-phosphazane isomerization reaction is described below:

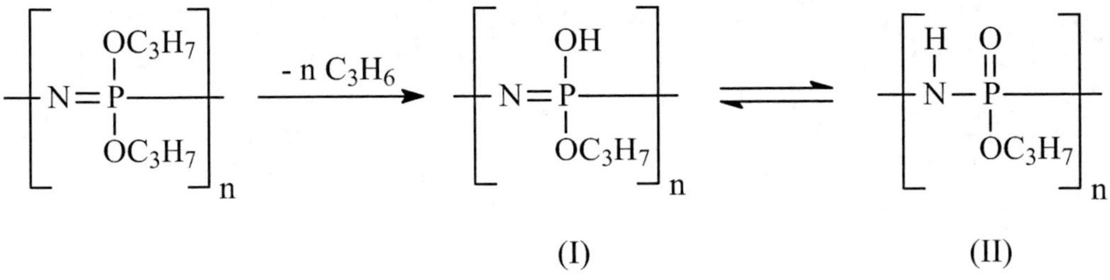

Figure 5

It is possible that during the final step of the synthesis of PDPP-PU small quantities of the isomeric compounds (I) or (II) are formed by the thermal degradation of PDPP. Since these species contain hydroxyl and secondary amino groups, respectively, the following reactions are possible:

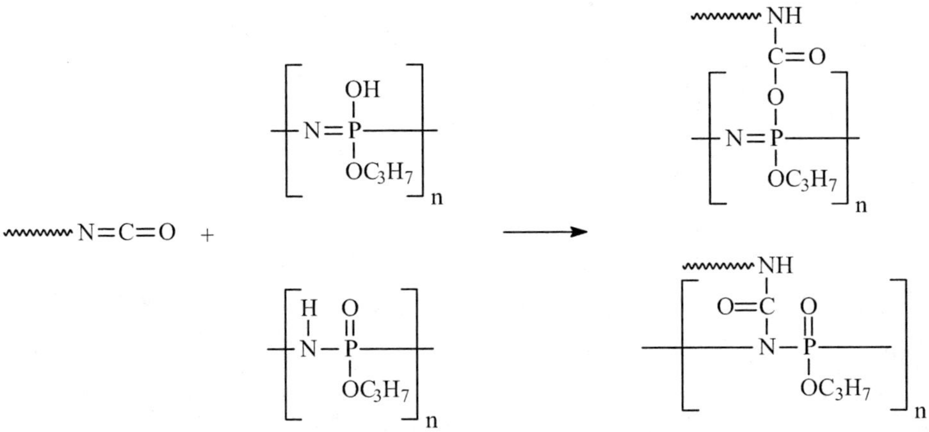

Figure 6

The addition of PDPP would accelerate urethane reactions and induce formation of micro-crosslinks. Consequently, the glass transition temperature (T_g) of PU increases and the rubbery plateau region is extended. From SEM observations, the PDPP additive phase separated from the PU matrix with good interfacial bonding. The presence of PDPP decreased both the ultimate tensile strength and elongation at break of PU[86].

Brackenridge[29] claimed to increase fire retardancy of polyesters by the addition of (halophenoxy)phosphazenes having the following structure:

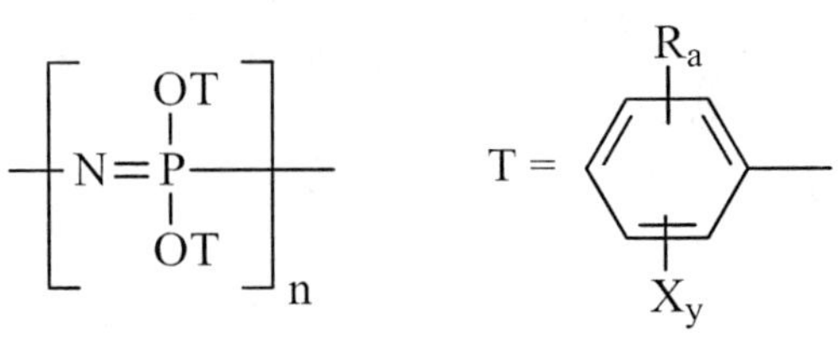

where: n=3 or more; R=alkyl group and a=1-3; X=Cl or Br

Figure 7

In fact, polyester resin, styrene, and a hexakis(4-bromophenoxy)cyclophosphazene were blended, cured overnight at 50°C with 1% benzoyl peroxide, and post-cured 3 h at 80°C. The LOI test for flame resistance was 21.6, compared with 18.4 for a similar blend containing no (halophenoxy)phosphazenes[29].

Chang[53] described a process for forming a low smoke, low flame spread composition, by blending a mixture of polybenzimidazole fibers, inorganic powder fillers, phosphazene polymers not containing F, F-containing POPs, curing agents with processing aids, and curing the blends.

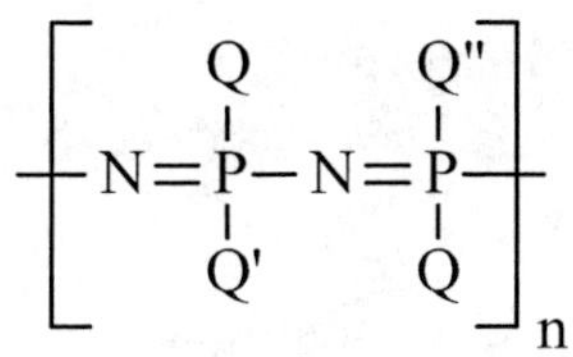

Non Fluorinated phosphazene polymers, where n=integer from 20 to 50,000
Q and Q' are the same or different and are C_1-C_6
Q" is a group having an olefinic unsaturation

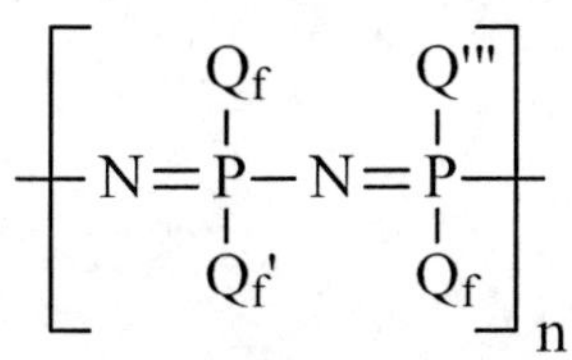

Fluorinated phosphazene polymers, where n=integer from 20 to 50,000
Q_f and Q_f' are the same or different and are $-OCH_2(CF_2)_xCF_3$ or $-OCH_2(CF_2)_nH$ with x=1-20;
Q''' is a group having an olefinic unsaturation

Figure 8

Foamed blends phosphazene polymers are fire-retarded and give little or no smoke emission on burning[28].

Useful compositions for electric wires and cables, comprising inorganic fire-retardant and powdered poly(organophosphazene), were prepared by Ishikawa[56]. For example a composition containing ethylene-ethyl acrylate copolymer, $Al(OH)_3$, a poly(organophosphazene), carbon black, and stabiliser parts, showed a LOI of 30 and a tensile strength of 1.2 kg/mm^2.

Blends for Dimensional Stabilization

Dimensional stabilization in phosphazene materials is an important topic for polymers substituted with short PEO units, e.g. MEEP. First synthesized by Allcock in 1984[107] this material proved to be extremely interesting because of its extraordinary high ionic conductivity at room temperature when doped with Li^+ or Ag^+ ions[108], thus suggesting application as polymeric electrolyte in the construction of solid state batteries[109,110]. However, the principal drawback of these derivatives is related to the poor dimensional stability and mechanical properties[111] which prevent the preparation of free standing films.

Attempts to solve this problem involved chemical reticulation of the polymer[112,113], UV[114-117] light or γ-ray[118-122]-induced crosslinking of this material, MEEP incorporation into three-dimensional silica networks by using the sol-gel technique[111,123-130], or enhancing the complexity of the ethylene oxide substituents by branching and functionalization processes[131].

Dimensional stabilization of MEEP, could be obtained also following another strategy based on the formation of blends between this polymer and PEO. Thus, Abraham[48-50] thoroughly investigated the ionic conductivity of new blends of MEEP and PEO doped with silver or lithium salts. In general these materials showed excellent dimensional stability, were completely miscible below 70% of MEEP content in the blend, still showing the presence of two crystalline phases (PEO and $LiBF_4$ complexed PEO) mixed with the MEEP-$(LiBF_4)_n$ component. These blends exhibit an electric conductivity at 60°C comparable to that of neat PEO$(LiX)_n$ at 100°C, with a ionic conductivity at room temperature 500 times higher than that of the doped PEO.

Similar results were obtained by blending MEEP with poly(propylene oxide) (PPO)[54]. In this case the two components were completely miscible over the whole range of compositions and the resulting blends were amorphous. However, the blends showed good mechanical properties and dimensional stability only when the MEEP content in the mixture was below 80%. The ionic conductivity of MEEP/PPO 80/20 blends doped with Li^+ salts was about 10^{-5} S cm^{-1} while that of MEEP/PPO 55/45 was considerably lower.

Blends between MEEP and PVPh, their miscibility in the whole range of compositions due to formation of hydrogen bonding between the two macromolecules, and their possible utilization in conductivity problems have been already discussed in another part of this article[6,57,60,66,125,126,128].

A very interesting approach to dimensional stabilization of MEEP has been proposed by Abraham[55] who reinforced MEEP polymer by photoreticulating this substrate in the presence of poly(vinyl pyrrolidone) or poly(ethylene glycol diacrylate)

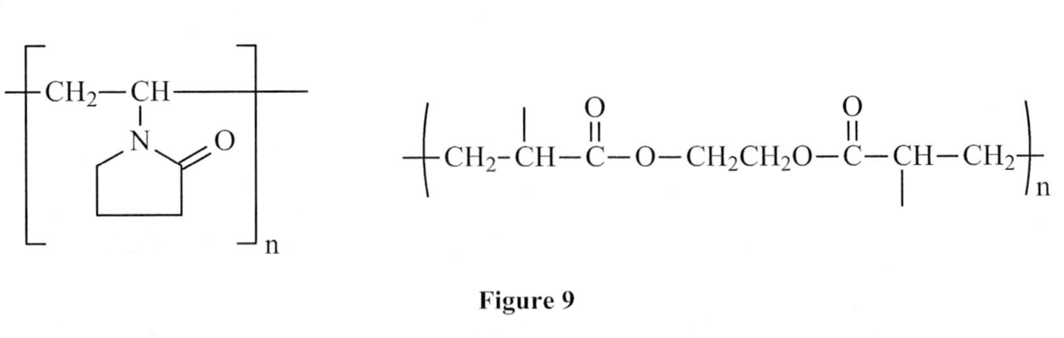

Figure 9

The resulting blends showed excellent dimensional stability and film homogeneity attributed to the presence of a second organic phase finely dispersed into the phosphazene matrix, together with a rather high ionic conductivity (6.7 x 10^{-5} S cm^{-1}).

Following a similar approach Gleria[39,89] recently reported the possibility of forming partially miscible blends of PEO and an intrinsically photosensitive phosphazene copolymer substituted with benzophenone and ethylene oxide groups (BzMEEP).

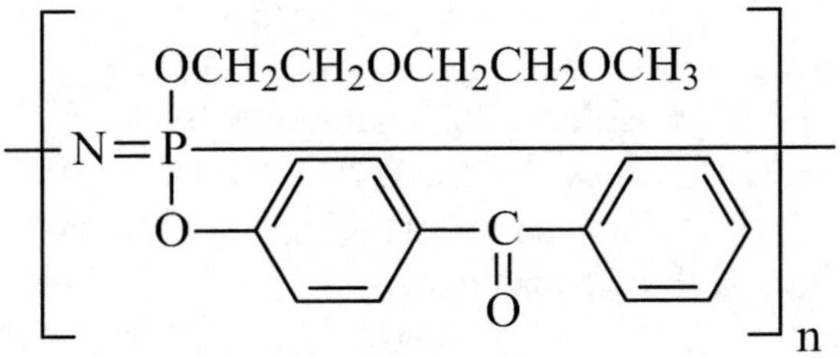

Figure 10

Miscible blends of these two materials could be obtained up to a BzMEEP content of 50% w/w due to the chemical similarity of these two compounds. Dimensional stability could be improved by controlled photocrosslinking of the blends by selective irradiation of the benzophenone moiety of BzMEEP. No information about the ionic conductivity of the blends is presently available.

Phosphazene Blends to Improve Ionic Conductivity of PEO

The increasing interest of new polymeric electrolytes for use as rechargeable solid-state lithium batteries directed research to obtain materials showing high ionic conductivity close to room temperature. PEO is one of the most appropriate candidates for the manufacture of solid polymeric membranes in secondary batteries, but its semicrystalline character and the increase in the glass transition temperature (T_g) resulting from the formation of polymer-salt complexes are two factors that diminish the ionic conductivity at room temperature. In fact, it is well known that conduction takes place predominantly in the amorphous phase and that the segmental mobility of the polymer is an important factor in determining the ionic mobility.

The problem of improving PEO ionic conductivity was faced by Acosta who studied blends of PEO with a phosphazene elastomer, poly(octofluoropentoxy trifluoroethoxy phosphazene) (PFP)[65]

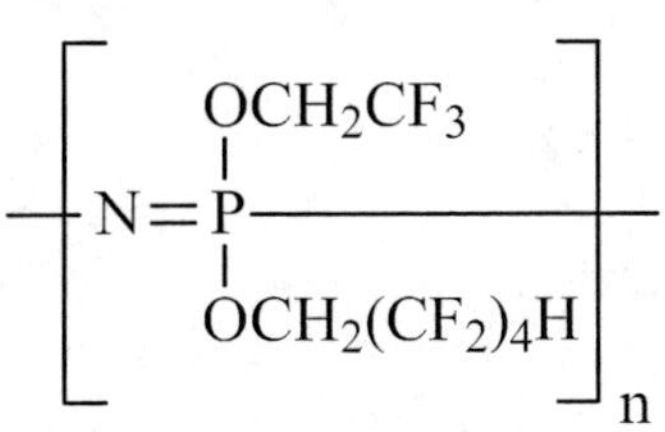

Figure 11

a material which possesses high thermal stability, elasticity at low temperature and good mechanical properties.

Studies on the isothermal crystallization measured by DSC demonstrated the partial compatibility of PEO/PFP blend. In fact, the crystallization rate increased up to 40 wt % of PFP, where the partially miscible polymer exerted a nucleating action and diminished above 40 wt % of PFP, where the dilution property became predominant. The system showed two T_g that differ from those of the pure components. According to the authors, this confirms that the blend is only partially compatible. Contradictory results were obtained by examining the crystallization kinetics using the Avrami or the Ozawa model.

Successively, attention was paid to the thermal behavior of PEO/PFP blends after solvation with lithium triflate, $LiCF_3SO_3$ (TfLi)[69]. The solvated PEO system presented two melting peaks: the first one corresponding to the T_m of unsolvated PEO (67°-68°C), the second one appeared at a higher temperature (150°-190°C), which was assigned to the solvated complex PEO/TfLi. The T_g of homopolymer PEO/TfLi was shifted toward higher values (up to 80°C) as a function of increasing TfLi content; on the contrary, T_g values of PFP/TfLi remained constant, suggesting that TfLi was not coordinated to the ether groups. Microstructural study[71] by FTIR spectroscopy showed broadening of the absorption band at 1108 cm^{-1}, when polyether oxygen was coordinated to TfLi.

Binary PEO/PFP blends containing vanadium oxides and carbon black were also investigated[70,72]. The blend compatibility was not affected by the presence of inorganic oxides, while electric conductivity was significantly enhanced (up to 10^{-3} S cm^{-1}).

Magnesium salts, rather than V_2O_5, were also considered as possible substituents for lithium salts in the preparation of light batteries[102]. Microstructural characterization by X-ray spectroscopy showed that complete coordination of magnesium triflate with the oxygen of the polyethers occurs.

Attempts were also carried out to physically compatibilize the binary blends of PEO/PFP, through the incorporation of a third polymeric component, e.g. poly(epichlorohydrin) (PECH)[68,73,74]. Isothermal and non-isothermal crystallization studies demonstrated that PEO/PECH was a compatible system, while the addition of PECH to the PEO/PFP binary system induced only a limited compatibility. T_g investigations confirmed these results. No improvement of electrical conductivity was observed for these ternary blends[81].

The effect of four different additives [e.g. ethylene carbonate (EC), propylene carbonate (PCA), γ-butyrolactone and ethyleneglycol dimethylether] on the thermal properties, ionic conductivity and electrochemical stability of polymer electrolytes based on blend PEO/PFP was also been investigated[80]. The obtained results indicate that only two species act as plasticizers (EC and PCA) decreasing the melting temperatures, enthalpy, and T_g values of systems with consequent enhancement of the ionic conductivity.

Composite solid polymers based on $[PEO/PFP]_8$ with $LiCF_3SO_3$[83,84] and $LiPF_6$[90,92] electrolytes added with finely dispersed solid fillers (γ-$LiAlO_2$) were recently studied with DSC, FTIR and X-ray spectroscopy. In some cases a plasticizer (e.g. PCA) was added to the composite to compensate for the rigidity introduced into the system by the filler. Limited interactions were found between the ceramic filler and the uncomplexed polymer, while the addition of the plasticizer decreased the crystallinity of the system, but did not influence the final morphology. When the filler content in the composite was above 20% by weight, the effect observed was an increase in the ionic conductivity, possibly due to the inhibition of PEO crystallization processes[84].

Phosphazene Blends for Biomedical Applications

To modulate the rate of biodegradability of materials used in implants a clever approach is to develop blends based on polymers that exhibits different behavior with respect to decay and interaction with the biological environment.

The mechanism of degradation, i.e. the bulk erosion process which usually leads to uncontrollable time for the loss of mechanical properties, is one of the drawbacks limiting the usefulness of commonly used biodegradable polymers. This is the case for glycolic and lactic acid (co)-polymers, which may be advantageously blended with different kind of polyphosphazenes[38,78,79]. The substitution of the chlorines in PDCP with p-methyl phenoxy and ethyl glycinate moieties represents a simple way to prepare interesting phosphazene copolymers. Ethyl glycinato-substituted polyphosphazenes have already been recognized to support the adhesion and growth of osteoblast-like cells[132,133]. Ibim[78] thus prepared polyphosphazenes with different glycinate content, blending them with commercial poly(lactide-co-glycolide), 50:50 (PLAGA). Blends were cast as films after mutual solvent dissolution and then dried. The intermediate T_g observed complies with the miscibility between PLAGA and phosphazene. The authors found useful changes in the degradation kinetics which basically proceeds by gradual surface erosion, thus avoiding an abrupt collapse of the mechanical integrity. Such behavior is suitable as well for drug delivery applications. To impart additional stability, IPN or semi-IPN can be prepared[38] where the crosslinking reaction may occur both ex-vivo or in-vivo.

Phosphazene Blends for Membrane Separation Processes

Nagai[40] employed spin casting and solvent casting methods in order to prepare blends of two polyphosphazenes as thin films for permeability measurements.

The selected polymers, PTFEP and adamantane amino/trifluoroethoxy polyphosphazene (AdAPTFEP)

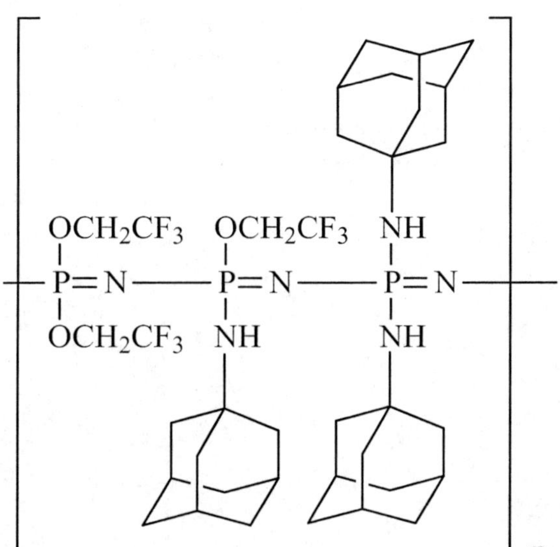

Figure 12

showed a quite dissimilar calorimetric behavior; therefore a DSC analysis revealed the unambiguous immiscible nature of this system. The two T_g values of the original members of the blend are observed to be unchanged.

PTFEP is known as a polymer with one of the highest permeabilities to gases[134,135]. In particular it is highly permeable to carbon dioxide because of the CO_2–philic nature of fluorine substituents which increase its solubility.

Blending the above mentioned homopolymer with a (50/50) AdAPTFEP resulted in a lower permeability composition with a modest increase in selectivity factors between the majority of tested gases which are behind about one order magnitude below the theoretical values.

Other Phosphazene/Organic Polymer Blends

A condis-mesophase may be developed in PE/POP blends although no mesogenic groups are present. "Condis" is an acronym for a conformationally disordered state proposed by Wunderlich[136] to distinguish such new types of mesophases. Borisenkova[51,52] observed that the introduction of a small amount of PTFEP into a high density polyethylene (HDPE) matrix results in substantial changes in the rheological behavior. Modification of the flow properties of ultra high molecular weight polyethylene or similar hardly processing polymers can be advantageously achieved by such compounding. The viscosity of HDPE (η_{HDPE}) at 225°C is two orders of magnitude higher than that of PTFEP. The addition of 30% wt of PTFEP is enough to depress the viscosity to near η_{PTFEP} values. Above 30% PTFEP the blend morphology consists of a HDPE matrix pervaded by fibrillar PTFEP microphase which in turn forms a skin around flowing PE.

Blends between Polyphosphazenes and Inorganic Polymers

Blending phosphazene polymers with inorganic macromolecules was attempted by Witner[24] immediately after the discovery of these materials[9-11], with the aim of improving mill processing, low temperature flexibility and volume swell in low aromatic content fluids of fluorinated phosphazene polymers substituted with trifluoroethoxy, octafluoropentoxy and of 2-allylphenoxy moieties. The inorganic partner selected for phosphazenes in these blends was polydimethylsiloxane containing silyl groups (SILASTIC 430®). Using this approach phosphazene polymers, showing improved processability, were obtained.

In the same period Dieck[26,31] showed that aryloxy-substituted phosphazene derivatives

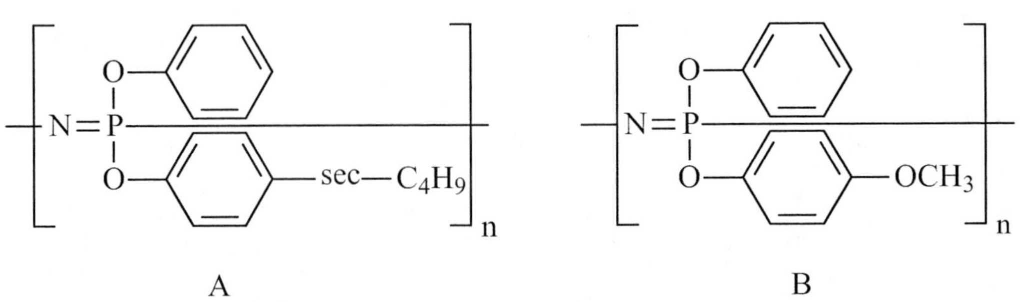

Figure 13

could be also blended with a silicone elastomer containing vinyl groups (SILASTIC 765U®) producing curable blends that contained 15-85% by weight of polyphosphazene. These formulations showed good flame retardancy, and may be used as low-smoke coating films and foams[26,31].

BLENDS BETWEEN DIFFERENTLY SUBSTITUTED POLY(ORGANOPHOSPHAZENES)

Considerable attention has been paid in the past to the possibility of obtaining blends of polyphosphazenes bearing different substituent groups[5,22,27,30,32,44-46]. The main target of these studies was basic research on crystallinity[32], polymer-polymer interactions and long-term miscibility[5,44], synthesis of new materials showing improved fire-retardant properties and/or low-smoke characteristics[27,30] and influence on the miscibility of different side substituents[45,46].

The first attempt in phosphazene/phosphazene blending was carried out by Connelly[22] who investigated the blending of PTFEP with poly[bis(heptafluorobutoxy)phosphazene] or poly[bis(octafluoropentoxy)phosphazene]. Thermomechanical analysis of these blends showed the presence of two separated phases in the polymer mixture, thus suggesting the substantial incompatibility of these materials.

Research performed by Dieck[27,30] showed successively that new fire-retardant, flexible phosphazene films and foams could be prepared simply by blending two phosphazene homopolymers or copolymers having rather different elasticity features. In these experiments elastomeric phosphazene substrates having a Young's Storage Modulus of 10^6-10^8 dyne/cm^2 and materials showing a Young's Storage Modulus of 10^8-10^{10} dyne/cm^2 were exploited. These differences in mechanical properties could be obtained by the introduction of different nucleophilic derivatives on the phosphazene skeleton of $(NPCl_2)_n$ starting from phenoxy up to 4-*tert*-butyl-phenoxy groups. The resulting blends showed only one T_g, thus indicating that the polymer blends approach true solutions. Furthermore, flammability tests carried out on these new materials gave values of 25.9 for L.O.I. and 95 Dm for smoke evolution, thus suggesting excellent properties for the flame resistance of the final substrates.

Schneider[32] studied a series of equiponderal blends of PBCPP with four differently substituted polyphosphazenes (i.e. phenoxy, 3- and 4-methylphenoxy, and 4-ethylphenoxy-substituted phosphazene macromolecules) cast from THF and analyzed by DSC and X-ray spectroscopy. The blends with poly[bis(4-ethylphenoxy)phosphazene] and poly[bis(3-methylphenoxy)phosphazene] exhibit two distinct T_g values (-24°C and 4°C in the first case, -17°C and 4°C in the latter) and therefore were incompatible in the amorphous phase, whereas those containing PBPP and poly[bis(4-methylphenoxy)phosphazene] have a T_g that overlaps those of the single components, and therefore no conclusions could be drawn on the compatibility of the two latter blends. From the observation of the melting transitions (broadened endotherms and lower temperatures compared to the parent components), it appears that some interaction exists among the crystalline phases, pointing to the possibility that cocrystallization occurs to a some extent.

Allcock[5,44] examined the 10:90 - 90:10 w/w binary blends based on the following polyphosphazenes: poly[bis(methylamino)phosphazene] (PBMAP), MEEP, PBPP, PTFEP and poly[bis(propyloxy benzoate)phosphazene]. The blends were prepared from solutions in suitable solvents and analyzed by DSC and TEM.

The only partially miscible pair was that consisting of PBMAP and MEEP containing more than 60% of the first component, probably as a result of hydrogen bonding interactions. Miscibility was revealed by a decrease in the value of T_g of PBMAP, as shown by the appearance in the thermogram of a second order transition at around -50°C, suggesting that two different regions are present in the blend: a PBMAP-rich phase and a totally miscible phase. The blend was analyzed again after five years, and the disappearance of the transition at -50°C suggested that phase separation slowly occurs in the material.

Semi-compatibility was observed among poly(fluoroalkoxyphosphazenes)[45,46]. Optical interferometry, rheological and mechanical analyses were performed, showing the presence of a miscibility zone between PTFEP and copolymers containing different amounts of trifluoroethoxy and dodecafluoroheptoxy side groups. As expected compatibility decreased with an increased amount of the second substituent and the viscosity showed positive deviations inside the miscibility window.

Crosslinking and Reactive Blending Processes

According to literature[1], a very simple method to blend polymers, to stabilize their morphology, and to form new materials is to promote the formation of interchain bonds between different macromolecules. This can be easily achieved by a variety of methods, for instance by using UV or high energy radiation (*e.g.* electron beam or gamma rays) to crosslink materials, or by the action of reactive processes involving suitable chemical functions present in the structure of the two partners of the blend.

Two examples of the first strategy are described below.

In one case blends of polyphosphazenes and ethylene-propylene-diene rubbers were prepared through accelerated electron beam irradiation of heterogeneous mixtures of these substrates[47,101] to obtain new materials showing stabilized morphology even during processing conditions. In the second case, the UV irradiation of partially miscible blends formed from phosphazene copolymers containing methoxyethoxyethoxy- and 4-benzoylphenoxy- moieties (BzMEEP)[39] and PEO leads to the formation of crosslinked, highly swellable soft hydrogels, which have potential application in the preparation of biomaterials for entrapment of drugs and their controlled release, or in the development of electroconductive materials (see above).

An example of the second approach involves the reaction between a fluorinated polyphosphazene and an organic fluoroelastomer functionalized with variable percentages of iodine atoms in the presence of organic peroxides[137,138] to form new, co-crosslinked polymeric compositions which show favorably improved mechanical properties, fuel oil resistance and reduced drawbacks at low temperature. More recently Allcock and co-workers[41,85] showed that new blends could be formed by treating poly[bis(4-carboxylatophenoxy)phosphazene] (a polymer containing free carboxylic groups) with PU (substrates that bear pending isocyanate moieties), according to the following general Scheme:

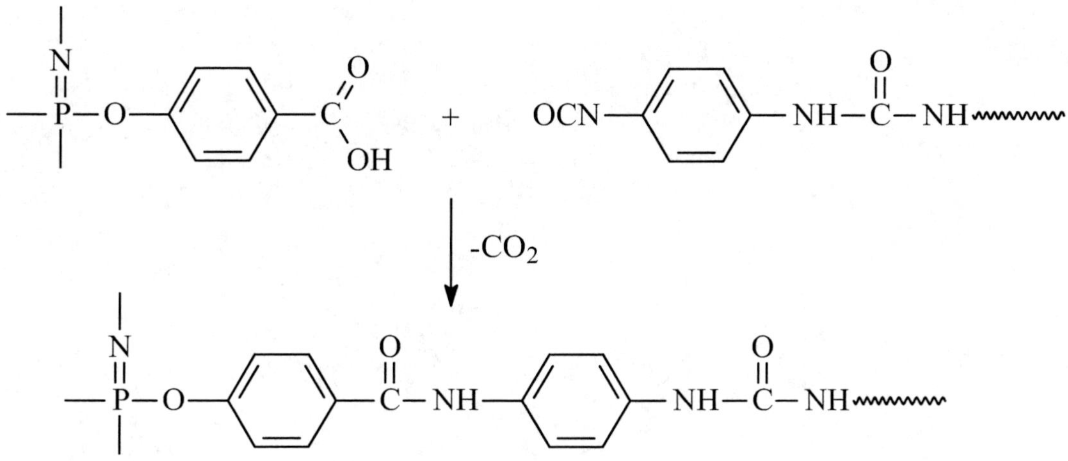

Figure 14

The thermal stability of the resultant foams were analyzed by thermogravimetric analysis (TGA). The char yields at both 400°C and 600°C were increased relative to the pure polyurethane precursor on increasing the content of polyphosphazene from 5 wt% to 20 wt%. With a higher phosphazene content, the char at 400°C remained the same, but the char at 600°C continued to increase. Flammability tests showed that the presence of 15 wt% or less of the polyphosphazene had little effect on the overall flammability, but a significant flame resistance was obtained at polyphosphazene concentrations higher than 20% w/w in the mixture. The improved properties are believed to originate from the crosslinking of the poly(organophosphazene), which released non-combustible gases and produced protective char.

In both cases described above, the chemical functions exploited to form blends were already present in the polymeric materials, however, this strategy works equally well if the chemical groups used to promote crosslinking reactions can be introduced into the phosphazene macromolecules through suitable functionalization reactions. For example, important modifications of phosphazene polymers have been based on the preparation of grafted phosphazene copolymers[20,139-141] which contain both low molecular weight substrates such as maleic anhydride[142-145] or variable types of maleates[82,146-148], and/or high molecular weight products [for instance, PS[149], PMMA[150], HALS-containing polyacrylates[151-153], poly(acrylic or methacrylic acids)[154], PVP[154], Poly(vinyl pyrrolidone)[155], poly(vinyl acetate)[156,157], PVA[156,157], poly(N,N-dimethylacryl amide)[158,159], PEO[39,89], dimethyaminoethyl methacrylate[160], etc.].

The general approach we used for these processes is described in the Scheme below:

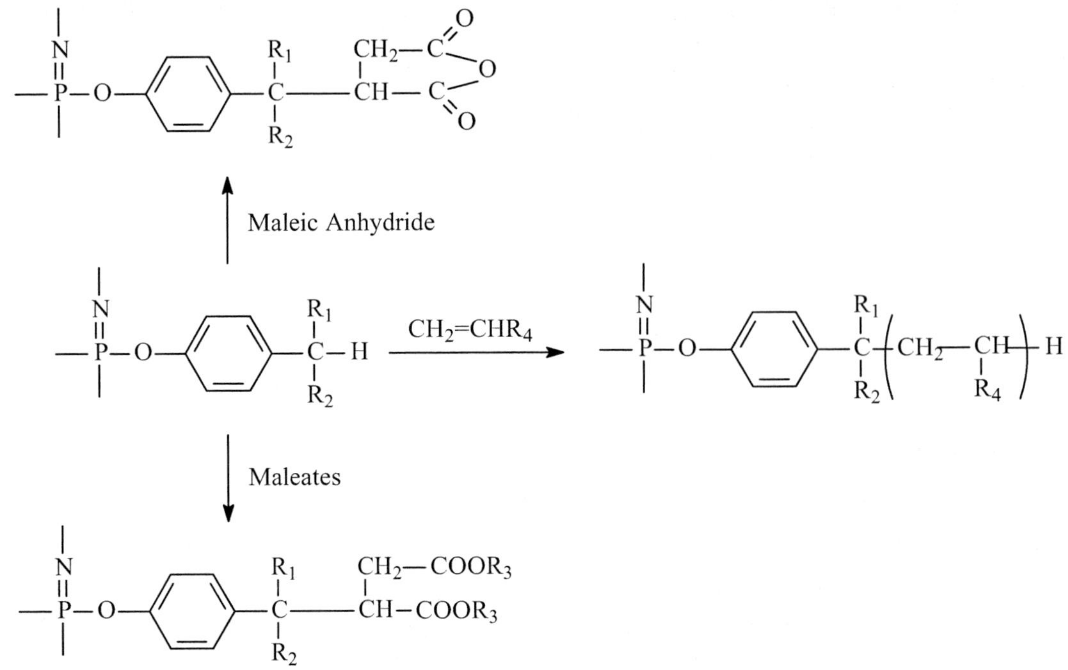

where:

$R_1 = R_2 = H$	Poly[bis(4-methylphenoxy)phosphazene][82,144,145,147,153,156,157];
$R_1 = H$; $R_2 = CH_3$	Poly[bis(4-ethylphenoxy)phosphazene][82,143,145-148,153,156,157];
$R_1 = R_2 = CH$	Poly[bis(4-isopropylphenoxy)phosphazene][145,147,149,150,153,158];
$R_1 = CH_3$; $R_2 = C_2H_5$	Poly[bis(4-secbutylphenoxy)phosphazene][82,142,145,147,153,156-158];
$R_1 = H$; $R_2 = C_6H_5$;	Poly[bis(4-benzylphenoxy)phosphazene][145,151-154,158];
$R_3 =$	methyl[146,147], ethyl[146-148] or butyl[146,147] groups, or an organic compound containing free oxazoline derivatives[82,162];

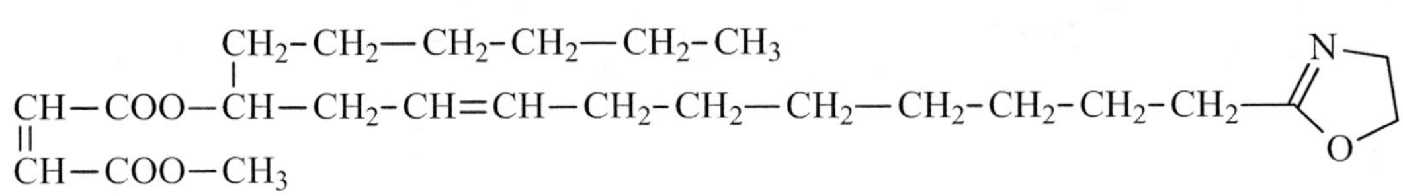

$R_4 =$	Phenyl[149], COOH[154], $COOCH_3$[158], $OCOCH_3$[156], OH[156], COO-HALS[151-153,157], pyridyl[154], etc.,

Figure 15

In other experiments, PBPP[160], PTFEP[155,158], and poly[bis(4-tert-butylphenoxy)phosphazene][145] have also been exploited.

In the above mentioned reactions, aryloxy-substituted POPs containing labile hydrogen atoms in the side substituent of the polyphosphazene material were treated with organic peroxides[82,142,144-146,148,149,161,162], UV light[147,150-154,156-158] or γ-rays[155,159,160,163] to split an hydrogen radical on the alkyl substituent of the phenoxy phosphazene substituents. The phosphazene macroradicals formed are able to interact with the unsaturated molecules present in the reaction mixture and to start grafting processes onto the polyphosphazene matrices.

Additional grafting processes of PS[164,165], PMMA[166] and polydimethylsiloxane[164,167,168] onto poly(alkyl/aryl)phosphazenes were almost simultaneously reported by Wisian-Neilson, who basically adopted anionic polymerization processes for the growth of the side organic or inorganic branches from the main polyphosphazene skeleton.

The new materials proved to be useful for a variety of practical applications, such as to improve adhesion of the functionalized POPs onto metal surfaces[169], or to change the surface polarity of polyphosphazene films[158], and showed potential for practical exploitation in the formation of blends between phosphazene polymers and organic commercial macromolecules[170].

For example, aryloxy-substituted polyphosphazenes functionalized with oxazoline moieties[82,162]

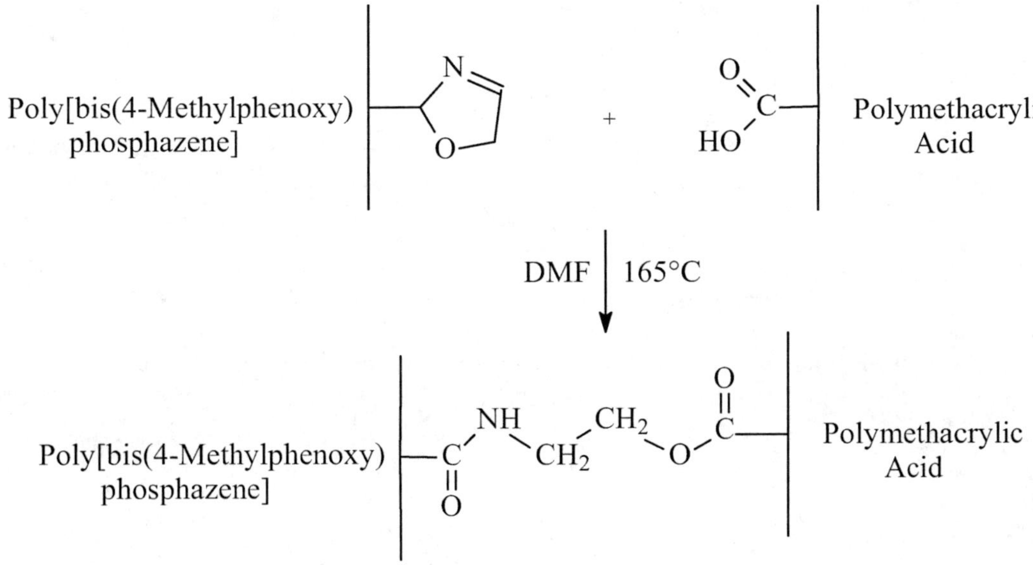

Figure 16

have be reacted at 165°C in dimethylformamide with poly(acrylic or methacrylic) acid forming new blends in which the phosphazene polymer and the organic macromolecules are linked together through amide and ester functionalities[171].

Cyclophosphazenes as Modifiers for Macromolecules

Cyclophosphazene derivatives are low molecular weight analogues of poly(organophosphazenes) that have recently assumed an increased importance in recent times as multifunctional modifiers for macromolecules.

These characteristics were first recognized by Inoue[172] in 1996, who synthesized a heteroarm, star-shaped cyclophosphazene copolymer bearing polystyrene and polyamide-6 (PA-6) chains as phosphorus substituents and used it to promote compatibilization between PPE and PA-6.

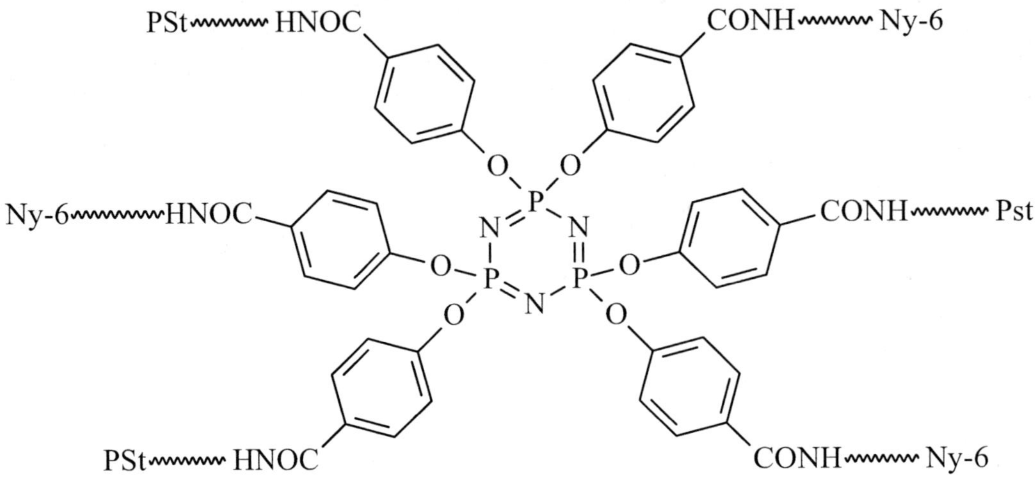

Figure 17

It is supposed that, during blend formation, the cyclophosphazene copolymer is located at the interface of PPE and PA-6 macromolecules, and that its side polystyrene arms strongly interfere with the PPE macromolecules, while the polyamide substituents of the cyclophosphazene interact with the PA-6 polymer. The resulting blends show a single T_g by DSC and exhibit improved mechanical properties, i.e. higher modulus, superior tensile strength and elongation at break.

Following the strategy reported above, Gleria recently developed the synthesis of a new cyclophosphazene, the hexakis(4-oxazolinophenoxy)cyclophosphazene [88,98,162] in which

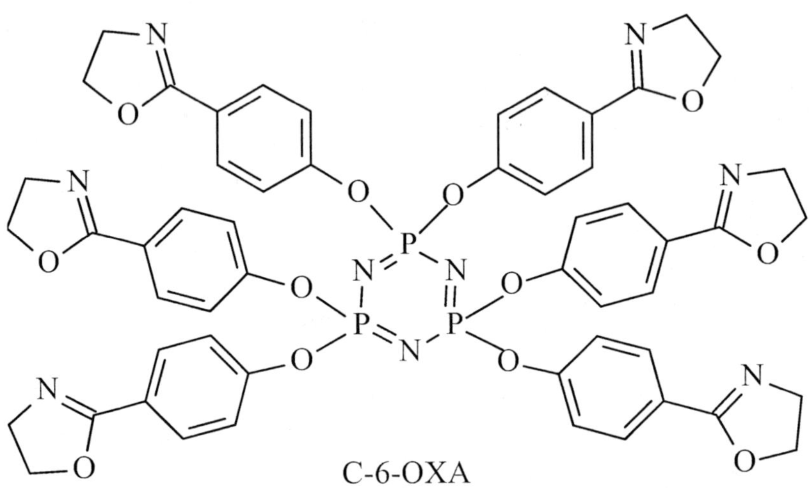

C-6-OXA

Figure 18

six free oxazoline groups are present in the side phosphorus substituents. These groups are reactive towards carboxylic[171] and (to a rather low extent) towards hydroxylic group-containing organic molecules[173], forming amide and ester groups, or amide and ether residues, respectively.

The cyclophosphazene therefore can be reacted with -COOH end-capped polyester substrates (e.g. PET) to act as a chain extender derivative[88,98,162] and forming star-shaped polymers, or as a compatibilizing agent[35,100] when reacted with polycarbonate and polyamide substrate.

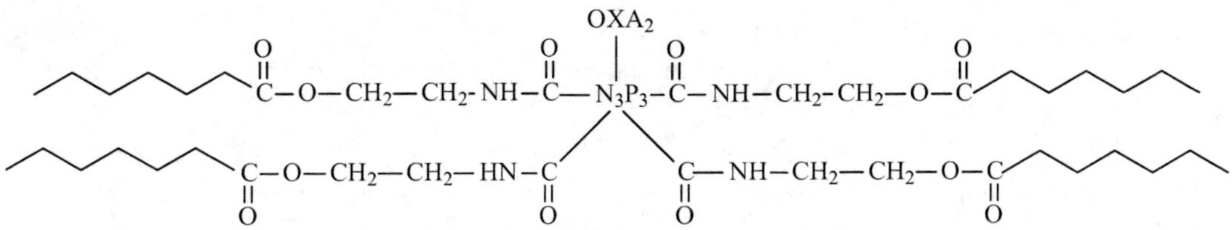

Figure 19

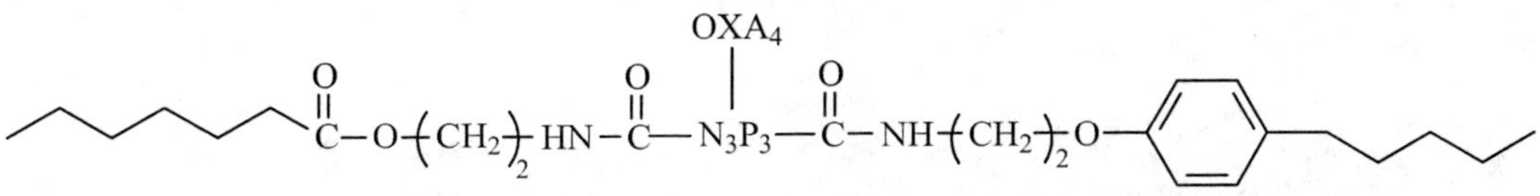

Figure 20

In the first case the recovering of molecular weight for PET is accompanied by the concurrent improvement of the mechanical properties of this material[88,98,162], with a possible utilization of the trimer in the recycling of organic polymers. In the second case a compatibilizing agent is generated *in situ* that bears polycarbonate and polyamide chains and is able to promote the formation of blends between these two, usually incompatible, macromolecules[35,100]. Even in this case, the resulting blends showed higher modulus values as compared with those of the original macromolecules.

An additional example of the utilization of cyclophosphazene derivatives was recently provided by Allcock[94-97] who synthesized cyclophosphazene substrates containing azide groups.

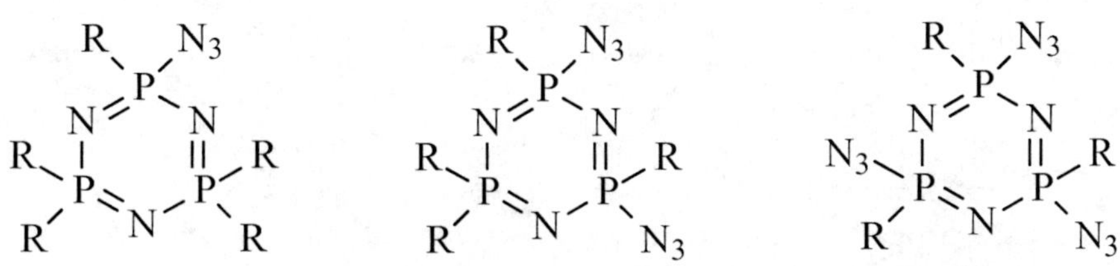

where: R may be C_6H_5O-; CF_3CH_2O- or $(CH_3)_2N$-

Figure 21

Due to the high reactivity of these functionalities in the trimers, a possible utilization of these materials for polyolefin modifiers could be envisaged.

CONCLUSION

In this article we have reviewed the work carried out during the last forty years in the field of phosphazene blends, using as partners for these substrates other members of the phosphazene series containing different phosphorus substituents[5,22,27,30,32,44-46], inorganic plastics (e.g. PDMS)[24,26,31] or selected organic polymers[5,6,23,25,28,29,33,35-41,44,47-93]. Interesting fundamental results and wide range of potential applications for these new materials resulted from these investigations.

First of all, it could be proved that POP polymers hardly compatibilize spontaneously both with similar or with different macromolecules[5,32], and that weak interactions (e.g. hydrogen bonding)[6,57,60,66] or strong covalent bonds[35,41,82,85,100] are necessary to induce blend formation and morphology preservation[47,101] of these substrates.

Secondly, it could be demonstrated that phosphazene blends show potential applications in different domains, such as:

(1) flame retardancy and thermal stability of blended materials[23,25,28,29,41,53,56,59,61,62,75-77,85-87,91], due to the intrinsic self-extinguishibility and heat resistance of POPs; these characteristics assumed particular relevance when phosphazene substrates bearing fluorinated substituents were exploited in the blend formation process;
(2) dimensional stabilization of polymers (e.g. MEEP) that did not show good mechanical properties[6,39,48-50,54,55,57,60,66,89], and, for these reason, cannot be used as ionic conductors in the construction of light batteries;
(3) improvement of the ionic conductivity of PEO[65,68-74,80,81,83,84,90,92,102] both by enhancing miscibility with this compound and by decreasing the crystallinity of PEO, acting as a plasticizer for the polymer;
(4) biomedical applications[38,78,79], by devising phosphazene blends that slowly degrade thus allowing the controlled delivery of drugs;
(5) membrane separation processes[40], controlling both the permeability and the selectivity of the final blends by appropriate composition of the system.

Finally, carefully functionalized, low-molecular weight cyclophosphazenes could be used as polymer modifiers to selectively crosslink organic macromolecules[94-97], as compatibilizers to facilitate blend formation between different organic polymers[35,88,100], and as chain extenders[88,98-100] for molecular weight recovery of functionalized organic macromolecules.

All of these investigations are extremely promising and may contribute to the industrial and technological development of phosphazene materials.

REFERENCES

1. Coleman, M. M.; Graf, J. F.; Painter, P. C. *Specific Interactions and the Miscibility of Polymer Blends*; Technomic: Lancaster PA, USA, 1991.
2. Allcock, H. R. *Chem. Eng. News* **1985**, *63*, 22.
3. *Chem. Week* **1983**, *March 1, 72.*
4. Garbassi, F. *CHEMTECH.* **1999**, *October*, 48.
5. Allcock, H. R.; Visscher, K. B. *Chem. Mater.* **1992**, *4*, 1182.

6. Landry, C. J. T.; Ferrar, W. T.; Teegarden, D. M.; Coltrain, B. K. *Macromolecules* **1993**, *26*, 35.
7. *Polymer Blends and Alloys*; Folkes, M. J.; Hope, P. S., Eds.; Blackie Academic & Professional: London, 1993.
8. De Jaeger, R.; Gleria, M. *Prog. Polym. Sci.* **1998**, *23*, 179.
9. Allcock, H. R.; Kugel, R. L. *J. Am. Chem. Soc.* **1965**, *87*, 4216.
10. Allcock, H. R.; Kugel, R. L.; Valan, K. J. *Inorg. Chem.* **1966**, *5*, 1709.
11. Allcock, H. R.; Kugel, R. L. *Inorg. Chem.* **1966**, *5*, 1716.
12. Allcock, H. R. *Phosphorus-Nitrogen Compounds. Cyclic, Linear, and High Polymeric Systems*; Academic Press: New York, USA, 1972.
13. Allen, C. W. *J. Fire Sci.* **1993**, *11*, 320.
14. Potin, P.; De Jaeger, R. *Eur. Polym. J.* **1991**, *27*, 341.
15. Allcock, H. R. In *Biodegradable Polymers as Drug Delivery Systems*; Chasin, M., Langer, R., Eds.; Marcel Dekker: New York, 1990; Chapt. 5, p 163.
16. Scopelianos, A. G., In *Biomed. Polym.*; Shalaby, S. W., Ed.; Hanser: Munich, 1994; p 153.
17. Schacht, E.; Vandorpe, J.; Lemmouchi, Y.; Dejardin, S.; Seymur, L., In *Frontiers in Biomedical Polymer applications*; Ottembrite, R. M., Ed.; Technomic: Lancaster, UK, 1998; Vol. 1; p 27.
18. Bortolus, P.; Gleria, M. *J. Inorg. Organomet. Polym.* **1994**, *4*, 1.
19. Bortolus, P.; Gleria, M. *J. Inorg. Organomet. Polym.* **1994**, *4*, 95.
20. Bortolus, P.; Gleria, M. *J. Inorg. Organomet. Polym.* **1994**, *4*, 205.
21. Critchley, J. P.; Knight, G. J.; Wright, W. W. *Heat-Resistant Polymers*; Plenum Press: New York, USA, 1983.
22. Connelly, T. M.; Gilham, J. K. *J. Appl. Polym. Sci.* **1976**, *20*, 473.
23. DeEdwardo, A. H.; Zitomer, F.; Stuetz, D.; Singler, R. E.; Macaione, D. *Org. Coat. Prep.* **1976**, *36*, 737.
24. Witner, J. F.; Kyker, G. S. U.S. US, 4,000,166 (1976), Chem. Abstr. 86, 6303y (1977), assigned to The Firestone Tire & Rubber Co.
25. DeEdwardo, A. H.; Zitomer, F.; Stackman, R. W.; Kramer, C. E. US Patent, 4,042,561 (1977), Chem. Abstr. 87, 118861f (1977), assigned to Celanese Corp.
26. Dieck, R. L.; Quinn, E. J. US Patent, 4,026,839 (1977), Chem. Abstr. 87, 24536b (1977), assigned to Armstrong Cork Co.
27. Dieck, R. L.; Quinn, E. J. US Patent, 4,055,520 (1977), Chem. Abstr. 88, 7903x (1978), assigned to Armstrong Cork Co.
28. Dieck, R. L.; Quinn, E. J. US Patent, 4,061,606 (1977), Chem. Abstr. 88, 51619h (1978), assigned to Armstrong Cork Co.
29. Brackenridge, D. R. US Patent, 4,079,035 (1978), Chem. Abstr. 89, 44646g (1978), assigned to Ethyl Corp.
30. Dieck, R. L.; Quinn, E. J. US Patent, 4,073,824 (1978), Chem. Abstr. 88,7903x (1978), assigned to Armstrong Cork Co.
31. Dieck, R. L.; Quinn, E. J. US Patent, 4,073,825 (1978), assigned to Armstrong Cork Co.
32. Beres, J. J.; Schneider, N. S.; Desper, C. R.; Singler, R. E. *Macromolecules* **1979**, *12*, 566.
33. Carriedo, G. A.; Garcia Alonso, F. J.; Gomez Elipe, P.; Gonzalez, P. A.; Marco, C.; Gomez, M. A.; Ellis, G. *J. Appl. Polym. Sci.* **2000**, *77*, 568.

34. Chiu, W. Y.; Wang, F. T.; Chen, L. W.; Don, T. M.; Lee, C. Y. *Polym. Degrad. Stab.* **2000**, *67*, 223.
35. Giannotta, G.; Po', R.; Fiocca, L.; Cardi, N.; Lucchelli, E.; Braglia, R.; Fambri, L.; Pegoretti, A.; Gleria, M., *Proceedings of the The 9th International Conference on Inorganic Ring Systems (IRIS)*, Saarbruecken, Germany, July 23-28, 2000, Poster No. P-107.
36. Gleria, M.; Minto, F.; Braglia, R.; Garbassi, F.; Giannotta, G.; Meda, L.; Po', R. *J. Inorg. Organomet. Polym.* **2000**, *10*, 23.
37. Gleria, M.; Minto, F.; Garbassi, F.; Giannotta, G.; Meda, L.; Po', R. *J. Inorg. Organomet. Polym.* **2000**, *10*, 61.
38. Laurencin, C. T.; Allcock, H. R.; Ibim, S. E. M.; Ambrosio, A.; Kwon, M. U.S. US, 6,077,916 (2000), Chem. Abstr. 133, 44341 (2000), assigned to Penn State Research Foundation, USA.
39. Minto, F.; Gleria, M.; Pegoretti, A.; Fambri, L. *Macromolecules* **2000**, *33*, 1173.
40. Nagai, K.; Freeman, D. B.; Cannon, A.; Allcock, H. R. *J. Membrane Sci.* **2000**, *172*, 167.
41. Reed, C. S.; Taylor, J. P.; Guigley, K. S.; Coleman, M. M.; Allcock, H. R. *Polym. Eng. Sci* **2000**, *40*, 465.
42. Gleria, M. *Phosphorus Res. Bull.* **1999**, *10*, 55.
43. *Japan Chemcal Week* **1985**, June 20.
44. Allcock, H. R.; Visscher, K. B. U.S. US, 5,457,160 (1995), Chem. Abstr. 124, 31250q (1996), assigned to The Penn State Research Foundation.
45. Avdeev, N. N.; Novikova, D. K.; Vasil'ev, G. B.; Kulichikhin, V. G.; Tur, D. R. *Vysokomol. Soedin. Ser. A Ser. B* **1999**, *40*, 1757, Chem. Abstr. 131, 116817z (1999).
46. Borisenkova, E. K.; Vasil'ev, G. B.; Kulichikhin, V. G.; Tur, D. R. *Vysokomol. Soedin. Ser. A Ser. B* **1999**, *40*, 1809, Chem. Abstr. 131, 117015y (1999).
47. Bohm, G. G. A.; Nelson, C. J. U.S. US, 4,220,512 (1980), Chem. Abstr. 89, 130820r (1978), assigned to The Firestone Tire & Rubber Co.
48. Abraham, K. M.; Alamgir, M.; Perrotti, S. J. *EIC Laboratories, Inc. Norwood, MA*, Report, available NTIS AD-A179925, **1987**, Chem. Abstr. 108, 41083w (1988).
49. Abraham, K. M.; Alamgir, M.; Perrotti, S. J. *J. Electrochem. Soc.* **1988**, *135*, 535.
50. Abraham, K. M.; Alamgir, M.; Reynolds, R. K. *J. Electrochem. Soc.* **1989**, *136*, 3576.
51. Antipov, E. M.; Borisenkova, E. K.; Kulichikhin, V. G.; Plate, N. A. *Makromol. Chem. Macromol. Symp.* **1990**, *38*, 275.286.
52. Borisenkova, E. K.; Tur, D. R.; Litvinov, V. M.; Antipov, E. M.; Kulichikhin, V. G.; Plate, N. A. *Vysokomol. Soedin. Ser. A* **1990**, *32*, 1505, Chem. Abstr. 113, 116448h (1990).
53. Chang, S. C. U.S. US, 4,966,937 (1990), Chem. Abstr. 114, 45169a (1991), assigned to Ethyl Corp.
54. Abraham, K. M.; Alamgir, M.; Moulton, R. D. *J. Electrochem. Soc.* **1991**, *138*, 921.
55. Abraham, K. M.; Alamgir, M. *Chem. Mater.* **1991**, *3*, 339.
56. Ishikawa, T.; Sunatsuka, H.; Matsura, M.; Asaba, S. Jpn. Kokai Tokkyo Koho JP, 03 281,553 [91, 281,553] (1991), Chem. Abstr. 116, 153156u (1992), assigned to Fujikura Ltd.
57. Landry, C. J. T.; Ferrar, W. T.; Teegarden, D. M. Eur. Pat. Appl. EP, 485,033 (1991), Chem. Abstr. 117, 192999m (1992), assigned to Eastman Kodak Co.

58. Braswell, L. *Annu. Tech. Conf. - Soc. Plast. Eng.* **1992**, *1*, 407, Chem. Abstr. 120, 271760s (1994).
59. Engelhardt, H.; Spilgiers, G.; Tuerck, G. Ger. Offen. DE, 4,026,883 (1992), Chem. Abstr. 116, 237298e (1992), assigned to Bayer Aktiengesellschaft.
60. Landry, C. J. T.; Ferrar, W. T.; Teegarden, D. M.; Coltrain, B. K. *ACS Polym. Prep.* **1992**, *33(2)*, 557.
61. Yamamoto, Y.; Tanmachi, M.; Takahata, N. Jpn. Kokai Tokkyo Koho JP, 04 154,852 [92 154,852] (1992), Chem. Abstr. 117, 173135k (1992), assigned to Hitachi Cable, Ltd.
62. Yamamoto, Y.; Hori, A.; Takahata, N. Jpn. Kokai Tokkyo Koho JP, 04 154,851 [92 154,851] (1992), Chem. Abstr. 117, 173488c (1992), assigned to Hitachi Cable, Ltd.
63. Borisenkova, E. K. *Int. J. Polym. Mater.* **1993**, *21*, 79.
64. Chen-Yang, Y. W.; Wu, T. T. *ACS Polym. Prep.* **1993**, *34(1)*, 272.
65. Herrero, C.; Acosta, J. L. *Polymer International* **1993**, *32*, 349.
66. Landry, C. J. T.; Ferrar, W. T.; Teegarden, D. M. U.S. US, 5,250,626 (1993), Chem. Abstr. 117, 192999m (1992), assigned to Eastman Kodak Co.
67. Chen-Yang, Y. W.; Lee, H. F.; Wu, T. T. In *Inorganic and Organometallic Polymers II. Advanced Materials and Intermediates*; Wisian-Neilson, P., Allcock, H. R., Wynne, K. J., Eds.; ACS Symposium Series: Washington, DC, USA, 1994; Chapt. 23, Vol. 572, p 295.
68. Herrero, C.; Acosta, J. L. *Polym. J. (Tokyo)* **1994**, *26*, 786.
69. Acosta, J. L.; Morales, E. *J. Appl. Polym. Sci.* **1996**, *60*, 1185.
70. Acosta, J. L.; Morales, E.; Paleo, M.; Jurado, J. R. *Eur. Polym. J.* **1996**, *32*, 13.
71. Acosta Luque, J. L.; Morales Bergas, E. *Solid State Ionics* **1996**, *91*, 75.
72. Acosta Luque, J. L.; Morales Bergas, E.; Jurado Egea, J. R. *J. Appl. Polym. Sci.* **1996**, *59*, 1173.
73. Del Rio, C.; Acosta, J. L. *Polym. Bull. (Berlin)* **1997**, *38*, 63.
74. Del Rio, C.; Martin-Alvarez, P. J.; Acosta, J. L. *Polym. Bull. (Berlin)* **1997**, *38*, 353.
75. Denq, B. L.; Chiu, W. Y. *Guoli Taiwan Daxue Gongcheng Xuekan* **1997**, *70*, 41, Chem. Abstr. 127, 346969x (1997).
76. Denq, B. L.; Chiu, W. Y.; Chen, L. W.; Lee, C. Y. *Polym. Degrad. Stab.* **1997**, *57*, 261.
77. Denq, B. L.; Hu, Y. S.; Chiu, W. Y.; Chen, L. W.; Chiu, Y. S. *Polym. Degrad. Stab.* **1997**, *57*, 269.
78. Ibim, S. E. M.; Ambrosio, A. M. A.; Kwon, M. S.; El-Amin, S. F.; Allcock, H. R.; Laurencin, C. T. *Biomaterials* **1997**, *18*, 1565.
79. Laurencin, C. T.; Ibim, S. E. M.; Allcock, H. R.; Ambrosio, A. M. A.; El-Amin, S.; Kwon, M. S. *Proc. 24th Int. Symp. Controlled Release Bioact. Mater.* **1997**, 971, Chem. Abstr. 127, 140370m (1997).
80. Morales, E.; Acosta, J. L. *Solid State Ionics* **1997**, *96*, 99.
81. Del Rio, C.; Martin-Alvarez, P. J.; Acosta, J. L. *J. Appl. Polym. Sci.* **1998**, *70*, 2181.
82. Gleria, M.; Minto, F.; Po', R.; Cardi, N.; Fiocca, L.; Spera, S. *Macromol. Chem. Phys.* **1998**, *199*, 2477.
83. Morales, E.; Acosta, J. L. *J. Appl. Polym. Sci.* **1998**, *69*, 2435.
84. Morales, E.; Acosta, J. L. *Solid State Ionics* **1998**, *111*, 109.
85. Allcock, H. R.; Coleman, M. M.; Reed, C. S.; Guigley, K. S. U.S. US, 5,965,627 (1999), Chem. Abstr. 131, 272704 (1999), assigned to The Penn State Research Foundation.
86. Chiu, W. Y.; Wang, P. S.; Don, T. M. *Polym. Degrad. Stab.* **1999**, *66*, 233.

87. Denq, B. R.; Hu, Y. S.; Chen, L. W.; Chiu, W. Y.; Wu, T. R. *J. Appl. Polym. Sci.* **1999**, *74*, 229.
88. Gleria, M.; Minto, F.; Bertani, R.; Tiso, B.; Po', R.; Fiocca, L.; Lucchelli, E.; Giannotta, G.; Cardi, N. *Phosphorus Res. Bull.* **1999**, *10*, 730.
89. Minto, F.; Pegoretti, A.; Fambri, L.; Gleria, M. *Phosphorus Res. Bull.* **1999**, *10*, 752.
90. Morales, E.; Villarreal, I.; Acosta, J. L. *J. Appl. Polym. Sci.* **1999**, *73*, 1023.
91. Ohno, Y.; Matsumoto, K.; Hirobe, K. Jpn. Kokai Tokkyo Koho JP, 11 181,268 [99 181, 268] (1999), Chem. Abstr. 131, 117059r (1999), assigned to Kanegafuchi Chemical Industry Co., Ltd., Japan.
92. Villarreal, I.; Morales, E.; Acosta, J. L. *Angew. Makromol. Chem.* **1999**, *266*, 24.
93. Wang, P. S.; Chiu, W. Y.; Chen, L. W.; Denq, B. R.; Don, T. M.; Chiu, Y. S. *Polym. Degrad. Stab.* **1999**, *66*, 307.
94. Allcock, H. R.; McIntosh, M. B.; Hartle, T. J. *Inorg. Chem.* **1999**, *38*, 5535.
95. Hartle, T. J.; McIntosh, M. B.; Allcock, H. R. *ACS Polym. Prep.* **1999**, *40(2)*, 908.
96. McIntosh, M. B.; Hartle, T. J.; Allcock, H. R. *J. Am. Chem. Soc.* **1999**, *121*, 884.
97. Hartle, T. J.; Sunderland, N. J.; McIntosh, M. B.; Allcock, H. R. *Macromolecules* **2000**, *33*, 4307.
98. Po', R.; Fiocca, L.; Giannotta, G.; Lucchelli, E.; Cardi, N.; Minto, F.; Fambri, L.; Gleria, M., *Proceedings of the The Ninth International Conference on Inorganic Ring Systems (IRIS-IX)*, Saarbrücken, Germany, July 23-28, 2000, Comm. O-50.
99. Gleria, M.; Minto, F.; Tiso, B.; Bertani, R.; Tondello, E.; Fiocca, L.; Lucchelli, E.; Giannotta, G.; Cardi, N. *Designed Monomers and Polymers*, in press.
100. Giannotta, G.; Po', R.; Fiocca, L.; Cardi, N.; Lucchelli, E.; Braglia, R.; Fambri, L.; Pegoretti, A.; Minto, F.; Gleria, M. *Phosphorus, Sulfur and Silicon Relat. Elem.*, in press.
101. Italian Patent, 1,093,258 (1985), assigned to The Firestone Tire & Rubber Co.
102. Acosta, J. L.; Morales, E. *Electrochim. Acta* **1998**, *43*, 791.
103. Couchman, P. R. *Macromolecules* **1978**, *11*, 1156.
104. Allcock, H. R.; McInstosh, M. B.; Klingenberg, E. H.; Napierala, M. E. *Macromolecules* **1998**, *31*, 5255.
105. Fox, T. G. *Bull. Amer. Phys. Soc.* **1956**, *1*, 123.
106. Ebdon, J. R., In *Polymeric Materials Encyclopedia*; Salamone, J. C., Ed.; CRC Press: Boca Raton, USA, 1996; Vol. 4; p 2397.
107. Blonsky, P. M.; Shriver, D. F.; Austin, P. E.; Allcock, H. R. *J. Am. Chem. Soc.* **1984**, *106*, 6854.
108. Chandrasekhar, V. *Adv. Polym. Sci.* **1998**, *135*, 139.
109. Allcock, H. R.; Austin, P. E.; Neenan, T. X.; Sisko, J. T.; Blonsky, P. M.; Shriver, D. F. *Macromolecules* **1986**, *19*, 1508.
110. Blonsky, P. M.; Clancy, S.; Hardy, L. C.; Harris, C. S.; Spindler, R.; Tonge, J. S.; Shriver, D. F. *ChemTech* **1987**, 758.
111. Coltrain, B. K.; Ferrar, W. T.; Landry, C. J. T.; Molaire, T. R.; Zumbulyadis, N. *Chem. Mater.* **1992**, *4*, 358.
112. Tonge, K. M.; Shriver, D. F. *J. Electrochem. Soc.* **1987**, *134*, 269.
113. Tonge, J. S.; Blonsky, P. M.; Shriver, D. F.; Allcock, H. R.; Austin, P. E.; Neenan, T. X.; Sisko, J. T. *Proc. Electrochem. Soc.* **1988**, *1*, 533.
114. Nelson, C. J.; Coggio, W. D.; Allcock, H. R. *Chem. Mater.* **1991**, *3*, 786.

115. Nelson, C. J.; Coggio, W. D.; Allcock, H. R. *Department of Chemistry, Pennsylvania State University, University Park, PA*, TR-3, Available NTIS AD-A241746 **1992**, Chem. Abstr. 118, 255465c (1993).
116. Allcock, H. R.; Nelson, C. J.; Coggio, W. D. *Polym. Mater. Sci. Eng.* **1993**, *68*, 76.
117. Allcock, H. R.; Nelson, C. J.; Coggio, W. D. U.S. US, 5,414,025 (1995), Chem. Abstr. 123, 61305h (1995), assigned to The Penn State Research Foundation.
118. Bennett, J. L.; Dembek, A. A.; Allcock, H. R.; Heyen, B. J.; Shriver, D. F. *Chem. Mater.* **1989**, *1*, 14.
119. Bennett, J. L.; Dembeck, A. A.; Allcock, H. R.; Heyen, B. J.; Shriver, D. F. *ACS Polym. Prep.* **1989**, *30(1)*, 437.
120. Nazri, G. A.; Meibuhr, S. G. *J. Electrochem. Soc.* **1989**, *136*, 2450.
121. Nazri, G. A.; Meibuhr, S. G. *Proc. Electrochem. Soc.* **1989**, *89-4*, 332.
122. Allcock, H. R.; Pucher, S. R.; Turner, M. L.; Fitzpatrick, R. J. *Macromolecules* **1992**, *25*, 5573.
123. Coltrain, B. K.; Ferrar, W. T.; Landry, C. J. T. PCT Int. Appl. WO, 90 11,323 (1990), Chem. Abstr. 114, 165903g (1991), assigned to Eastman Kodak Co.
124. Coltrain, B. K.; Ferrar, W. T.; Landry, C. J. T.; Molaire, T. R. *ACS Polym. Prep.* **1991**, *32(3)*, 477.
125. Landry, C. J. T.; Ferrar, W. T.; Coltrain, B. K. PCT Int. Appl. WO, 91 09,080 (1991), Chem. Abstr. 115, 234535y (1991), assigned to Eastman Kodak Co.
126. Landry, C. J. T.; Ferrar, W. T.; Coltrain, B. K. U.S. US, 5,104,602 (1992), Chem. Abstr. 115, 234535y (1991), assigned to Eastman Kodak Co.
127. Coltrain, B. K.; Ferrar, W. T.; Landry, C. J. T.; Molaire, T. R.; Schildkraut, D. E.; Smith, V. K. *ACS Polym. Prep.* **1993**, *34(1)*, 266.
128. Landry, C. J. T.; Ferrar, W. T.; Coltrain, B. K. U.S. US, 5,190,819 (1993), assigned to Eastman Kodak Co.
129. Brusatin, G.; Guglielmi, M.; De Jaeger, R.; Facchin, G.; Gleria, M.; Musiani, M. In *Syntheses and Methodologies in Inorganic Chemistry. New Compounds and Materials*; Italy ed.; Daolio, S., Tondello, E., Vigato, P. A., Eds.: Bressanone, 1996; Vol. 6, pp 315.
130. Brusatin, G.; Guglielmi, M.; De Jaeger, R.; Facchin, G.; Gleria, M.; Musiani, M. *J. Mater. Sci.* **1997**, *32*, 4415.
131. Allcock, H. R.; Napierala, M. E.; Olmeijer, D. L.; Cameron, C. G.; Kuharcik, S. E.; Reed, C. S.; O'Connors, S. J. M. *Electrochim. Acta* **1998**, *43*, 1145.
132. Laurencin, C. T.; Norman, M. E.; Elgendy, H. M.; El-Amin, S. F.; Allcock, H. R.; Pucher, S. R.; Ambrosio, A. M. A. *J. Biomed. Mater. Res.* **1993**, *27*, 963.
133. Laurencin, C. T.; El-Amin, S. F.; Ibim, S. E. M.; Willoughby, D. A.; Attawia, M.; Allcock, H. R.; Ambrosio, A. M. A. *J. Biomed. Mater. Res.* **1996**, *30*, 133.
134. Drioli, E.; Zhang, S. M.; Basile, A.; Golemme, G.; Gaeta, S. N.; Zhang, H. C. *Gas Separation & Purification* **1991**, *5*, 252.
135. Hirose, T.; Kamiya, Y.; Mizoguchi, K. *J. Appl. Polym. Sci.* **1989**, *38*, 809.
136. Wunderlich, B.; Grebowicz, J. *Adv. Polym. Sci.* **1984**, *60/61*, 2.
137. Tatemoto, M.; Tomoda, M.; Ueta, Y. U.S. US, 4,260,698 (1981), Chem. Abstr. 93, 27580w (1980), assigned to Daikin Kogyo Co., Ltd.
138. Tatemoto, M.; Tomoda, M.; Ueta, Y. Italian Patent, 1,125,443 (1986), assigned to Daikin Kogyo Co., Ltd.

139. Gleria, M.; Minto, F.; Bortolus, P.; Facchin, G.; Bertani, R.; Scoponi, M.; Pradella, F. *ACS Polym. Prep.* **1993**, *34(1)*, 270.
140. Gleria, M.; Minto, F.; Bortolus, P.; Facchin, G.; Bertani, R. In *Inorganic and Organometallic Polymers II: Advanced Materials and Intermediates*; USA ed.; Wisian-Neilson, P., Allcock, H. R., Wynne, K. J., Eds.; ACS Symposium Series: Washington DC, 1994; Vol. 572, pp 279.
141. Minto, F.; Gleria, M.; Bortolus, P.; Scoponi, M.; Pradella, F.; Fambri, L. In *Current Trends in Polymer Photochemistry*; USA ed.; Allen, N. S., Edge, M., Bellobono, I. R., Selli, E., Eds.; Ellis Horwood: New York, 1995; pp 165.
142. Gleria, M.; Minto, F.; Scoponi, M.; Pradella, F.; Carassiti, V. *Chem. Mater.* **1992**, *4*, 1027.
143. Gleria, M.; Minto, F.; Fontana, G.; Bertani, R.; Facchin, G. *Macromolecules* **1995**, *28*, 4399.
144. Fontana, G.; Minto, F.; Gleria, M.; Facchin, G.; Bertani, R.; Favero, G. *Eur. Polym. J.* **1996**, *32*, 1273.
145. Minto, F.; Fontana, G.; Bertani, R.; Facchin, G.; Gleria, M. *J. Inorg. Organomet. Polym.* **1996**, *6*, 367.
146. Gleria, M.; Minto, F.; Doriguzzi, F.; Bertani, R.; Facchin, G.; Tondello, E. *Macromolecules* **1997**, *30*, 4310.
147. Minto, F.; Gleria, M.; Favero, G. *J. Macromol. Sci., Pure Appl. Chem.* **1997**, *A34*, 2177.
148. Gleria, M.; Minto, F.; Bertani, R.; Facchin, G.; Doriguzzi, F.; Tondello, E. In *Synthesis and Methodologies in Inorganic Chemistry. New Compounds and Materials*; Italy ed.; Daolio, S., Tondello, E., Vigato, P. A., Eds., 1997; Vol. 7, pp 318.
149. Gleria, M.; Bolognesi, A.; Porzio, W.; Catellani, M.; Destri, S.; Audisio, G. *Macromolecules* **1987**, *20*, 469.
150. Minto, F.; Scoponi, M.; Fambri, L.; Gleria, M.; Bortolus, P.; Pradella, F. *Eur. Polym. J.* **1992**, *28*, 167.
151. Gleria, M.; Minto, F.; Scrima, R.; Borzatta, V. *J. Appl. Polym. Sci.* **1996**, *61*, 1675.
152. Minto, F.; Borzatta, V.; Gleria, M. *J. Inorg. Organomet. Polym.* **1996**, *6*, 171.
153. Minto, F.; Borzatta, V.; Bertani, R.; Gleria, M. *J. Appl. Polym. Sci.* **1997**, *65*, 217.
154. Minto, F.; Gleria, M.; Bortolus, P.; Fambri, L.; Pegoretti, A. *J. Appl. Polym. Sci.* **1995**, *56*, 747.
155. Pezzin, G.; Palma, G.; Lora, S.; Carenza, M. *Chem. Ind. (Milan)* **1991**, *73*, 185.
156. Minto, F.; Fambri, L.; Gleria, M. *Macromol. Chem. Phys.* **1996**, *197*, 3099.
157. Fambri, L.; Minto, F.; Gleria, M. *J. Inorg. Organomet. Polym.* **1996**, *6*, 195.
158. Minto, F.; Scoponi, M.; Gleria, M.; Pradella, F.; Bortolus, P. *Eur. Polym. J.* **1994**, *30*, 375.
159. Lora, S.; Palma, G.; Carenza, M.; Caliceti, P.; Pezzin, G. *Biomaterials* **1994**, *15*, 937.
160. Lora, S.; Carenza, M.; Palma, G.; Pezzin, G.; Caliceti, P.; Battaglia, P.; Lora, A. *Biomaterials* **1991**, *12*, 275.
161. Gleria, M.; Minto, F.; Fontana, G.; Bertani, R.; Facchin, G. *Macromolecules* **1995**, *28*, 4399.
162. Gleria, M.; Minto, F.; Galeazzi, A.; Po', R.; Cardi, N.; Fiocca, L.; Spera, S. *Phosphorus, Sulfur and Silicon Relat. Elem.* **1999**, *144-146*, 201.
163. Lora, S.; Carenza, M.; Palma, G.; Caliceti, P. *Radiat. Phys. Chem.* **1990**, *35*, 117.
164. Wisian-Neilson, P.; Schaefer, M. A.; Islam, M. S. *ACS Polym. Prep.* **1989**, *30(1)*, 185.

165. Wisian-Neilson, P.; Schaefer, M. A. *Macromolecules* **1989**, *22*, 2003.
166. Garcia-Alonso, F. J.; Wisian-Neilson, P., *Proceedings of the Workshop on the Status and Future of Polyphosphazenes*, Raleigh, NC, USA, March 15-18th, 1992, Poster No. 10.
167. Wisian-Neilson, P.; Islam, M. S. *Macromolecules* **1989**, *22*, 2026.
168. Wisian-Neilson, P.; Islam, M. S.; Schaefer, M. A. *Phosphorus, Sulfur, and Silicon* **1989**, *41*, 145.
169. Gleria, M.; Minto, F.; Scoponi, M.; Pradella, F.; Carassiti, V. Italian Patent, 1,258,815 (1996), assigned to Consiglio Nazionale delle Ricerche.
170. Gleria, M.; Minto, F.; Bertani, R.; Facchin, G.; Tondello, E. *Gazz. Chim. Ital.* **1997**, *127*, 343.
171. Frump, J. A. *Chem. Rev.* **1971**, *71*, 483.
172. Miyata, K.; Watanabe, Y.; Itaya, T.; Tanigaki, T.; Inoue, K. *Macromolecules* **1996**, *29*, 3694.
173. Inata, H.; Matsumura, S. *J. Appl. Polym. Sci.* **1985**, *30*, 3325.

List of Abbreviations

AdAPTFEP	Adamantane amino/trifluoroethoxy polyphosphazene copolymer
BzMEEP	Benzophenone/methoxyethoxyethoxy phosphazene copolymers
DETA	Diethylenetriamine
DGEBA	Diglycidyl ether of bisphenol A
EC	Ethylene carbonate
HDPE	High density polyethylene
HALS	Hindered amine light stabilizers
INP	Interpenetrating Polymer Networks
MEEP	Poly[bis(methoxyethoxyethoxy)phosphazene]
MEEEP	poly[bis(methoxyethoxyethoxyethoxy)phosphazene]
OXA	2-Oxazoline
PA-6	Polyamide-6
PBCPP	Poly[bis(p-chlorophenoxy)phosphazene]
PBMAP	Poly[bis(methylamino)phosphazene]
PBPP	Poly[bis(phenoxy)phosphazene]
PBPPP	Poly[bis(4-phenoxyphenoxy)phosphazene]
PCA	Propylene carbonate
PDPP	Poly(dipropoxyphosphazene)
PDCP	Poly(dichlorophosphazene)
PECH	Poly(epichlorohydrin)
PEO	Poly(ethylene oxide)
PET	Poly(ethylene terephthalate)
PFP	Poly(octofluoropentoxy trifluoroethoxy phosphazene)
PLAGA	Poly(lactide-co-glycolide)
PMMA	Poly(methyl methacrylate)
POP	Poly(organophosphazenes)
PPE	Poly(2,6-dimethylphenylene ether)

PPO	Poly(propylene oxide)
PS	Polystyrene
PTFEP	Poly[bis(2,2',2"-trifluoroethoxy)phosphazene]
PTHF	Poly(tetrahydrofuran)
PU	Polyurethane
PVA	Poly(vinyl alcohol)
PVP	Poly(4-vinylpyridine)
PVPh	Poly(vinyl phenol)
THF	Tetrahydrofuran

In: Applicative Aspects of Poly(organophosphazenes)
Editors: M. Gleria and R. De Jaeger, pp. 169-188
ISBN 1-59454-025-X

Chapter 8

IONICALLY CONDUCTIVE PHOSPHAZENE POLYMERS

Kenzo Inoue
Department of Applied Chemistry
Faculty of Engineering
Ehime University
Matsuyama, 790-8577
Japan

INTRODUCTION

Since the pioneering work on ionically conducting polyether-metal salt complexes has been reported in 1970's, solid polymer electrolytes are the topic of considerable current research both for academic interests and for applications.[1-3] A number of polymer electrolytes, especially polymers containing ethyleneoxy units, have been synthesized and studied as host materials.[4-9] The ionic conductivity (σ) is known to be given simply by

$$\sigma = ne\mu$$

where "n", "e", and "μ" are the total number of carrier ions, the elementary electric charge, and the carrier mobility, respectively. The generally accepted model for ionic conduction is that the ionic conductivity occurs via random walk of ions from one transient coordination site to another through the amorphous region, facilitated by the segmental motions of the polymers. This implies that two basic requirements are needed to construct solid polymer electrolytes having a high conductivity: a number of donor atoms capable of complexing metal ions and high degree of segmental motion to pave the road for fast ion migration, which is associated with a glass transition temperature (Tg) of polymers. The crystalline domains, often formed in the polymers, for example, more than 60% of poly(ethylene oxide) (PEO) is crystalline at room temperature, act as barriers to the migration of ions.

In the last decade, the use of phosphazene-based polymer electrolytes has been studied extensively since well-designed phosphazene polymers may meet most requirements for polymer electrolytes. These polymers can be divided into three categories: (1) linear polyphosphazenes with donor atoms in the side chains, (2) carbon-based polymers with pendant oligo(ethyleneoxy)cyclotriphosphazenes, and (3) star-shaped polymer electrolytes with cyclotriphosphazenes as core. In this chapter, the ionic conductivity of such phosphazene-based polymer electrolytes is described.

1. Linear Polyphosphazenes with Pendant Donor Atoms

1.1 Polyphosphazenes with Oligo(Ethyleneoxy) Side Groups

Poly(organophosphazene)s composed of alternating phosphorus and nitrogen atoms continue to attract the attention of chemists, not only from synthetic points of view, but also with respect to the wide spectrum of chemical and physical properties of their materials.[10-15] Although there are now several methods for the preparation of such poly(organophosphazene)s, most of them were prepared *via* a thermal ring-opening polymerization of hexachlorocyclotriphosphazene (**1**). The substituents introduced into the poly(dichlorophosphazene) (**2**) backbone become the primary determining factor in creating desired functional polymers.

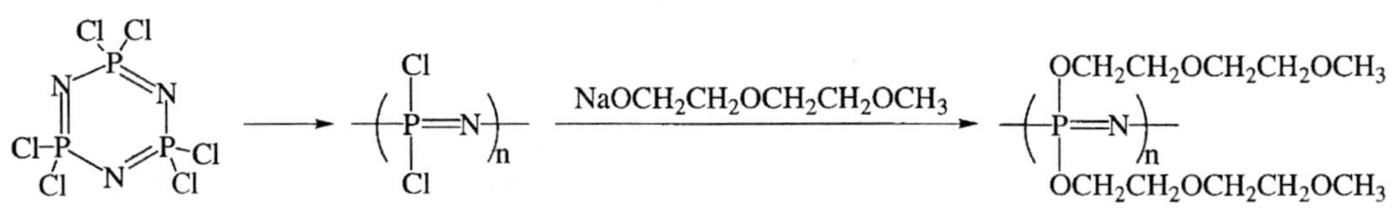

Scheme 1

One of characteristic features of such polyphosphazenes is the conformational flexibility of the backbone.[16] The polyphosphazene carrying $-(OCH_2CH_2)_2OCH_3$ (**3**, MEEP) was first synthesized and studied as a polymer electrolyte by Shriver, Allcock, and their co-workers.[17,18] The reaction of **2**, obtained by the ring-opening polymerization of **1**, with sodium salt of 2-(2-methoxyethoxy)ethanol gave the fully substituted, stable polymer MEEP. MEEP was a completely amorphous with glass transition temperature of –84 °C, which is prerequisite to achieve a high ionic conductivity. The Tg values of the polymers increased with increasing the concentration of $AgSO_3CF_3$ as shown in Figure 1. This behavior can be attributed to the restriction of segmental motion caused by the polymer-salt interaction.

The conductivity of an amorphous polymer follows the Vogel-Tammann-Fulcher(VTF) equation

$$\sigma = AT^{-1/2} \exp[-B/(T-T_0)]$$

where A, B and T_0 are empirical constants. A and B are related to the number of carrier ions and the pseudo-activation energy, respectively. T_0 is closely related with the glass transition temperature.

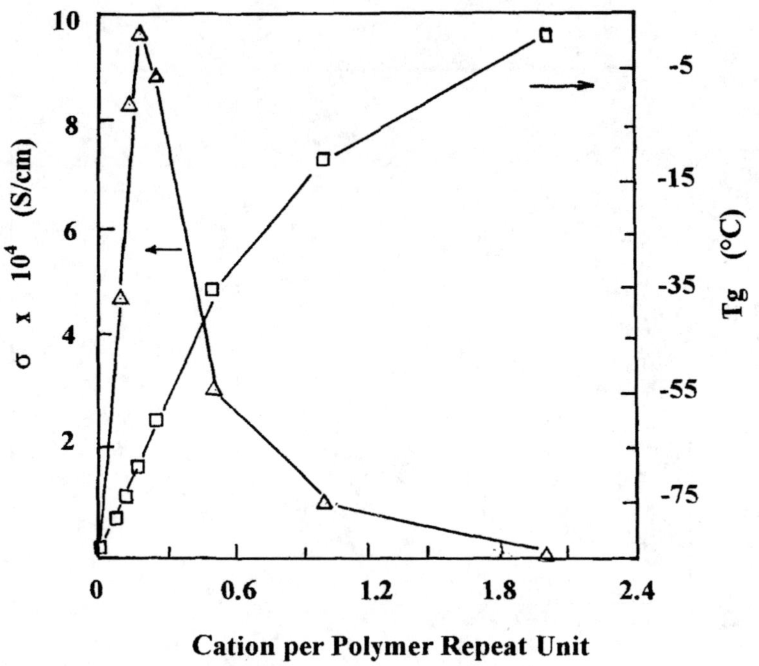

Figure 1 Left axis: △, conductivity at 70 °C *vs.* compositions of $AgSO_3CF_3$-MEEP complexes. Right axis: □, Tg (extrapolated to 0 °C/min heating rate) *vs.* composition of $AgSO_3CF_3$-MEEP complexes

As shown in Figure 2, the plots of ln $\sigma T^{1/2}$ vs. 1/T gave gentle curves, as expected for amorphous polymer electrolytes. Among them, the MEEP complexed with 17.1 mol% $LiSO_3CF_3$ and 10.9 mol% of $AgSO_3CF_3$ showed the maximum conductivities of 2.9 x 10^{-5} S cm^{-1} and 1.68 x 10^{-4} S cm^{-1} at 25 °C, respectively. These values are 2 - 3 orders of magnitude higher than those of PEO-Li^+ salt complexes. The high conductivity is largely due to the highly flexible backbone and relatively rich contents of coordination site per monomer units. For all the polymer systems, the conductivity increased steeply with increasing the salt concentrations, reached a maximum, and then decreased. The behavior is observed for every polymer electrolyte systems. The initial enhancement of conductivity is ascribed to the increase of the number of carrier ions. The decrease of conductivity for further addition of salts is believed to be due to following reasons: (a) the decrease of free coordination sites, (b) the decrease of segmental motion by ionic crosslinking between different side chains, and (c) the ion-pair association.

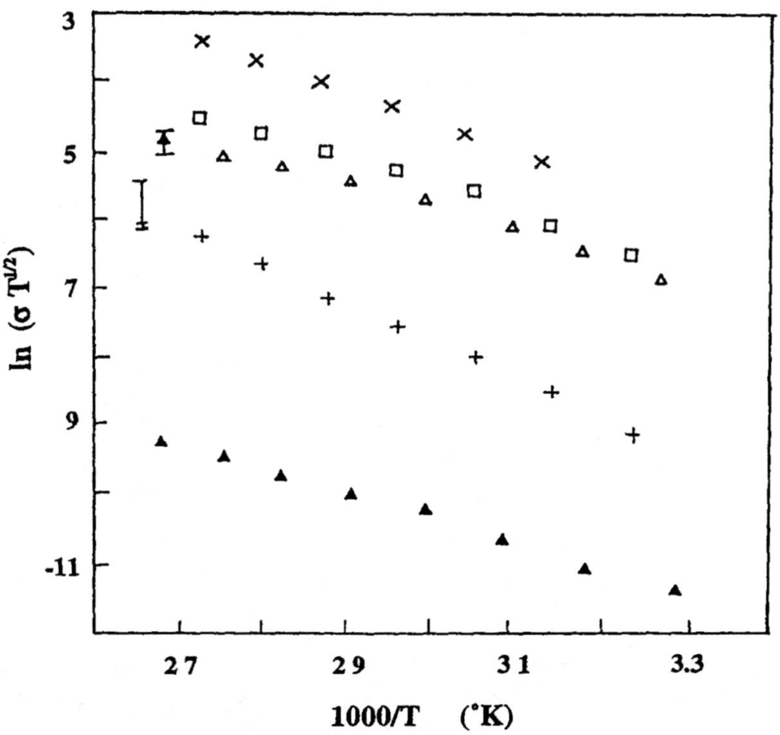

Figure 2 Temperature dependence of ionic conductivity for MEEP (▲), MEEP-$Sr(SO_3CF_3)_2$ (+), MEEP-$NaSO_3CF_3$ (△), MEEP-$LiSO_3CF_3$ (□), and MEEP-$AgSO_3CF_3$ (×) complexes.

Infrared and Raman spectroscopies are useful tools for understanding of the interaction of the ions with the polymers, as well as ion-ion interactions. The ions such as $CF_3SO_3^-$ and ClO_4^- might act as probes and their vibrational spectra make it possible to determine whether the species exist as free ions, ion pairs or highly associated ions.[19] The Raman spectra of the 40: 1 MEEP- $LiSO_3CF_3$ complex suggest that most of the triflate ions exist as a contact ion pair (88 %), together with free ion (12 %).[20] The triple ions, $[Li_2SO_3CF_3]^+$, were not formed at such a dilute metal salt concentration. As expected, the amounts of free ions increased with increasing of ethyleneoxy chain length, e.g., for the polyphosphazene with $-(OCH_2CH_2)_5OCH_3$, the percentage of free ion reached to 24 %. In contrast, for the 10:1 MEEP-$LiSO_3CF_3$ complex, triple ions (96 %) are dominant species, and no measurable amounts of free ion are formed. In these systems, it is believed that the backbone nitrogen atoms of polyphosphazenes with ethyleneoxy side chains do not directly participate in the Li^+ ion conduction and that the oxygen atoms of the P-O bonds are little-used coordination sites.[21]

A series of poly(organophosphazene)s bearing longer ethyleneoxy side chains (**3** – **8**) or branched oligo(ethyleneoxy) (**9** - **14**) side chains were synthesized and the effects of these chains on the ionic conduction and physical properties such as glass transition temperature, crystallinity, and dimensional stability were studied.[22] The results are summarized in Table 1.

Table 1 Maximum Ionic Conductivities of $LiSO_3CF_3$-Polymer 3-14 Complexes

Polymers	$-P(R)_2=N-$ R	T_g (°C)	Li^+ (mol %)	Conductivity 10^5 S/cm (25°C)
4	$-OCH_2CH_2OCH_3$	-81	6.2	2.6
3	$-O(CH_2CH_2O)_2CH_3$	-84	20.0	2.7
5	$-O(CH_2CH_2O)_3CH_3$	-80	11.1	3.2
6	$-O(CH_2CH_2O)_4CH_3$	-80	24.6	3.7
7	$-O(CH_2CH_2O)_6CH_3$	-76 (Tm=-25)	41.5	4.8
8	$-O(CH_2CH_2O)_8CH_3$	-75 (Tm=0)	40.5	4.5
9	$-OCH_2CHOCH_3$ | CH_2OCH_3	-79	16.5	0.16
10	$-OCH_2CHOCH_2CH_2OCH_3$ | $CH_2OCH_2CH_2OCH_3$	-81	14.5	1.2
11	$-OCH_2CHO(CH_2CH_2O)_2CH_3$ | $CH_2O(CH_2CH_2O)_2CH_3$	-81	19.0	3.7
12	$-OCH_2CHO(CH_2CH_2O)_3CH_3$ | $CH_2O(CH_2CH_2O)_3CH_3$	-82	21.5	3.9
13	$-OCH_2CHOCH_2CH_2O(CH_2)_3CH_3$ | $CH_2OCH_2CH_2O(CH_2)_3CH_3$	-86	20.6	0.12
14	$-OCH_2CHOCH_2CH_2OCH(CH_3)_2$ | $CH_2OCH_2CH_2OCH(CH_3)_2$	-77	8.4	0.064

The ionic conductivities of polymers, **3** – **8**, complexed with $LiSO_3CF_3$ varied from 2.6 x 10^{-5} S cm^{-1} to 4.8 x 10^{-5} S cm^{-1} at 25 °C. The maximum conductivity increases slightly as the lengths of the ethyleneoxy side chains increase. The polyphosphazenes with longer side groups (**7** and **8**) have a tendency to form crystalline domain as observed in other comb-shaped polymer systems. The glass transition temperature and ionic conductivity of polyphosphazenes bearing branched side chains were comparable to those of their corresponding linear analogues. In contrast to the low temperature crystallite formation for **7** and **8**, the branched polymers are amorphous over a wide temperature range. Interestingly, the branched polymers have significantly higher dimensional stability than the polymers **3** - **8**. The development of stability might be due to the enhancement of the interdigitated entanglement of side chains. The relatively bulky group at the terminal position in the side chains of **14** affects the conductivity in a striking manner, suggesting that short oligo(ethyleneoxy) chains with aliphatic group can not serve as effective coordination sites probably due to the steric hindrance.

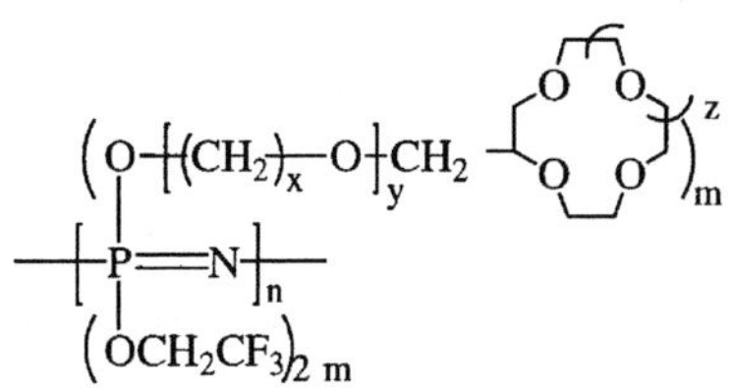

15a x = 0 y = 0 z = 1

15b x = 3 y = 1 z = 1

15c x = 6, y = 1 z = 1

16a x = 0 y = 0, z = 2

16b x = 3, y = 1, z = 2

16c x = 6, y = 1, z = 2

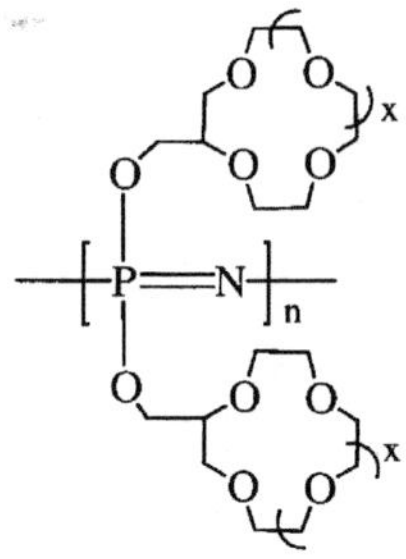

17a x = 1

17b x = 2

17c x = 3

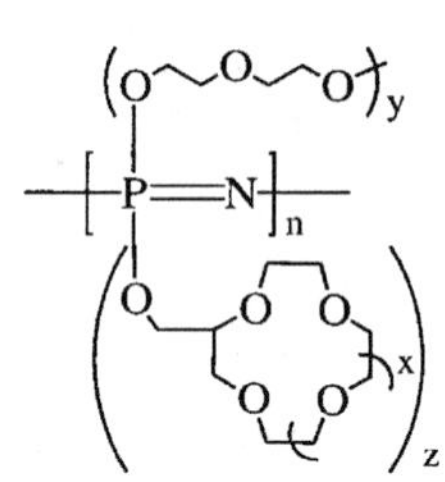

18a x = 1, y = 1.50 z = 0.50

18b x = 2, y = 1.51, z = 0.49

18c x = 3, y = 1.54 z = 0.46

Chart 1

Cowie and co-workers[23-25] synthesized polyphosphazenes with pendant 12-crown-4 ether (**15**) and 15-crown-5 ether (**16**), where these units are spaced from the backbone by zero, three, and six methylene units, starting from polybis(trifluoroethoxy)phosphazene. A 50 – 80 % of crown ether units was introduced into the backbone by the displacement of trifluoroethoxy group with the sodium salt of crown ether alcohols. For the 12-crown-4 ether substituted polyphosphazene, the highest value, 10^{-5} S cm^{-1} at room temperature, was obtained for the polyphosphazene with six methylene spacer (**15c**) complexed with $LiBF_4$. When $NaBF_4$ and KBF_4 were used, the conductivity decreased in the order of 10^{-7} S cm^{-1}. The behavior could be rationalized in terms of well-known complex stability between metal ion and crown ether. The lithium salt is more effectively ionized by the crown ether than $NaBF_4$ and KBF_4. For such a system, the migration of anions rather than cations is considered to be mainly responsible for conduction. A significant different result was reported for the polyphosphazenes (**17**) prepared by the displacement of **2**, -(N=P(Cl)$_2$)-, with lithium salt of hydoxymethyl crown ethers.[26] In this system, no measurable ionic conductivity was observed when complexed with $LiSO_3CF_3$ or $LiClO_4$, indicating that both cation and anion could not migrate efficiently through the polymers (**17**) bearing fully substituted crown ethers. When both 2-(2-methoxyethoxy)ethoxy and 15-crown-5 ether groups in a 3: 1 ratio were introduced into polyphosphazene backbone (**18b**), the Li^+ salt complex showed ionic conductivity of 1 x 10^{-5} S cm^{-1}, which was somewhat higher than that of the **18a** –Li^+ salt system (σ = 6.6 x 10^{-6} S cm^{-1}). The introduction of crown ether units enhances the degree of ion separation of added salts but not contribute to the increase of the number of net carrier ions due to the stable complex formation. From the difference of the conduction behavior between polyphosphazenes partial (**18**) and fully substituted with crown ethers (**17**), it is suggested that the major portion of conduction for the MEEP-metal salt complexes is due to the migration of cations.

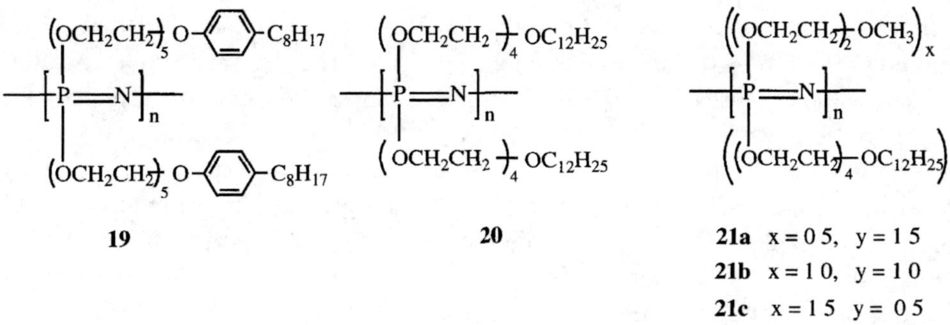

Chart 2

Chandrasekhar et al.[27] have prepared linear polyphosphazene (**19**) appending surfactant Igepal 520, which is insoluble in water. The main motivation of introduction of long alkyl chains is to improve the cumbersome purification process for the synthesis of MEEP, the separation of by-product NaCl. The polymer was a completely amorphous with glass transition temperature of –28.8 °C. The highest conductivity of 5.8 x 10^{-6} S cm^{-1} at 60 °C was obtained for an $[CH_2CH_2O]/[Li^+]$ = 8. Likewise, cosubstituent polyphosphazenes containing various side groups such as 2-(2-methoxyethoxy)ethoxy and polyethyleneoxy lauryl ether units (**20** and **21**) have been studied as hosts for lithium salts by Allcock and co-workers.[28] The polymers have low glass transition temperatures (-53.6 °C - -68.0 °C), but they have at least one melting transition. The complex of the polymer (**20**) bearing $(OCH_2CH_2)_4OC_{12}H_{25}$ and $LiSO_3CF_3$, for example, has a conductivity of 3.3 x 10^{-6} S cm^{-1}. Thus, the polymers with long alkyl chains have poor ionic conductivities.

As a series of cosubstituted polyphosphazenes, the synthesis and conductivity of mixed poly(phosphazene)s bearing etheric $-((OCH_2CH_2)_2OCH_3)$ and alkoxy ($-O(CH_2)_nCH_3$, n = 2-9, **22**) substituents as repeating polymer units have been reported.[29] The latter group was introduced into the backbone with the intention of increase of free volume around coordination sites, which would assist ion mobility. The polymers were amorphous over a wide temperature range, and the glass transition temperatures (-95 - -85 °C) were roughly 10 °C below compared to the Tg value of MEEP, indicating that these polymers are highly flexible. The maximum conductivity of $LiSO_3CF_3$-complexed polyphosphazenes with R = -$O(CH_2)_2CH_3$ (**22a**) had 1.29 x10^{-5} S cm^{-1} at 25 °C while for the polymer with the longest alkyl chain (**22h**) the conductivity showed 1.96 x 10^{-6} S cm^{-1}. Thus, the increase in the length of alkoxy groups brought about the decrease in conductivity. These results indicate that the introduction of noncoordination sites interferes the ion migration even though the segmental mobility is enhanced, in other word, the existence of abundant donor atoms is the most important for fast ion transport.

Aside from these polyphosphazenes bearing pendant ethyleneoxy chains, the conductivity of PEO-polyphosphazene triblock copolymer (**23**) was reported.[30] The copolymer was synthesized via a phosphoramine intermediate, prepared from the reaction of NH_2-PEG-

$CH_2CH_2NH_2$ (M_n =3400) and $(CF_3CH_2O)_2BrP{=}NSiCH_3$. The block copolymer complexed with $LiSO_3CF_3$ exhibited the conductivity of 2 x 10^{-5} S cm^{-1} at room temperature.

Recently, Chen-Yang et al.[31] reported the ionic conductivity of poly[bis(amino)phosphazene] (**24**)-$LiClO_4$ complexes. The glass transition temperature of the polymer with pentylamino group (**24a**) was significantly low compared to that of pendant hexylamino analogue (**24b**). The conductivity (σ = 1.5 x 10^{-7} S cm^{-1} at room temperature) for the Li^+ salt complex system with **24a** was almost the same as that of the PEO system. Reflecting of relatively high segmental mobility of **24a**, the conductivity of the **24a**-Li^+ system was roughly 10 times higher than that of the **24b**- Li^+ complex. FTIR and NMR spectra suggested that Li^+ ions in these polyphosphazene systems interact predominantly with nitrogen atoms in the side chains. They also prepared cosubstituted polyphosphazenes (**25**) bearing $CH_3OCH_2CH_2OCH_2CH_2O$- and alkyl amino side $CH_3(CH_2)_mNH$ chains by a two-step substitution reaction of polydichlorophosphazene.[32] The ratio of the two side chains in each polyphosphazene substituted unit was controlled by the feed ratio of the corresponding alcohol and amine. DSC analysis showed that all of the polymers were amorphous with glass transition temperatures of –50 – -68 °C, depending on the length and nature of the side groups. The polymer-$LiClO_4$ complexes gave translucent freestanding films with satisfactory dimensional stability. The ionic conductivity was significantly affected by the degree of substitution of ethyleneoxy units and length of amino chains. The Li^+ salt complex of the polymers with the higher contents of ethyleneoxy units gave a higher conductivity. The maximum conductivity was 2.2 x 10^{-5} S cm^{-1} for the Li^+ salt complex of cosubstituted polymer **25c**. This value was higher than that of the complex of poly(bis(alkylamino)phosphazene-$LiClO_4$ system. The VTF plots of the complexes exhibit straight lines when the standard temperature is defined as Tg - 50 °C, suggesting that the segmental motion of ethyleneoxy chains controls the ion transport.

Table 2 Maximum Ionic Conductivities of $LiSO_3CF_3$-22 Complexes

Polymer	$-P(R)[(OCH_2CH_2)_2OCH_3]{=}N-$	T_g (°C)	Li^+ (mol %)	Conductivity 10^5 S/cm (25°C)
22a	$-O(CH_2)_2CH_3$	-88	3.5	1.29
22b	$-O(CH_2)_3CH_3$	-82	14.4	1.06
22c	$-O(CH_2)_4CH_3$	-72	22.1	0.91
22d	$-O(CH_2)_5CH_3$	-72	15.9	0.72
22e	$-O(CH_2)_6CH_3$	-75	16.5	0.54
22f	$-O(CH_2)_7CH_3$	-78	17.3	0.37
22g	$-O(CH_2)_8CH_3$	-63	25.8	0.21
22h	$-O(CH_2)_9CH_3$	-64 (T_m^1=-36; T_m^2=-29)	18.6	0.19

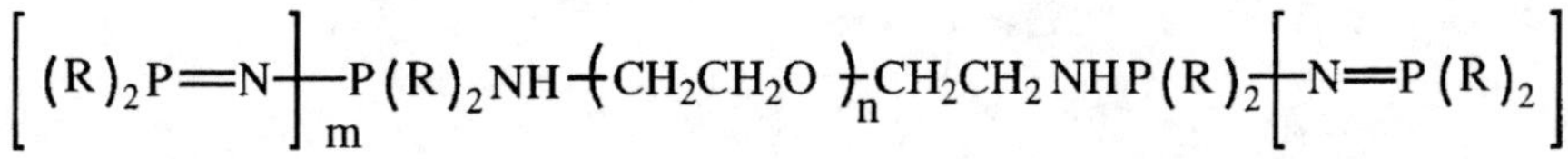

23 $R = -OCH_2CF_3$

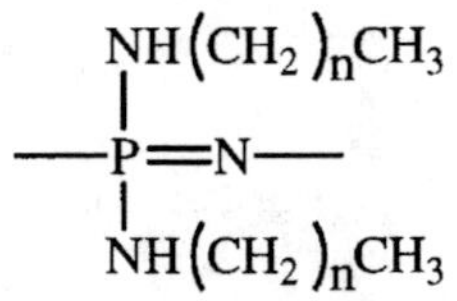

24a n = 4

24b n = 5

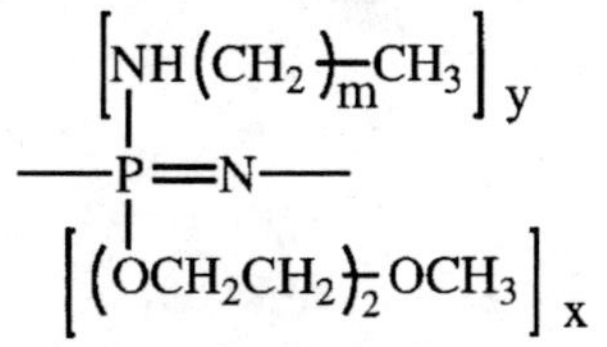

25a m = 2, x = 1 0, y = 1 0

25b m = 3 x = 1 0 y = 1 0

25c m = 4 x = 0 8, y = 1 2

25d m = 5, x = 1 0, y = 1 0

Chart 3

1.2. Poly(Phosphazene)-Based Single Ion Conductor

The anions of lithium salt are known to enhance the degradation of electrodes. Attempts to minimize such anion migration in the polymer electrolytes have been reported by Shriver and coworkers.[33,34] They have studied the conductivity of polyphosphazene with sulfonate and oligo(ethyleneoxy) side chains (**26**) as a single ion conductor. The polymer was synthesized from the partial displacement of parent $-(N{=}P(Cl)_2)-$ with $NaOC_2H_4SO_3Na$ in the presence of 15-crown-5 ether, followed by substitution of remaining P-Cl group with sodium salt of 2-(2-methoxyethoxy)ethanol or poly(ethylene glycol methyl ether). The polymers with high content of sulfonate were rigid solids whereas the polymers with 10 mol% of sulfonate group were elastomers at room temperature. The ionic conductivity of amorphous $[NP[O(CH_2H_4O)_{7.22}CH_3]_{1.8}(OC_2H_4SO_3Na)_{0.2}]_n$ was 1.7 x 10^{-6} S cm^{-1} at 80 °C. This value was somewhat lower than the complex of MEEP with $NaCH_2{=}CHCH_2SO_3$ and was two orders of magnitude lower than that of the MEEP-$NaSO_3CF_3$ complex. These results indicate that most of Na^+ cation form ion-pair with sulfonate anion.

As the parent polymer of an anionic conductor, they also prepared the polyphosphazene bearing dimethyl- or diethylethanolamine units and $CH_3OCH_2CH_2OCH_2CH_2O$- groups (**27**).[34] The mobile anions such as Cl^-, Br^-, I^- were introduced by the quaternization of pendant amino group with haloalkanes or hydrochloric acid. This treatment brought about the elevation of the glass transition temperature of the polymer. When bromide and iodide ions were used as mobile ion, *e, g.*, the polymer (**28**, R' = CH_3-, R"X = C_2H_5Br, x = 0.19) showed high conductivity on the order of 10^{-4} S cm^{-1} at 80 °C, indicating that large halide ions form loose ion pair.

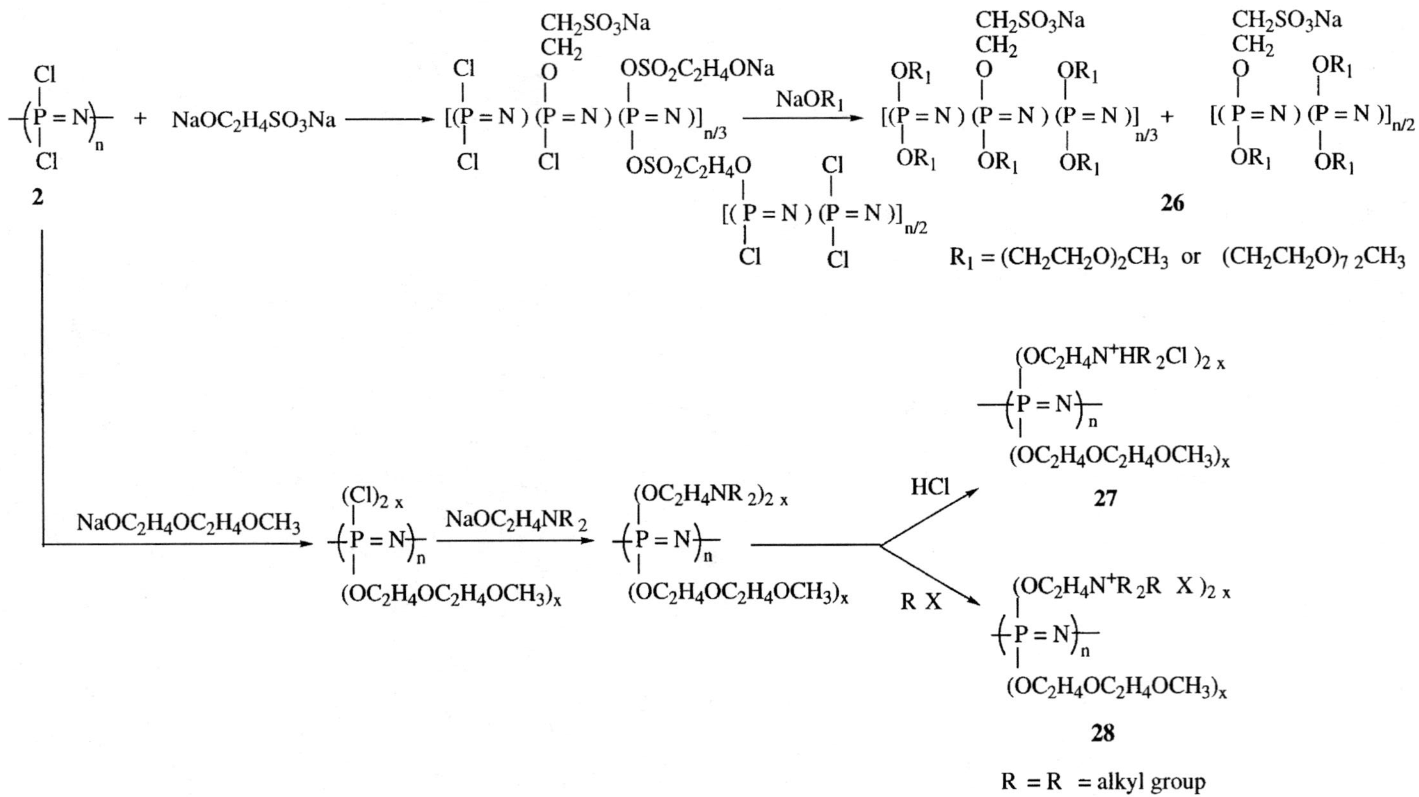
2
NaOC2H4SO3Na
NaOR1
26
R1 = (CH2CH2O)2CH3 or (CH2CH2O)7 2CH3
NaOC2H4OC2H4OCH3
NaOC2H4NR 2
HCl
R X
27
28
R = R = alkyl group

Scheme 2

1.3 Modified Linear Polyphosphazenes with Oligo(Ethyleneoxy) Side Chains

Although the MEEP-metal salt complexes have a high ionic conductivity, the drawback of MEEP is poor dimensional stability, and hence several attempts to overcome this problem have been reported.[35-41] The dimensional stability of the system can be improved by the chemical or radiation cross-linking without a significant decrease in the conductivity.

One approach was chemically cross-linking of linear polymer. Cross-linked polyphosphazenes were prepared from the reaction of poly(dichlorophosphazene) with sodium salts of methoxyethoxyethanol and polyethylene glycol.[35] The conductivity of cross-linked polyphosphazene-$LiSO_3CF_3$ complexes (σ = 4.1 x 10^{-5} S cm^{-1} at 30 °C) was comparable to the parent linear polyphosphazene system. As expected, the flow of polymer was not observed even at 140 °C for long periods. Thus, the cross-linking polyphosphazenes exhibit a high ionic conductivity, yet they possess improved dimensional stability. A similar result was obtained for the cross-linked MEEP – Li^+ salt complex by γ-radiation.[36]

The ionic conductivity of Li^+ complexed MEEP/polysiloxane hybrid materials has been reported, which were prepared by the sol-gel processing of tetraethyl orthosilicate (TEOS) in the presence of MEEP and $LiSO_3CF_3$.[37-39] When hydroxylated MEEP (**29**) and TEOS were used, the homogenous, transparent and mechanically stable materials (**30**) were obtained.[38] The conductivity of **30** was unexpectedly high (σ = 3 x 10^{-6} S cm^{-1}), however, probably due to the occurrence of proton transport ascribed to absorbed water. The Li^+-doped materials, which were treated at 120 °C, showed a maximum conductivity of 5 x 10^{-6} S cm^{-1}. For the materials obtained from MEEP and TEOS, the highest conductivity was found to be 4.6 x 10^{-5} S cm^{-1} for the complex containing 10 wt% TEOS and 15 wt% of $LiSO_3CF_3$.[39] The increase of TEOS caused the decrease in conductivity, although the glass transition temperature was not significantly affected by the amount of TEOS. This implies that the decrease in conductivity is not due to the restriction of segmental mobility. It is suggested that the phase separation between MEEP and TEOS networks is responsible for the decrease in the conduction. These hybrid materials are mechanically stable owing to the pseudo-crosslinking of etheric side groups in MEEP with silicate network.

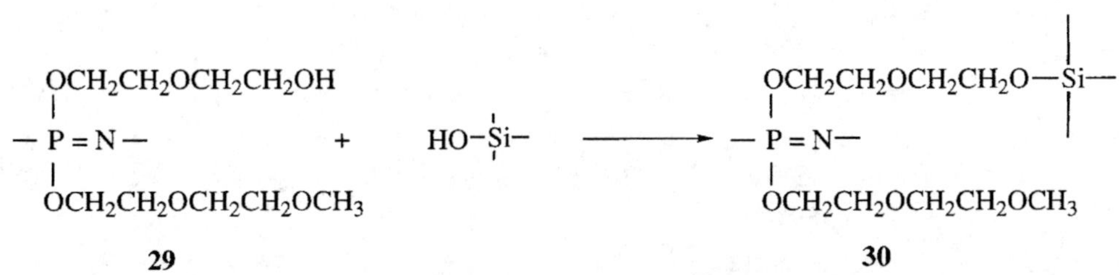

29 **30**

Scheme 3

The blend of MEEP and PEO also gave a highly mechanically stable material.[40,41] A conductivity of 6.7 x 10^{-5} S cm^{-1} at 25 °C was found for the fully amorphous 55: 45 MEEP-PEO when $LiSO_3CF_3$ was added. In this blend, MEEP seems to act as a network around PEO-salt complex to afford dimensionally stable solid electrolytes.

2. Polyelectrolytes with Pendant Oligo(Ethyleneoxy)Cyclotriphosphazenes

The polymers with pendant chlorocyclophosphazene units may provide a new type of polymers, since the cyclophosphazenes can also serve as modification sites of polymers, leading to new functional materials. So far, the synthesis and radical polymerizations of vinyloxy-, 3- and 4-(α-methylethenylphenyl)-, methacryloyl-, and styryl derivatives with halocyclophosphazenes have been reported.[42-49] From an experimental point of view, the synthesis of vinyl monomers with pendant pentachlorocyclotriphosphazene is quite straightforward. The monomer (**31**), for example, was synthesized in a moderate yield by the reaction of 4-vinylphenol and **1** in the presence of triethylamine.[42] The radical polymerization of these monomers led to moderate molecular weights. The side cyclotriphosphazene units in **31** could be easily modified by the displacement of P-Cl group with sodium salt of oligo(ethylene glycol) methyl ether. The monomer (**32**) also underwent a facile polymerization with radical initiators to give the corresponding multi-armed polymers, in spite of having sterically large side group. It should be noted here that the complete substitution of P-Cl group in the vinyl polymers with nucleophiles was rather difficult and a few percent of P-Cl groups often remained without any suffering displacement. The polymers with pendant penta(oligooxyethylene)cyclotriphosphazene have various features as a polymeric electrolyte host: (i) the introduction of many short ethyleneoxy chains into a phosphazene ring brings about a low glass transition temperature; (ii) the rigid and large groups of the main and/or side chains impart a greater free volume around ethyleneoxy chains. If the ethyleneoxy chains form a continuous conducting phase around the backbone, carrier ions could move in the phase independent of the nature of the polymer backbone and a high conductivity would be expected.

The polymers, **33**, **34**, and **35** exhibited the low glass transition temperatures (-65 - -60 °C), indicative of a very high segmental mobility of side chains.[42, 50-52] When $LiClO_4$ was added, the Tg values of the polymers increased monotonically with an increase in the amount of dissolved $LiClO_4$ in the systems. As with other systems discussed in this chapter, the behaviors can be attributed to the restriction of segmental motion caused by the polymer-salt interaction. For all of the complexes, the plots of log σ vs. 1/T showed convex curves. The VTF plots of the complexes exhibit straight lines when the standard temperature is taken as Tg-50 °C, suggesting that the ions migrate through amorphous phase facilitated by the segmental motion of ethyleneoxy chains. The maximum conductivity of 1.8×10^{-5} S cm^{-1} at 30 °C was achieved for the **33b**-Li^+ salt complex of $[Li^+]/O = 0.05$. A similar value was obtained for polybiphenyl derivatives **34b**-Li^+ complexes. The observed values were 10-10^4 times higher than those of most of the complexes based on polyacrylates and poly(styrene-*alter*-maleic anhydride) and comparable to those of MEEP-$LiSO_3CF_3$ complexes.[8,17,18] These results suggest that the conductivity is not significantly affected by the mobility of the main chain, *i.e.*, the flexibility of the backbone is not the critical factor required to achieve high conductivities. An additional interesting behavior of these multi-armed polymer electrolytes is that the temperature dependence of conductivity is rather small compared to that of simple comb-shaped polymer-Li^+ salt complexes reported in the literatures.

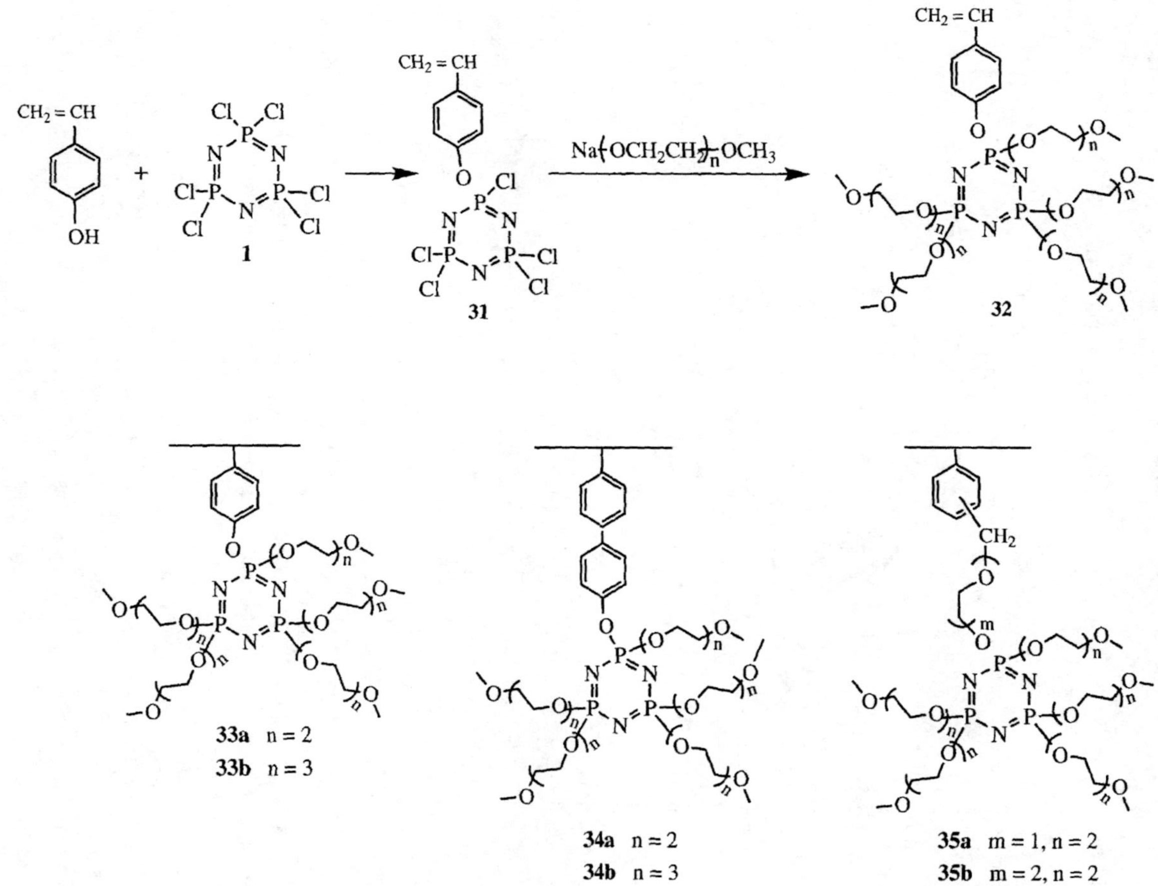

Scheme 4

The conductivity values of Li^+ salt complexes of polystyrenes with a $-CH_2(OCH_2CH_2-)_m$ spacer (**35**) between the polymer backbone and (methoxyethoxyethoxy)cyclotriphosphazene moiety have been studied.[52] The extents of Tg elevation of these polymers are almost the same with that of **33b**. In Figure 4, the conductivities of **33a**, **34a**, **35a**, and **35b**-$LiClO_4$ complexes at 60 °C were plotted as a function of $[Li^+]/[O]$. For the **35a**- and **35b**-Li^+ salt complexes, the ratio of $[Li^+]/[O]$ giving rise to the maximum conductivity shifts to a higher value of 0.08 from 0.067 in the **34a**-Li^+ system and 0.047 in the **33a**-Li^+ system. The maximum conductivity at 60 °C increased in the order **35b** > **33b** = **34b** > **35a** = **33a** > **34a.**

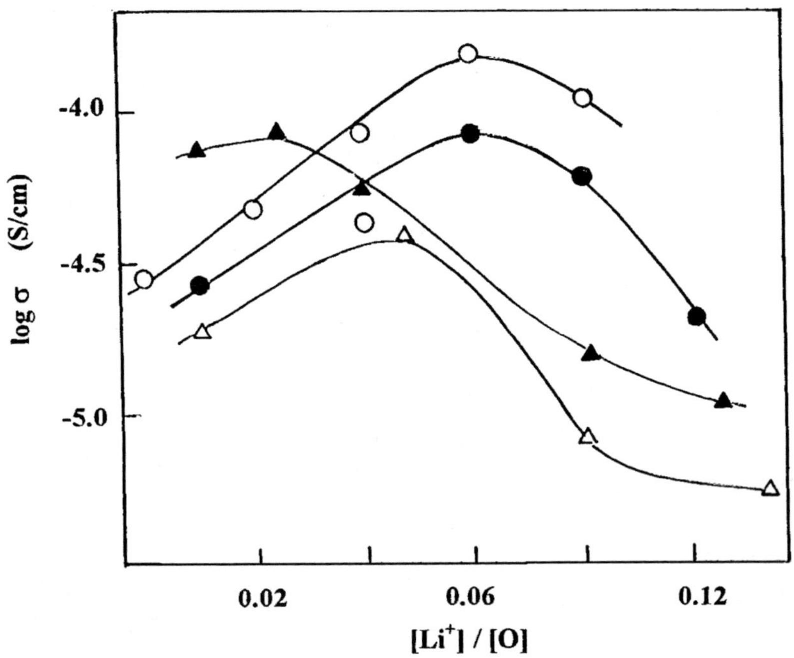

Figure 3 Ionic conductivity *vs.* Li^+ salt concentration for the complexes of $LiClO_4$-polymers with pendant cyclotriphosphazenes at 60 °C: (▲), 33a; (△), 34a; (●), 35a; (○), 35b.

Some speculations need to explain the results. For these polymers, the various complexes exist in the system. Some of them might be related structurally to the complex Li^+-octopus molecule.[53] Two typical conceptual models of complexes are a wrapping form and a cross-linking form (Figure 4). The wrapping structure is formed by three oligo(ethyleneoxy) chains attached to the same phosphazene ring, while the Li^+ mediated cross-linking is formed by side chains between neighboring phosphazene rings. It seems likely that the Li^+ ion in the wrapping complex moves with difficulty (a tight complex), whereas the cross-linking complex is relatively loose and the Li^+ ion moves with ease. However, this interchain-Li^+ salt interaction adversely effects on the segmental mobility. From the behavior of Tg elevation, it appears that the spacer acts as the part of coordination sites for the formation of the wrapping complex. This implies that higher concentrations of Li^+ ions are required to attain high conductivities. The shift of the $[Li^+]/[O]$ ratio for **35** might be related with the increase of such additional place for the complex formation.

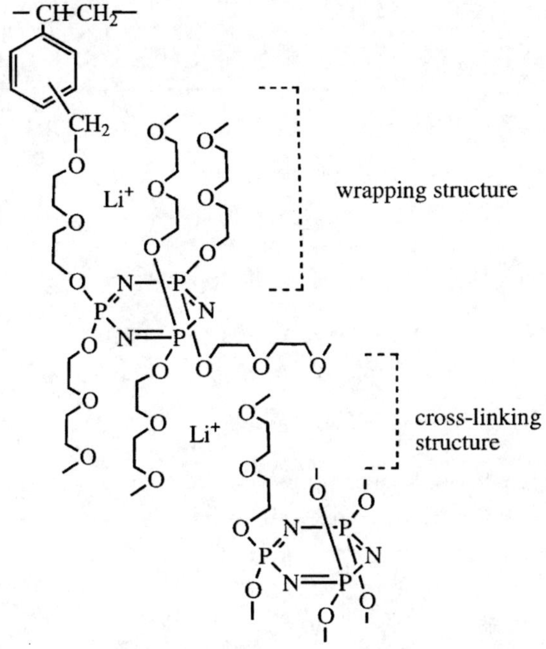

35b

Figure 4 Schematic representation of 35b-Li$^+$ complexes.

Allcock and co-workers[54] have reported the ionic conductivity of polynorbornenes bearing pendant cyclotriphosphazenes with oligo(ethyleneoxy) side groups. The monomer, norbornenyl cyclotriphosphazene (**36**), was synthesized from the reaction of sodium salt of 5-norbornene-2-methanol with cyclotriphosphazene with penta(oligoethyleneoxy) side chains ($-O(CH_2CH_2O)_nCH_3$, n = 1 - 3). The ring-opening metathesis polymerization with a ruthenium-based initiator gave polymers (**37**) with M_n = 1.6 - 14 x 10^4. The glass transition temperatures of these polymers were in the range -49.0 - -66.5 °C, depending on the length of ethyleneoxy chains. The polymers produced with monomer-to-initiator ratios of less than 100: 1 were mobile liquids, but polymers formed from ratios greater than 200: 1 were solid or gums. The conductivities of the **37c**-$LiN(SO_2CF_3)_2$ and the **37c**-$LiSO_3CF_3$ complexes were 2.2 x 10^{-5} and 0.9 x 10^{-5} S/cm at 30 °C, respectively. The increase in Tg was less pronounced for the former system. These results imply that the larger anions induce better ion-separation and contribute to the increase the free volume.

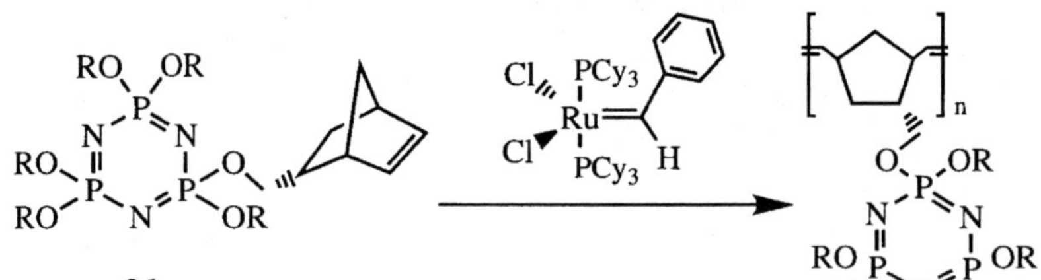

36 37

36a 37a R = $-CH_2CH_2OCH_3$

36b, 37b = $-CH_2CH_2OCH_2CH_2OCH_3$

36c 37c = $-CH_2CH_2OCH_2CH_2OCH_2CH_2OCH_3$

Scheme 5

The additions of small molecule additives into polymeric electrolytes have been demonstrated to be one of the effective methods to improve the conductivity. These additives could contribute to the increase in carrier ions due to effective cation solvation and the decrease of microviscosity of polymer matrix, although their addition to the polymer electrolyte system often leads to a loss of dimensional stability.

The addition of 18-crown-6 ether (18CE6) and 12-crown-4 ether (12CE4) to the poly(**33b**)-Li^+ system did not affect the extents of Tg elevation, whereas the increase in Tg values was considerably suppressed when 1,4,8,11-tetrakis(2-methoxyethyl)-1, 4, 8, 11-tetraazacyclotetradecane (TETA) is used.[55] By the addition of 18CE6 to poly(**33b**), the conductivity of the complex with $Li^+/O = 0.03$ increased slightly from $\sigma = 1.8 \times 10^{-4}$ S cm^{-1} to $\sigma = 2.7 \times 10^{-4}$ S cm^{-1} at 30 °C, and, for 12CE4, the increase in the conductivity was a factor of 2.5 compared to the **33b**-$LiClO_4$ complex with the same Li^+/O ratio. The addition of TETA was more effective to improve the conductivity; the conductivities of the complexes with $Li^+/O = 0.04$ have been raised to 10^{-4} S cm^{-1} at 30 °C and 10^{-3} S cm^{-1} at 90 °C, suggesting the occurrence of effective ion separation by TETA. It is suggested that Li^+ ions could not appreciably contribute to the conduction, and hence the anions might be mainly responsible for conduction. Such anions are considered to be loosely associated to the dipoles of the polymer, the transport of which might be possible in tune with the segmental motion.

The norbornene-based polymers are known to absorb various organic liquid to form soft elastomers. Allcock et al.[56] prepared lightly cross-linked polynorbornene by using oligo(ethyleneoxy)cyclotriphosphazenes with one and two norbornyl groups. The conductivity of the cross-linked polymer-$LiN(SO_3CF_3)_2$ complex was reported to be 5×10^{-5} S cm^{-1} at 30 °C. The conductivities increased by the addition of propylene carbonate and the maximum value of 2×10^{-3} S cm^{-1} at 30 °C was achieved for the Li^+ salt complex containing 50 wt% of the additive. Interestingly, the film of polynorbornene-Li^+-propylene carbonate complexes is highly dimensionally stable.

3. Polyelectrolytes with Phosphazene Core

The cyclophosphazenes with six or more reactive sites are also useful to prepare star-shaped polymers, hyperbranched polymers, and dendrimers.[57-60] The number of end groups and low melt viscosities are examples of unique properties of branched polymers. The approaches represent a great potential for the preparation of new functional polymers including ionic conductive materials.[61]

For the Li^+ salt complexes with star-shaped polymer electrolytes based on cyclotriphosphazene core, the conductivities depend on the number and length of ethyleneoxy units $-(CH_2CH_2O)_n-CH_3$ (**38a**, n = 7.2 and **38b**, n = 11.8).[62] The Tg elevation of the polymer-Li^+ complexes was significantly suppressed compared to that of linear counterparts. The maximum conductivity of 4×10^{-5} S/cm at 30 °C was than those of the polystyrenes carrying pendant oligo(ethyleneoxy)cyclotriphosphazene-Li^+ salt system.[42] This improvement is observed for the **38b**-Li^+ salt complexes. The observed values are somewhat higher ascribed to the high segmental mobility of ethyleneoxy chains, as judged from the extent of Tg elevation.

Hyperbranched poly(ethyleneoxy) attached to cyclotriphosphazene core (**39** - **41**) was also used as ion transport assistance in the MEEP -Li^+ system.[63] The addition of **41** to the MEEP-$LiSO_3CF_3$ system improved the conductivity, suggesting that the hyperbranched cyclotriphosphazenes contribute to decreasing the degree of ionic crosslinking between metal ion and ethyleneoxy chains in MEEP.

Conclusion

The design and synthesis of polymer-based electrolytes that have a high ionic conductivity and are stable mechanically as well as electronically are one of the major challenges of current polymer science. As described in this chapter, phosphazene-based polymer electrolytes offer several important advantages over poly(ethylene oxide) and traditional polymer-based electrolytes. These promising polymers are of potential interest for various applications such as batteries, data storage, sensors, and electrochromic devices. The maximum conductivities at room temperature for most of these materials, however, were reported as in the order of 10^{-5} -10^{-4} S/cm. In order to meet the criteria for practical applications, further improvement of conductivity and dimensional stability is required. The virtue of the facile introduction of functional groups into the phosphazene as well as the flexibility of P=N backbone might provide a route to meet the requirement for practical polymer electrolytes.

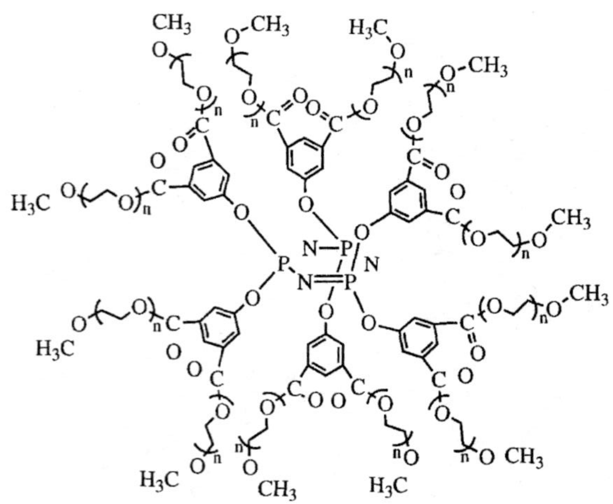

38a n = 2

38b n = 3

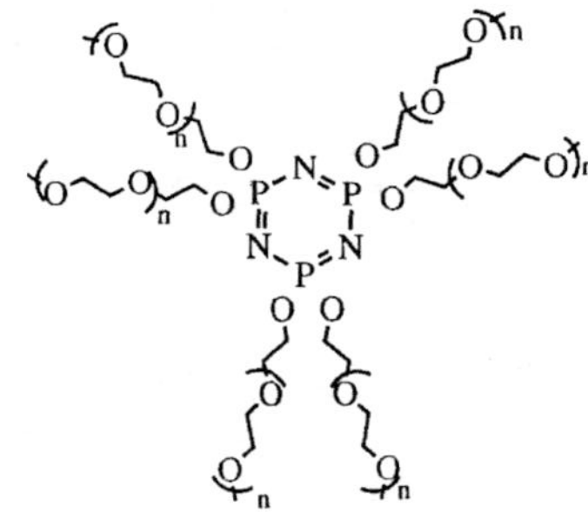

39a n = 1

39b n = 6

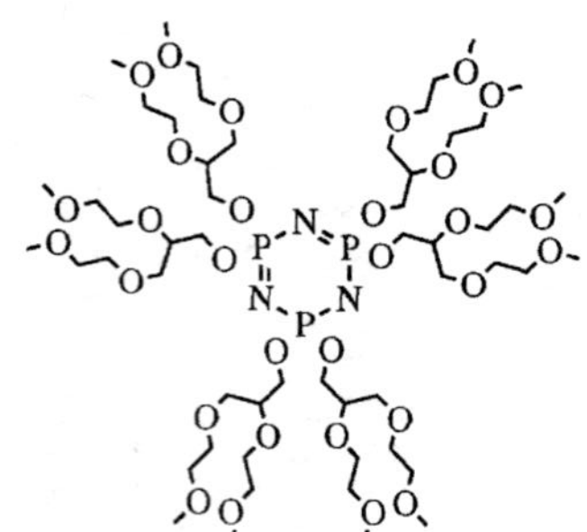

40

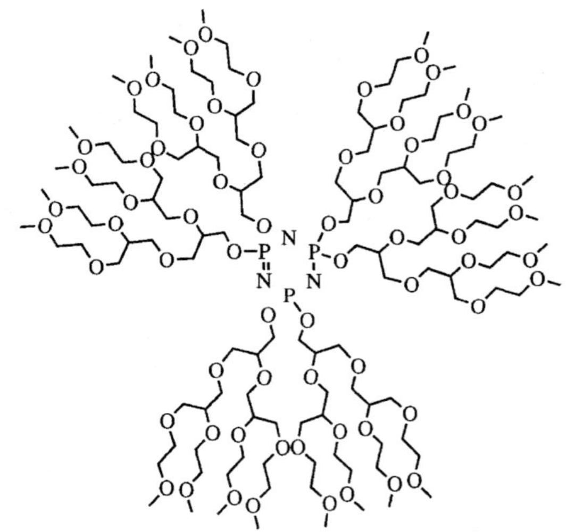

41

Chart 4

REFERENCES

1. D. E. Fenton, J. M. Parker, and P.V. Wright, *Polymer*, **14**, 589 (1973).
2. M. B. Armand, J. M. Chabagno, and M. J. Duclot, *In Second International Conference on Solod Electrolytes,* St Andrews, Scotland, 1978.
3. M. Armand, *Solid State Ionics*, **69**, 309 (1994).
4. M. A. Ratner and D. F. Shriver, *Chem. Rev*., **88**, 109 (1988).
5. M. B. Armand, *Adv. Mater.*, **6** & **7**, 278 (1990).
6. P. V. Wright, *J. Makromol. Sci. Chem*., A26, **2** & **3**, 519 (1989).
7. F. B. Dias, L. Plomp, and J. B. J. Veldhuis, *J. Power Sources*, **88**, 169 (2000).
8. R. Koksbang, I. I. Olsen, and D. Shackle, *Solid State Ionics*, **69**, 320 (1994).
9. V. Chandrasekhar, *Adv. Polym. Sci*., **135**, 139 (1998).
10. R. De Jaeger and M. Gleria, *Prog. Polym. Sci*., **23**, 179 (1998).
11. G. A. Carriedo, F. J. G. Alonso, P. A. González, and J. L. Garica-Alvarez, *Macromolecules*, **31**, 3189 (1998).
12. H. R. Allcock, W. R. Laredo, C. R. deDenus, and J. P. Taylor, *Macromolecules*, **32**, 7719 (1999).
13. H. R. Allcock, S. D. Reeves, J. M. Nelson, and I. Manners, *Macromolecules*, **33**, 3999 (2000).
14. Y. Cho, H. Baek, and Y. S. Sohn, *Macromolecules*, **32**, 2167 (1999).
15. M. L. White and K. Matyjaszewski, *Macromol. Chem. Phy*., **198**, 665 (1997).
16. H. R. Allcock, *Phosphorus-Nitrogen Compounds*, Academic Press, 1972, New York.
17. P. M. Blonsky, D. F. Shriver, P. Austin, and H. R. Allcock, *J. Am. Chem. Soc.*, **106**, 6854 (1984).
18. P. M. Blonsky, D. F. Shriver, P. Austin, and H. R. Allcock, *Solid State Ionics*, **18** & **19**, 258 (1986).
19. S. Schantz, L. M. Torell, and J. R. Stevens, *J. Appl. Phys*., **64,** 2038 (1988).
20. S. York, E. C. Kellam III, H. R. Allcock, and R. Freh, *Electrochim. Acta*, **46**, 1553 (2001).
21. H. R. Allcock, M. E. Napierala, D. L. Olmeijer, S. A. Best, and K. M. Merz, Jr., *Macromolecules*, **32**, 732 (1999).
22. H. R. Allcock, S. J. M. O'Connor, D. L. Olmeijer, M. E. Napierala, and C. G. Cameron, *Macromolecules*, **29**, 7544 (1996).
23. J. M. J. Cowie and K. Sadaghianizadeh, *Makromol. Chem. Rapid Commun*., **9**, 387 (1988).
24. J. M. J. Cowie and K. Sadaghianizadeh, *Polymer Commun*., **29**, 126 (1988).
25. M. Andrei, J. M. J. Cowie, P. Prosperi, *Electrochim. Acta*, **37**, 1545 (1992).
26. H. R. Allcock, D. L. Olmeijer, and S. J. M. O'Connor, *Macromolecules*, **31**, 753 (1998).
27. I. Selvaraj, S. Chaklanobis, and V. Chandrasekhar, *J. Electrochem. Soc*., **142**, 3434(1995).
28. H. R. Allcock, S. E. Kuharcik, C. S. Reed, and M. E. Napierala, *Macromolecules,* **29**, 3384(1996).
29. H. R. Allcock, M. E. Napierala, C. G. Cameron, and S. J. M. O'Connor, *Macromolecules*, **29**, 1951 (1996).

30. J. M. Nelson, A. P. Primrose, T. J. Hartle, and H. R. Allcock, *Macromolecules*, **31**, 947 (1998).
31. Y. W. Chen-Yang, J. J. Hwang, and F. H. Chang, *Macromomlecules*, **30**, 3825 (1997).
32. Y. W. Chen-Yang, J. J. Hwang, and A. Y. Huang, *Macromomlecules*, **33**, 1237 (2000).
33. S. Ganapathiappan, K. Chen, and D. F. Shriver, *Macromolecules*, **21**, 2299 (1988).
34. S. Ganapathiappan, K. Chen, and D. F. Shriver, *J. Am. Chem. Soc.* **111**, 4091 (1989).
35. J. S. Tonge and D. F. Shriver, *Electrochem. Soc.*, 269 (1987).
36. G. A. Nazri and S. G. Meibuhr, *J. Electrochem. Soc.*, **136**, 2450 (1989).
37. B. K. Coltrain, W. T. Ferrar, C. J. T. Landry, T. R. Molaire, and N. Zumbulayadis, *Chem. Mater.*, **4**, 358 (1992).
38. G. Brusatin, M. Guglielmi, R. De Jaeger, G. Facchin, M. Gleria, and M. Musiani, *J. Mater. Sci.*, **32**, 4415 (1997).
39. C. Kim, and J. S. Kim, Myong-Hoon Lee, *Synthetic Metals*, **98**, 153 (1998).
40. K. M. Abraham, M. Alamgir, and S. J. Perrotti, *J. Electrochem. Soc.*, **105**, 535 (1988).
41. K. M. Abraham and M. Almgir, *J. Power Sources*, **195**, 43 (1993).
42. K. Inoue, Y. Nishikawa, and T. Tanigaki, *J. Am. Chem. Soc.*, **113**, 7609 (1991).
43. K. Inoue, M. Nakano, M. Takagi, and T. Tanigaki, *Macromolecules*, **22**, 1530 (1989).
44. K. Inoue, H. Nakamura, S. Ariyoshi, M. Takagi, and T. Tanigaki, *Macromolecules*, **22**, 4466 (1989).
45. K. Inoue and T. Itaya, *Bull Chem Soc. Jpn.*, **74**, 1381 (2001).
46. J. C. Shaw and C. W. Allen, *Inorg. Chem.*, **25**, 4632 (1986).
47. C. W. Allen, J. C. Shaw, and D. E. Brown, *Macromolecules*, **21**, 2653 (1988).
48. C. W. Allen and R.P. Bright, *Macromolecules*, **19**, 571 (1986).
49. Dez and R. D. Jaeger, *Macromolecules*, **30**, 8262 (1997).
50. K. Inoue, Y. Nishikawa, and T. Tanigaki, *Macromolecules*, **27**, 3464 (1991).
51. K. Inoue, Y. Nishikawa, and T. Tanigaki, *Solid State Ionics*, **58**, 217 (1992).
52. K. Inoue, T. Tanigaki, and Z. Yuan, *Adv. Polym. Technol.*, **4**, 74 (1993).
53. F. Vögtle and E. Weber, *Angew. Chem., Int. Ed. Engl.*, **13**, 814 (1974).
54. H. R. Allcock, W. R. Laredo, E. C. Kellam, III and R. V. Morford, *Macromolecules*, **34**, 787(2001).
55. K. Inoue, K. Takiue, and T. Tanigaki, *J. Polym. Sci., Polym. Chem. Ed.*, **34**, 1331 (1996).
56. H. R. Allcock, W. R. Laredo, and R. V. Morford, *Solid State Ionics,* **139**, 27 (2001).
57. Anne-Marie Caminade and J. P. Majoral, *Chem. Rev.*, 94, 1183 (1994).
58. N. Launay, Anne-Marie Caminade, and J. P. Majoral, *J. Organomet. Chem.*, **529**, 51 (1997).
59. F. Sournies, F Crasnier, M. Graffeuil, J. -P. Faucher, R. Lahana, M. C. Labarre, and J. F. Labarre, *Angew. Chem., Int. Ed. Engl.*, **34**, 578 (1995).
60. K. Matyjaszewski, P. J. Miller, J. Pyun, G. Kickelbick, and S. Diamanti, *Macromolecules*, **32**, 6526 (1999).
61. C. J. Hawker, F. Chu, P. J. Pomery, and D. J, T, Hill, *Macromolecules*, **29**, 3831(1996).
62. K. Inoue, H. Miyamoto, and T. Itaya, *J. Polym. Sci., Polym. Chem.Ed.*, **35**, 1839 (1997).
63. H. R. Allcock, R. Ravikiran, and S. J. M. O'Connor, *Macromolecules*, **30**, 3184 (1997).

In: Applicative Aspects of Poly(organophosphazenes)
Editors: M. Gleria and R. De Jaeger, pp. 189-206
ISBN 1-59454-025-X

Chapter 9

POLYPHOSPHAZENE MEMBRANES

Gianni Golemme and Enrico Drioli
Dipartimento di Ingegneria Chimica e dei Materiali, Università della Calabria, and
Istituto per la Tecnologia delle Membrane, ITM-CNR
Via P. Bucci, I-87030 Rende (CS) Italy
<ggolemme@unical.it>

ABSTRACT

The 90's marked a growing interest in phosphazene based materials for membranes preparation, but probably the great capabilities of this class of materials have not yet been explored enough. The progress in the reproducible and cheaper preparation of new phosphazene based materials can give a new stimulus to the development of polyphosphazene membranes with tailored properties. The first part will be focused on the new available phosphazene materials. Then a survey will cover the literature of the last years on linear and cyclolinear polyphosphazene membranes with an outline of the possible future developments in the field.

Keywords: Phosphazenes; Membranes; Gas separation; Pervaporation; Vapor permeation; Proton Conduction; Ultrafiltration; Nanofiltration

INTRODUCTION

Membrane science and technology have introduced significant beneficial innovation in both processes and products. Membranes are used today for different applications: in the agro-food, pharmaceutical, chemical and electronic industries; for water desalination and potabilization, medical applications, waste water and flue gas treatment, gas separation, chloro-soda production, fuel cells and lightweight batteries, and more. About four decades ago the breakthrough was the discovery by Loeb and Sourirajan of an effective method for significantly increase the permeation flux of cellulose acetate membranes with no significant loss in selectivity[1]. The preparation of asymmetric membranes opened the way to the large scale desalination of brackish water and sea water by reverse osmosis (RO) and to several other molecular separations in different industrial areas. In 1999 the water production of RO desalination units worldwide (10^7 m^3/day) already exceeded the capacity of plants using thermal methods[2], and a total membrane area of more than 250,000 m^3 was installed for whey and milk ultrafiltration (UF)[3]. From the 1970s commercial membrane gas separation found its way and today more than 10^9 Nm^3 of pure gases are produced via selective permeation through composite asymmetric polymeric membranes. The combination of a chemical reaction with molecular separation in membrane reactors gives new opportunities in biotechnology, in the pharmaceutical and in the chemical industry. Membrane contactors and new production routes are made possible by the availability of rugged, thermally and chemically resistant membranes.

The nature of membrane operations is the key to understand the reasons of their success: they are modular, therefore simplify scale-up and are able to fit any desired plant size; their concept and operation is simple; usually they are athermal, do not require phase changes or addition of chemicals; their energy needs are low; minimise the use of raw materials, maximise their recovery and re-use, reduce the formation of by-products and wastes. When compared with current technologies, membrane operations offer in many cases better quality products with reduced investment and production costs, as a result of a smaller size of the plants, lower energy consumption and waste production and a better environmental impact[4]. The sales for different membrane operations in 1998 are listed in Table 1[5].

Polymeric materials are so popular for membrane preparation due to the low cost of most polymers employed and the ease of fabrication of asymmetric membranes with thin separating layers in flat, tubular or capillary configurations. Ceramic membranes are often more performant because they withstand high temperatures, aggressive chemicals and/or organic solvents, but the high cost and the low packing density limit the province of their applications in those cases where resistance is a must. Organo-inorganic polyphosphazenes and phosphazene based materials lend themselves to bridge the gap between the organic polymers and the ceramics thanks to their outstanding chemical, thermal, and solvent resistance[6,7]. If the cost of polyphosphazenes will decrease, their chemical versatility will impose them as starting materials for most diverse applications, and for sure also in the preparation of membranes, something similar to the explosion of the applications of siloxane polymers in the last decades.

Table 1. 1998 Worlwide Sales of Membranes and Membrane Modules in Selected Membrane Processes[5]

Membrane process	Sales (millions USD)	Yearly percent growth
Microfiltration	900	8
Ultrafiltration	500	10
Reverse Osmosis	400	10
Gas Separation	230	15
Electrodialysis	110	5
Electrolysis	70	5
Pervaporation	>10	not determined
Miscellaneous	30	10

The applications of polyphosphazenes in membrane science up to 1995 were covered by a previous review[8]. This paper examines the new production appeared in the literature in the meantime.

RECENT PROGRESS IN THE PREPARATION OF PHOSPHAZENES, PHOSPHAZENE BASED MATERIALS AND PHOSPHAZENE MEMBRANES

Linear (Fig. 1a), cyclolinear (Fig. 1b) and cyclomatrix (Fig. 1c) are the three main classes of polyphosphazenes. They are characterized by the presence of the inorganic phosphazene alternating single and double bonds between phosphorus and nitrogen atoms. Unlike carbon-carbon double bonds there is practically no energetic barrier to the rotation of substituents around it, and no conjugation exists for alternating double bonds with no electronic conduction through them. Therefore linear polyphosphazenes, like polysiloxanes, have extremely flexible backbones, and both classes of polymers are characterized by low glass transition temperatures[6,7]. However, the range of glass transition temperatures of linear polyphosphazenes spans an interval of almost 300°C, from –107°C of poly[bis(n-butoxy)phosphazene][9] to 180°C of poly[(2,2,2-trifluoroethoxy)(1-adamantylamino)-phosphazene][10], depending on the nature of the substituents. With a similarity taken from the world of movies, we can say that the phosphazene system is the Zelig of the world of polymers: the kind of substituents linked to P atoms can be extremely diverse, and they impart their own properties (optical, electrical, solubility, etc.) to the macromolecules. There's no other synthetic polymer system with the same versatility. In addition to this, an interesting feature of several semicrystalline polyphosphazenes is the presence of a liquid crystalline phase beyond a typical T(1) temperature[11], which makes them more permeable[12,13].

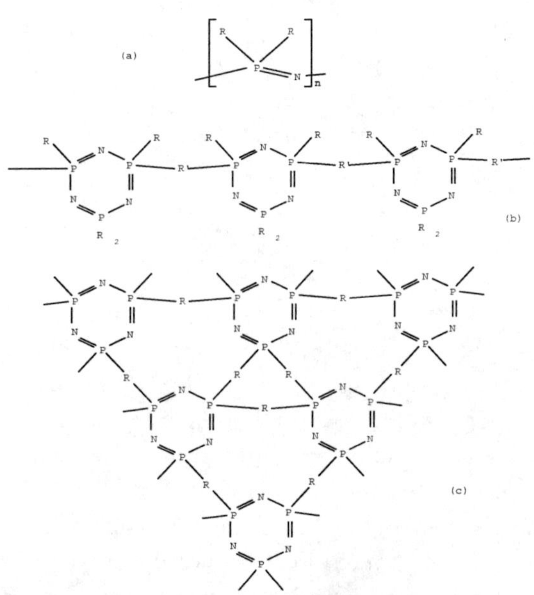

Fig. 1: Scheme of the three main classes of polyphosphazenes: linear (a), cyclolinear (b) and cyclomatrix (c) polymers

The research efforts of the last years have made the synthesis of linear polyphosphazenes more reproducible, with control over the molecular weight and with the possibility to obtain block co-polymers, also with units of different nature (e.g. siloxane). The exhaustive survey of these achievements is beyond the scope of this contribution and the reader is referred to other chapters of this book. However, some results are summarized here because they are relevant for the preparation of membranes with selected properties and functionalities.

The first route to linear polyphosphazenes is due to the pioneering work of Allcock, and involves the thermal polymerisation of hexachlorocyclotriphosphazene $(NPCl_2)_3$ to linear polydichlorophosphazene $(NPCl_2)_n$, followed by the substitution of the chloride ions by suitable nucleophiles to give stable and soluble linear poly(organophosphazene)s[14]. In the resulting polymers the substituents are bound to the inorganic nitrogen-phosphorus backbone via P-O, P-N or P-S bonds.

The thermal condensation of $Cl_3P{=}N{-}P(O)Cl_2$ yields high molecular weight polydichlorophosphazene characterised by no branching[15]. The thermal or ionically catalysed condensation of organophosphoranimines $ClRR'P{=}NSiMe_3$ yields polyphosphazenes of the generic repeat unit $(NPRR')_n$, where R and R' are alkyl and/or aryl groups directly linked to the P atom of the skeleton via a P-C bond[16,17]. An other route to the $(NPRR')_n$ class of polymers – not too much popular – is the denitrogenation of (explosive) phosphine azides $RR'PN_3$[18].

The new approach to linear polyphosphazenes, which enlarges greatly the already vast number of structures potentially available, is the room temperature living cationic condensation of phosphoranimines such as $Cl_3P{=}NSiMe_3$, $Cl_2RP{=}NSiMe_3$ and $ClRR'P{=}NSiMe_3$[19-21]. The phosphoranimine polymerises in presence of PCl_5 or Lewis acid initiators at room temperature in solution by elimination of Me_3SiCl to yield $(NPCl_2)_n$, $(NPClR)_n$ or $(NPRR')_n$, where the halogen atoms can be substituted in the usual way by O, N or S containing organic nucleophiles. The polymers have a narrow molecular weight distribution and the molecular weight can be controlled by variations in the initiator-to-monomer ratio. In addition to this, the living polymerisation mechanism offers the opportunity to employ multifunctional monomers giving rise to multi-arm or star-like polymers[22-24], and block copolymers with organic[25] and siloxane[26] units.

In synthesis, the variability of phosphazene polymers can be obtained at different levels. The first is the kind of skeleton desired: linear, branched, cyclolinear, etc.; the second is the eventual presence of other skeletal co-polymeric units; the third is the kind of substituent on phosphorus atoms: alkyl/aryl, alkoxy, amino groups, etc., or mixtures thereof; the fourth is the possibility to further modify the substituents by suitable reactions (e.g. sulfonation) to impart different properties and functionalities; the fifth is the crosslinking or the formation of hybrids with organic and inorganic substances in order to modify the swellability, the mechanical and the transport properties of the resulting materials. The several degrees of freedom give one the possibility to tailor the properties of polyphosphazenes (hydrophilicity/hydrophobicity, solubility, chemical affinity, etc.) for specific needs in different membrane operations. In the following the most important examples of novel polyphosphazenes for membrane fabrication will be highlighted.

Poly{bis[2-(2-methoxyethoxy)-ethoxy]phosphazene} or MEEP[27] is the best known polymer of a family containing ethylene oxide moieties as side groups. MEEP is water soluble, has a very low Tg (-75°C) thanks to the flexible polyether chains, like poly(ethylene oxide) is a polymeric electrolyte, is amorphous and not dimensionally stable. The purification of the crude polymer after synthesis from the salts dissolved in water, that used to be performed with a tedious dialysis procedure, can be simplified to a large extent by heating the aqueous solution above the lower critical solubility temperature of the polymer, and recovering the precipitated MEEP[28]. Several new polyphosphazenes, similar to MEEP in that they contain a variety of different alkyl ether and alkoxy groups, were prepared and studied as solid electrolytes for lithium triflate after γ ray crosslinking[29]. The presence of alkoxy groups of different length was found to be detrimental for the ion conduction. The ionic conductivity was improved to some extent by increasing the length of the ethyleneoxy side chains $–O(CH_2CH_2O)_nCH_3$ up to a value of n of six, to level off for longer chains. Polyphosphazenes containing branched alkyl ether side groups showed similar conductivity as the polymer with linear side chains, but two beneficial effects appeared: lack of crystallinity in polymers with long substituents, and dimensional stability without crosslinking.

γ ray crosslinked MEEP and poly[bis(propyl-4-oxybenzoate)phosphazene] were used to prepare interpenetrating polymer networks (IPN) through swelling with metal complexing monomers followed by polymerisation and crosslinking of the latter[30]. The IPN structures were intended for ion selective resins and eventually ion selective membranes.

Several crosslinked sulfonated poly[bis(aryloxy)phosphazene] have been prepared by Pintauro and co-workers for application in direct methanol fuel cells, obtaining membranes of comparable conductivity as Nafion 117 and reduced methanol mobility. The reader is referred to the chapter of this book for details on the preparation and properties of such membranes[31].

Phosphonated poly[bis(aryloxy)phosphazene] have been prepared by Allcock and co-workers[32] with different methods. In the first, a THF solution of poly[bis(phenoxy)phosphazene] (PPOP) was treated with $OPCl_3$, then reacted with a stoichiomentric amount of water, evaporated until viscous, cast on a Teflon plate, dried to produce a thin film, and then repeatedly soaked in water up to constant pH. No identification data on the real product are given in the patent, but evidence of the formation of a phosphonic acid polymer is given by the high conductivity of the material ($2.8\ 10^{-3}$ S/cm, identical to a Nafion 117 membrane). The polymer also resulted less permeable to methanol than Nafion 117 in permeation experiments. The second and third methods give rise to poly[bis(aryloxy)phosphazene]s containing pendant phosphonic ester groups[33], which can be hydrolysed to phosphonic acid groups. Some of these polymers might be the polymers used to prepare phosphonic acid functionalised poly[aryloxyphosphazene] membranes[34] with comparable ion exchange capacity (IEC) with respect to Nafion 117 and sulfonated poly[bis(aryloxy)phosphazene]. In an aqueous methanol solution (50% v/v) they were 40 times less permeable than Nafion 117, and 10–20 times less than sulfonated polyphosphazene membranes, with methanol diffusion of about $2.5\ 10^{-8}\ cm^2/s$.

Many water soluble polyphosphazenes (e.g. bearing acid (sulfonic, carboxyl), basic (amines) or ethylene oxide groups) lend themselves to the preparation of hydrogels when properly crosslinked[35]. Thermally sensitive hydrogels can be prepared by using those polymers which exhibit a lower critical solubility temperature (MEEP and analogs): an increase of temperature beyond the LCST may cause a hydrogel to shrink and squeeze its water content out, working as a thermally switched water pump[36]. pH sensitive hydrogels are available by crosslinking polyphosphazenes containing ethylene oxide groups and arylcarboxylic acids and their salts[37,38]. In water the anionic form of the polymer swells due to its net negative charge: upon acidification the conjugated neutral acid is formed, its lower affinity for water makes the hydrogel shrink.

The formation of poly(organophosphazene) (POP)/metal oxide composite materials has been tested by some authors by means of different strategies. In order to improve gas separation properties, Duval et al.[39] tried to prepare silicalite-1 (MFI) filled PPOP membranes by casting of a silicalite-1 suspension in a PPOP solution, followed by evaporation. SEM observations indicated a poor adhesion of the silicalite to the polymeric phase, probably due to the semicrystalline nature of the polymer whose T_g is close to 0°C. The authors did not try to use coupling agents. The hydrolysis and condensation of different metal alkoxides (Si, Ti, Zr and Al) in the presence of MEEP gives rise upon drying to clear glassy films with interesting properties.[40] The MEEP polymer mobility was not reduced by the intimate contact with the inorganic phase, as indicated by the unchanged glass transition temperature of –75°C. Also, the Li triflate doped composite maintained an appreciable ionic conductivity when compared to the doped MEEP, with much better mechanical properties. When subjected to swelling/drying cycles a MEEP/silica composite may fail and develop crazes. The optimal catalyst for improving the mechanical properties of hybrid MEEP/silica (2:3 mol/mol) nanocomposites was tetrabutylammonium fluoride, a salt in which the F^- ion is less prone to the formation of ion pairs and therefore is completely available to catalyse the hydrolysis of tetraethoxysilane[41].

POP/metal oxide composite materials with covalent bonds between the two phases can be prepared by hydrolysis of a metal oxide precursor (an alkoxide) and co-condensation with a phosphazene containing pendant hydroxyl groups. Hydroxylated MEEP (MEEP-OH) and analogs can be prepared by reacting MEEP with trimethylsilyl iodide $ISiMe_3$. Up to 40% of the terminal methoxy groups of the side chains of MEEP could be replaced by hydroxyls, and MEEP-OH, in turn, was co-condensed with tetraethoxysilane to yield a phosphazene-silica hybrid material with improved thermal stability of the phosphazene part[42]. These composite materials in which an intimate contact exists between the phosphazene and the metal oxide phase might prove interesting for several membrane applications, such as gas separation and pervaporation. A hybrid material obtained by acid hydrolysis of tetraethoxysilane and co-condensation with minor amounts of poly[bis(4-hydroxyphenoxy)phosphazene] (5.65 mol %) was deposited on TiO_2 porous supports (10 nm pore size)[43]; the phosphazene polymer improved slightly the He/N_2 selectivity of the hybrid membranes with respect to the poor ones obtained with pure silica, probably by acting as a binder and limiting the formation of cracks during the drying of the inorganic oxide.

Researchers from Idaho found out that the less hindered hydroxyl of t-butylhydroquinone is much more reactive than the other one in the nucleophilic substitution of hexachlorocyclotriphosphazene[44]: hexa-substituted trimers with six terminal sterically hindered hydroxyls can be produced neatly with high yields, and crosslinking reactions among phosphazene rings do not occur, as opposed to syntheses reported using simple hydroquinone. The same reaction on poly(dichlorophosphazene) should be tested in order to verify the possibility to prepare the analogous polymeric compound, poly[bis(4-hydroxy-3-t-butylphenoxy)phosphazene]. The availability of the hexa-(t-butylhydroquinone) cyclotriphosphazene (tBuHQCP) gives one the opportunity to prepare new composite materials, e.g. by co-condensation of hydrolysed metal alkoxides or reaction with bi-functional molecules (e.g. containing two acyl chloride groups). When tBuHQCP was reacted with an excess of succinyl chloride, adipoyl chloride or dodecanoyl chloride in certain ratios, soluble organic-inorganic hybrid cyclolinear polyesters were produced[45]. The nmr data on the polymers suggested that the ester moiety was attached either in a bidentate fashion, connecting two different phosphazene trimers, and in a monodentate fashion to a trimer ring; the free end in this case was hydrolysed by water to a carboxylic acid group during the isolation and purification of the polymer. The resulting polyesters therefore contain a variable amount of carboxylic and hydroxyl groups, and might undergo different further modifications, functionalisations and crosslinking reactions.

Gas Separation (GS)

The separation of gas mixtures with membranes is a technically well consolidated operation which is expanding to a growing number of applications. The separation of the components of air, the dehydration of gas streams, the separation and recovery of CO_2 from biogas, the upgrading of natural gas, the separation of hydrogen from industrial streams, can be considered now as traditional applications (Table 2). Table 3 lists the main research fields for gas separation.

Table 2. Current Membrane Processes for Gas Separation

Gas	separated from
H_2	NH_3 and CH_3OH synthesis purge gas
H_2	Petrochemical process streams
CO_2	mixtures with hydrocarbons in Enhanced Oil Recovery (EOR)
CO_2, H_2S, H_2O	natural gas, biomass and landfill gases
H_2O	air
VOCs	air
N_2	air

Table 3. Membrane Gas Separations Currently Studied

Volatile organic compounds from air
Oxygen from air
Higher alkanes and water from natural gas
Olefines from alkanes (e.g. propylene/propane)
Desulfurization of combustion flue gases
Sequestration of CO_2 from combustion flue gases
Upgrading of natural gas, biomass and landfill gases
High temperature separations

Commercial polymeric membranes for gas separation, either as hollow fibers or as plane sheets assembled in spiral wound modules, are made of a thin (down to 50 nm) dense separating layer on a highly porous support[46]. The driving force for the passage of penetrants in dense polymeric media is a difference in pressure between the feed and the permeate (downstream) sides, and the solution-diffusion mechanism adequately describes transport[47-49]. In this model, permeability is the product of solubility and of diffusion of the penetrant in the polymeric matrix. The rule of thumb is that highly permeable polymers are not very selective, and conversely highly selective polymers are not very permeable. When comparing selectivity for a certain gas pair vs. the permeability of the most permeable species in the pair for a large number of polymers, an upper bound is confining the permselectivity data in the lower left part of the graph[50].

The literature on GS polyphosphazene membranes up to 1995[8] indicated that their capability for the separation of poorly sorbing species is modest, and better results can be obtained in the selective separation of rather condensable gases, expecially acid (CO_2, SO_2, H_2S), from permanent gases. The permeability and selectivity of some selected polyphosphazenes and organic polymers are given in Table 4.

Table 4. CO_2 permeability (Barrer, 10^{-10} cm^3 cm / cm^2 s cmHg) and ideal separation factors of some polymeric membranes at room temperature (25-35°C)

Polymer	T_g	Selectivity O_2/N_2	Selectivity CO_2/N_2	Selectivity CO_2/CH_4	P_{CO2}	Ref.
$-[NP(O\,CH_2\,CH_2\,CH_2\,CH_3)_2]_n-$	-108	2.29	11.6	3.30	648	51
$-\{NP[(O\,CH_2\,CH_2)_2\,O\,CH_3]_2\}_n$	-81	1.8	63	22.7	250	52
$-[NP(O\,CH_2\,CF_3)_2]_n-$	-65	2.2	11.1	9.6	595	53
$-[NP\,(O\,Ph)_2]_n-$	-6	4.2	27	11	13.5	53
$-[NP(O\,C_6H_4\,Br\text{-}p)_2]_n-$	10	4.9	25.6	12.7	4.60	53
$-[NP(OPhBr\text{-}p)_{0.45}(OPhSiMe_2Ph\text{-}p)_{1.20}(OPh)_{0.32}]_n-$	10	3.6	22.2	8.4	49	53
$-[NP(OPhBr\text{-}p)_{0.07}(OPhSiMe_2Ph\text{-}p)_{1.80}(OPh)_{0.10}]_n-$	24	3.4	13.8	5.3	123	53
$-\{[NP(CH_3)(Ph)]_{0.46}\text{-}[NP(Ph)(CH_2\,SiMe_3)]_{0.27}\text{-}$ $[NP(CH_2\,Si\,Me_2\,((CH_2)_2(CF_2)_8F))(Ph)]_{0.27}\}_n-$	30	2.6	13.2	8.9	92.1	54
$-\{[NP(CH_3)(Ph)]_{0.4}\text{-}[NP(CH_2\,SiMe_3)]_{0.6}\}_n-$	50	4.6	15	10.2	16.9	54
$-\{[NP(O\,CH_2\,CF_3)\,\eta^2\text{-}(C_5H_4)_2Fe\,NP\,(O\,CH_2\,CF_3)]$ $-\,[NP(O\,CH_2\,CF_3)_2]_{0.4}\}_{n/3}-$	61	3.4	14.9	13.0	67.5	53
$-[O\,SiMe_2]_n-$	-123	2.0	11	3.4	2700	55
Bisphenol-A polysulfone	186	5.6	22	22	5.6	56
Cellulose acetate	187	5.5	32	32	4.75	57
Poly(2,4-dimethyl-1,4-phenylene oxide)	210	4.4	15.6	15.4	52.1	58
$-[(Me_3Si)C{=}C(CH_3)]_n-$	>250	1.5	4.9	2.1	33100	59
PMDA-4,4'-ODA polyimide	420	6.1	27	46	2.7	56

Poly[(n-butylamino)(di-n-hexylamino)phosphazene] and poly[(n-butylamino)(di-allylamino)phosphazene] were prepared and crosslinked by means of agents containing oxirane rings[60] or upon irradiation by UV light, electron beams and γ rays[61,62]. Crosslinking improved the mechanical properties of the membranes, and in the case of the first polymer oxygen permeability increased too.

Poly[bis(2,2,2-trifluoroethoxy)phosphazene] (PTFEP, Tg –62°C) is a semicrystalline polymer characterised by an unusual behavior: its crystalline domains seem to be permeable to gases[63], in this respect similar to few other polymers, such as poly(4-methyl-1-pentene). Its gas permeability is among the highest ever measured for polyphosphazenes[53], therefore it represents a good starting material to manipulate in order to improve permeability and selectivity. One of such attempts has been pursued by Nagai et al.[10], who studied poly[(2,2,2-trifluoroethoxy)(1-adamantylamino)phosphazene] (PAdATFEP). The amorphous polymer (Tg = 180°C) was powder-like and too brittle to be cast in free standing membranes, therefore it was necessary to cast PTFEP/PAdATFEP blends containing different amounts of the copolymer (30 and 40 wt.%, i.e. 13.8 and 18.8 mol% adamantane units, respectively) from THF solutions. The heterogeneous blends in the resulting membranes was characterized by almost unchanged crystallinity with respect to homopolymeric PTFEP membranes (31.4-31.8 % vs. 34.7%). In addition to this, the unchanged position of the crystalline peaks in the XRD spectrum and the DSC experiments indicated that there was no significant mixing of the polymers, that the crystallization of PTFEP was not much influenced by the presence of PAdATFEP, and that the latter polymer in the blend was phase segregated in the amorphous regions of the material.

The impossibility to cast membranes from the pure PAdATFEP co-polymer did not allow a thorough analysis of the gas permeation experiments through the blend membranes, however some conclusion could be drawn. First of all PAdATFEP decreased the permeability at 35°C of all gases, and caused a modest increase of the selectivity of PTFEP among light permanent gases (H_2, O_2 and N_2). Permeability decreased with the increasing size of such molecules for all of the membranes studied, suggesting that the order of permeability was governed by their diffusion coefficient, in both PTFEP and in the blend membranes. The situation was more complicated with CO_2 and hydrocarbons (CH_4, C_2H_6 and C_3H_8), suggesting a more important role of solubility in the transport properties of these more condensable species: CO_2 was by far the most permeable gas, more than the small H_2 molecule. The order of permeability was $C_2H_6 > CH_4 > N_2 > C_3H_8$ in all cases (PTFEP and blends), with small differences due to the different content in the adamantly group, and indicated that both solubility and diffusion were important in determining the permeation behavior of the different penetrants. The slight increase in selectivity introduced by the adamantly group, however, was not enough to make the new material interesting for any practical gas separation, also because it was accompanied by a decrease in permeability.

Orme et al.[52] studied the gas permeation properties of selected rubbery amorphous polyphosphazenes containing variable amounts of o-allylphenoxy (used for crosslinking), p-methoxyphenoxy (for dimensional stability and film forming properties), and 2-(2-methoxyethoxy)ethoxy (MEE) groups, including the water soluble MEEP. All the membranes but the MEEP one were cast on asymmetric anodic alumina supports (20 nm pore size) from THF solutions containing benzoyl peroxide as a radical initiator, and crosslinked by heating at 130°C. The MEEP membrane, instead, was cast from THF solutions on the same support and crosslinked with UV light. The gas permeability of He, H_2, CO_2, O_2, N_2 and CH_4 was measured through the crosslinked composite membranes at 30°C. Carbon dioxide resulted by far the most permeable gas, and the permeability of this gas through the different co-polymers studied was proportional to the percentage of the MEE substituent on the backbone. The highest CO_2 permeability of 250 Barrers corresponded to the largest CO_2/CH_4 separation factor of 22.7, on the upper bound limit in the Robeson's selectivity vs. permeability plot for the above pair[50]. For all of the gases but CO_2 the gas transport data could be interpreted in a satisfactory way by considering the size of that molecules and the different condensability of the gases. The authors evidenced that the basic difference between carbon dioxide and the other gases was the strong intermolecular interaction of this gas with the MEE substituent. Comparable results were found by other researchers in the separation of humid CO_2/CH_4

mixtures with supported liquid membranes made of triethylene glycol: CO_2/CH_4 separation factors of 20-30 and CO_2 permeability in excess of 100 Barrer are reported[64]. A recent work on the gas separation properties of a mixed matrix membrane obtained by dispersing polyethylene glycol in silicon rubber (PDMS) reports selectivities in excess of 4000 for polar gases (such as SO_2 and CO_2) over N_2[65].

Pervaporation (PV) and Vapor Permeation (VP)

Polymeric membranes for Pervaporation and Vapor permeation (Fig. 2) are dense, and, as in gas separation, transport can be described by a solution-diffusion model[66,67]. The driving force for the passage of penetrants through the membrane is usually given by vacuum on the permeate side. Due to the large affinity with the species to separate, VP and PV membranes are usually asymmetrically swollen during the separations. If the separating layer of the membrane is made of a glassy polymer, the face in contact with the feed is usually plasticized, whereas the face on the vacuum side is in the glassy state.

The applications of PV and VP are in the separation of water from organic/water mixtures, in the removal of volatile organic compounds from aqueous and gas streams, and in the separation of organic/organic mixtures[3].

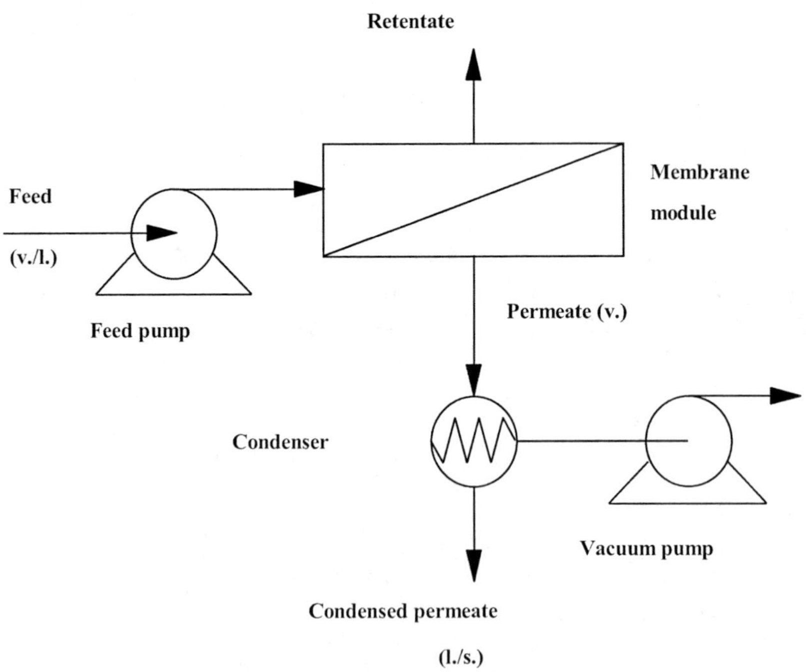

Fig. 2: Essential scheme of Vapor Permeation or Pervaporation. In VP the feed is in the gas phase, in PV the feed is a mixture of liquids.

Phosphazene homopolymer membranes for pervaporation and vapor permeation have been studied by several researchers in the last years. Sun et al. studied the sorption of alcohols, ketones and aromatic hydrocarbons in PTFEP and PPOP[68], and the data were analysed by the solution-diffusion model. In some cases the penetrant/air permselectivity of the phosphazene membranes was better than for poly(dimethylsiloxane), therefore worth to be explored, however with lower permeability. In a similar study of vapor sorption of water and ethanol in PTFEP membranes, Sun et al. found that water was more soluble than the alcohol, faster and therefore more permeable[69]. In VP experiments with water/methanol and water/ethanol mixtures through PTFEP membranes, Uragami et al. found that water was selectively transported with respect to the alcohols[70]. In PV experiments, instead, the selectivity switched in favor of the alcohols, unless the ethanol concentration in the feed was high and at low temperature. This behavior was explained by assuming that water transport was favored by the swelling of the polymer, which is larger for higher alcohol concentrations in the feed and at low temperatures. In all cases selectivity was too low for any practical applications.

The separation of benzene from mixtures with cyclohexane is energy expensive because the two substances form an azeotrope (60% benzene) and the difference in the boiling points is 0.6°C. A thorough picture of the pervaporation of benzene/cyclohexane mixtures with PPOP membranes was outlined by Sun et al.[71]. The maximum benzene/cyclohexane selectivity was 3 at 35°C with a 35-50% benzene in the feed.

The research group of the Idaho National Engineering and Environmental Laboratory has been very active in the search for innovative phosphazenes and phosphazene membranes for different separations (gas separation, pervaporation, vapor permeation, etc). An interesting application of polyphosphazenes is in Membrane Introduction Mass Spectroscopy (MIMS), in which an analyte in a liquid solution is selectively separated by a membrane and is desorbed in vacuum, ready to be ionised for mass spectroscopy. The membrane can enrich the analyte on the permeate side, therefore MIMS is able to increase the detection limits of MS, with PDMS being the routinely used polymer. PDMS is a very lipophilic polymer, however its water permeability is very high, and at the same time its sensitivity to polar analytes is not satisfactory. In a study of the capability of PTFEP and PPOP for MIMS, Stone et al.[72] found that the MIMS sensitivity with these polymers was about one order of magnitude lower than with PDMS when toluene, trichloroethylene, 2-propanol, THF and MEK were analysed. The more permeable PTFEP yielded the best results.

In the last years the search for water selective polymers indicated in the 2-(2-methoxyethoxy)ethoxy (MEE) group a substituent able to increase the sorption capacity of polyphosphazenes. In order to improve the dimensional stability and the film-forming capability of MEEP (a free flowing gum), finely tune the solubility properties of the resulting polymer and make crosslinking easier via radical initiators, the MEE group was co-substituted on polydichlorophosphazene with such hydrophobic and crosslinkable substituents as diacetone-D-glucofuranose[73], p-methoxy-phenol and o-allylphenol[74-76]. As an example, a water selective composite membrane prepared by casting a 60% MEE / 40% p-methoxy-phenol phosphazene on a ceramic support failed in long term pervaporation of water / ethylene glycol mixtures, probably due to high swelling and intrusion into the pores of the support. The durability was increased to a great extent by introducing, besides MEE and p-methoxy-phenol, a certain amount of o-allylphenol on the PN backbone, and crosslinking thermally with benzoyl peroxide or azo-bis(cyclohexane)carbonitrile: a membrane made of one of such co-polymers was able to withstand long term pervaporation tests with water / dye mixtures, aqueous ethylene glycol and aqueous 20% ethylene diamine. In the latter case the permeate fluxes at 35 – 38°C were 0.11 – 0.14 l/m^2 hr, with ethylene diamine rejection of 84-

92%[74,75]. In a similar study, Orme et al. have studied the pervaporation of water-dye, alcohol-dye, and water-alcohol mixtures by using a membrane prepared by casting a 48% MEE / 48% p-methoxy-phenol / 4% o-allylphenol phosphazene solution, containing benzoyl peroxide, on an anodic alumina support (200 nm pore size), and crosslinking at 130°C[76]. In the repeated water/dye, methanol/dye and 2-propanol/dye PV experiments between 25 and 54°C no dye was detected in the permeate, and the flux of solvent increased with temperature: the Arrhenius plot of fluxes indicated apparent activation energies in the order H_2O>2-propanol>MeOH. The order of permeability was instead reversed: H_2O<2-propanol<MeOH; at around 40°C the fluxes were 0.24, 0.73 and 4.7 kg/ m^2 hr, respectively. The performance of the membrane in the pervaporation of water/alcohol mixtures was instead unsatisfactory, due to the very low alcohol/water separation factor, below 7 in all cases. In the same paper the hard-to-find Hansen solubility parameters for four phosphazene polymers - MEEP, poly[bis(2-allylphenoxy)phosphazene], poly[bis(4-methoxyphenoxy)phosphazene] and the co-polymers used to prepare the membranes – were derived from group contribution methods, on the basis of solubility and swelling experiments, and by proper assumptions on the values of the group contribution parameters for N and P in the polyphosphazene polymers.

The concentration of dyes in aqueous waste streams in the textile industry by pervaporation is the subject of a patent filed by Stone[77]. Several linear polyphosphazenes and blends of polyphosphazenes were cast as solutions in the inside of ceramic tubes (200 nm pore size), and the resulting membranes were tested at temperatures between 22 and 70°C. The best results in terms of fluxes of water and retention of dyes were shown by poly[bis(trifluoroethoxy)phosphazene] (PTFEP), poly[bis(4-tetradecylanilino)phosphazene], and poly[(2-(2-methoxyethoxy)ethoxy)(cyclohexyloxy)phosphazene]. A membrane polymer with polyethylene glycol substituents, of repeat unit $-\{NP[(O\ CH_2\ CH_3)_mO\ CH_2\ CH_3)]_2\}_n-$, with $m \geq 1$, was only stable when blended with equal amounts of Atochem™ Z-1002-A, a crosslinkable polyaryloxyphosphazene with pendant allyl groups. A 33% 4-aminophenol / 67% phenol substituted polyphosphazene and poly[bis(phenoxy)phosphazene]s containing variable amounts of carboxylic acid groups in the para position were not stable, and spoiled even if cast as blends with Atochem™ Z-1002-A. Poly[bis(4-methoxyphenoxy)phosphazene], instead, was not permeable enough.

A concept similar to the blending with a cross-linkable polymer used in the previous patent was employed by Bac et al.[78] to prepare ten siloxane-phosphazene co-polymers with enhanced mechanical stability with respect to poly(dimethylsiloxane) (PDMS). The authors hydrosilylated an amorphous polyaryloxy phosphazene containing 4-methoxyphenoxy, 2-allylphenoxy and 4-(2-butyl)-phenoxy side groups (Atochem™ AMF 100) with a PDMS containing Si-H bonds; the content of PDMS in the copolymers varied from 20 to 95 wt %. The resulting membranes were made of linear polyphosphazene grafted and crosslinked by long PDMS chains. DSC experiments and TEM observations both indicated a fair phase separation between the phosphazenic and the siloxane phases. The membranes were used for the concentration of ethyl acetate from a saturated aqueous solution at 30°C. When the PDMS content was larger than 70% in the membrane, the permeate composition was 97-98% ethyl acetate with a 26-27 kg / m^2 h flux (normalized for a thickness of 5 μm), i.e. its behaviour was comparable to that of a pure PDMS membrane, but with better mechanical properties. Similar siloxane-phosphazene co-polymers were prepared by Roizard et al.[79] for the separation of 1-butanol from a saturated aqueous solution at 40°C. When passing from 40 to

50 wt % PDMS in the co-polymers, the continuous tridimensional phosphazene network disappeared and gave rise to a continuous tridimensional PDMS network embedding isolated phosphazene domains. This morphological change caused a marked increase in permeation rates and in the 1-butanol/water selectivity, and a weight fraction of 60% PDMS was enough to obtain results similar to those given by a PDMS film.

Ultrafiltration and Nanofiltration

The selectivity of Ultrafiltration membranes relies on a sieving mechanism, for Nanofiltration both on a solution – diffusion mechanism and on molecular sieving. The presence of fixed charges on NF membranes can discriminate ions on the basis of charge (Donnan exclusion) and induces preferential rejection of bi- or multivalent ions of the same charge[80].

Ultrafiltration and Nanofiltration are rather mature membrane operations. No wonder if the number of studies on such operations with polyphosphazene membranes is not large.

In a work of Rafik et al. an UF membrane was prepared by casting a solution of cyclolinear poly(p-oxydianiline-tetramorpholinocyclotriphosphazene) in methylcellosolve over a PVDF microporous support (IRIS, TechSep). The resulting membrane (cut-off 515 kD, water flux 380 l/m^2 day at a pressure difference of 300 torr) was tested for the decoloration of raw sugar solutions. The ultrafiltration of a 34.2 g/l raw sugar solution led to a 90% decoloration, a 3-7% saccharose retention and a flux of 150 l/m^2 day[81].

Peterson et al.[82] used the concept of dynamically adjustable membranes introduced by Guizard et al.[83], consisting in the modulation of the rejection properties of a pressure-sensitive swollen polymeric membrane. Three membranes (PPOP, Atochem AMF100, PDMS) were cast on 20 nm anodic alumina supports and tested for the permeation of 2-propanol/dye mixtures. The dye rejection changed from 100 to 0% when increasing the transmembrane pressure, and the transport could be described in terms of a combination of viscous flow and of a solution – diffusion mechanism.

Finally, tritiated water could be enriched of a few units percent in the feed by permeating contaminated water through dense PPOP and carboxyated PPOP membranes in an operation which could be indicated as Reverse Osmosis, since both the feed and the permeate are liquids[84-86]. The mechanism at the basis of this separation is not known.

PERSPECTIVES

The versatile chemical and physical properties of polyorganophosphazenes and their excellent chemical and thermal resistance are the key point to exploit in the development of new membranes. There is not much drive in the preparation of polyphosphazene porous membranes for MF and UF because the pore size and the pore size distribution are their main feature of these separations, and UF and MF membranes of excellent properties in rugged materials – polysulfone, polypropylene, poly(vinylidene difluoride), polytetrafluoroethylene – already exist. Membrane contactors (devices in which two phases are intimately contacted through membranes)[80,87] made of hydrophobic microporous polyphosphazenes, instead, might be of interest if we consider the ease of surface modification of these polymers. Some other specific application might be in the preparation of temperature or pH switchable membranes.

The chemical diversity is better exploited in cases where the transport mechanism is of the solution-diffusion type, such as gas separation, reverse osmosis, nanofiltration, pervaporation and vapor permeation. Nanofiltration and reverse osmosis are mature technologies, and excellent high flux membranes are on the market. Polyphosphazene membranes might find a way through such strong competitors only in niche areas (non aqueous, oxidant and aggressive systems, high temperatures) where other materials can hardly survive.

The concepts of today's technology for the manufacture of gas separation membranes were developed more than 20 years ago, and improvements and gains to date have largely been evolutionary and incremental in nature, and appear to be nearing their limits[46]. Next generation's gas separation membranes will require new concepts, such as liquid and solid facilitated transport membranes[88], in which one species in a mixture (e.g. olefins, CO_2) is exceedingly more soluble than the others[89]; molecular sieving ceramic or carbon membranes, or mixed matrix membranes[65] in which the separation characteristics of inorganic molecular sieves are combined with the ease of processing of polymeric membranes. The intrinsically basic nature of polyphosphazenes favors the solution of acidic species, therefore they are potentially interesting as scaffolds of immobilized liquid membranes or polymeric components of mixed matrix membranes for natural gas sweetening, for the separation of SO_2 from permanent gases, etc.

The ease in the tuning of the solubility properties of polyphosphazenes makes them very attracting as starting materials for pervaporation and vapor permeation, as demonstrated by the large number of reports of the last years. As is the case for other polymeric systems, the preparation of mixed matrix membranes might prove determining in enhancing permeability and selectivity. Proton conducting and ion exchange membranes, prepared for different purposes, are appealing candidated for selected liquid separations.

The ability to phosphonate polyphosphazenes gives an additional tool for the preparation of new proton conducting and ion exchange membranes. Some new options to explore are the introduction of different proton conducting groups, such as imidazole rings[90]; the modulation of the acidic strength of the substituent; the formation of mixed matrix membranes with inorganic superacid, such as layered zirconium phosphonates and sulfonates[91].

Polymer/ceramic hybrid membranes deserve more attention. The mechanical properties can be modulated to a large extent, and flexible but dimensionally stable hybrids can be produced by covalent crosslinking and intimate contact between the two phases. Proton and ion conductive membranes can take advantage from this, but also pervaporation in which excessive swelling or solubility of penetrants is a primary concern.

Polyphosphazene membranes containing inorganic catalysts, enzymes, living cells, etc., did not appear yet, nonetheless they may offer distinct advantages – thermal, chemical and solvent stability, ease of functionalization – over other polymer systems. For example, the availability of several hydrogel systems[35-37] could be exploited for the fabrication of catalytic membranes used in biphasic membrane reactors, in which one product could be extracted selectively from the reacting system, eventually with advantages in the reaction rates and overcoming limitations imposed by thermodynamic equilibria[92].

CONCLUSIONS

In the last years most of the research work with phosphazene membranes has been devoted to pervaporation and gas separation. The amount of studies on this subject is still modest, but the continuos advances in the control of the properties of polyphosphazenes and in the synthesis of new fascinating materials offer excellent opportunities to the membrane scientists. The chemical versatility of polyphosphazenes meets the need of tough materials with a wide range of properties for the preparation of tailored membranes. The operations where the chemical nature of the membrane is of paramount importance (pervaporation, vapor permeation, gas separation, ion conductive membranes, catalytic membranes) are those where polyphosphazenes may play a major role. Membrane contactors, Ultrafiltration and Nanofiltration however may also take advantage from the outstanding properties of phosphazene materials for non conventional applications where usual organic membranes cannot operate.

ACKNOWLEDGEMENTS

This work has been supported by MIUR (Italian Ministry of Education, University and Research) - PRIN 2002 (Polymer based mixed matrix membranes for gas, vapor and liquid separations).

REFERENCES

1. S. Loeb, *Desalination & Water Reuse*, **3** (2) 20-4 (1993).
2. D. H. Furukawa, *Reverse Osmosis process status* – 1999, Proc. ICOM'99, Toronto (CND), June 1999.
3. E. Drioli, M. Romano, *Ind. Eng. Chem. Res.*, **40** 1277-300 (2001).
4. A. I. Stankiewicz, J. A. Mouljin, *Chem. Eng. Progr.*, **96** 22 (2000).
5. H. Strathmann, Membrane processes for a sustainable industrial growth, paper presented at the Conference "*New frontiers for catalytic membrane reactors and other membrane systems*", Ravello (I), May 23-27, 1999.
6. H. R. Allcock, "*Phosphorous Nitrogen Compounds*", Academic Press: New York (NY, USA), 1972.
7. J. E. Mark, H. R. Allcock, R. West, Eds., "*Inorganic Polymers*", Prentice Hall: Englewood Cliffs (NJ, USA), 1992.
8. G. Golemme, E. Drioli, *J. Inorg. Organom. Polym.*, **6** (4) 341-65 (1996).
9. V. S. Papkov, M. N. Il'ina, V. P. Zhukov, D. Ja. Tsvankin, D. R. Tur, *Macromolecules*, **25** 2033-44 (1992).
10. K. Nagai, D. B. Freeman, A. Cannon, H. R. Allcock, *J. Membrane Sci.*, **172** 167-76 (2000).
11. E. M. Antipov, V. G. Kulichikhin, N. A. Platé, *Polym. Eng. Sci.,* **32** (17) 1188-203 (1992).
12. F. Suzuki, D. Onozato, H. Yaegashi, T. Masuko, *J. Appl. Polym. Sci.*, **34** 2197 (1987).
13. K. Mizoguchi, Y. Kamiya, T. Hirose, *J. Polym. Sci. B Polym. Phys.*, **29** 695 (1991).

14. H. R. Allcock, R. L. Kugel, *J. Am. Chem. Soc.*, **87** 4216-7 (1965).
15. P. Potin, R. De Jaeger, *Eur. Polym. J.*, **27** 341 (1991).
16. R. H. Neilson, P. Wisian-Neilson, *Chem. Rev.*, **88** 541-62 (1988).
17. R. A. Montague, K. Matyjaszewski, *J. Am. Chem. Soc.*, **112** 6721-3 (1990).
18. K. Matyjaszewski, U. Franz, R. A. Montague, M. L. White, *Polymer*, **35** 5005-11 (1994).
19. C. H. Honeymann, I. Manners, C. T. Morrissey, H. R. Allcock, *J. Am. Chem. Soc.*, **117** 7035-6 (1995).
20. H. R. Allcock, C. A. Crane, C. T. Morrissey, J. M. Nelson, S. D. Reeves, C. H. Honeymann, I. Manners, *Macromolecules*, **29** 7740-7 (1996).
21. H. R. Allcock, J. M. Nelson, S. D. Reeves, C. H. Honeymann, I. Manners, *Macromolecules*, **30** 50-6 (1997).
22. J. M. Nelson, H. R. Allcock, *Macromolecules*, **30** 1854-6 (1997).
23. H. R. Allcock, S. D. Reeves, J. M. Nelson, C. A. Crane, I. Manners, *Macromolecules*, **30** 2213-5 (1997).
24. J. M. Nelson, H. R. Allcock, I. Manners, *Macromolecules*, **30** 3191-6 (1997).
25. J. M. Nelson, A. P. Primrose, T. J. Hartle, H. R. Allcock, I. Manners, *Macromolecules*, **31** 947-9 (1998).
26. R. Prange, H. R. Allcock, *Macromolecules*, **32** 6390-2 (1999).
27. P. M. Blonsky, D. F. Shriver, P. E. Austin, and H. R. Allcock; *J. Am. Chem. Soc.*, **106** 6854-5 (1984).
28. M. K. Harrup, F. F. Stewart, *J. Appl. Polym. Sci.*, **78** 1092-9 (2000).
29. H. R. Allcock, M. E. Napierala, D. L. Olmeijer, C. G. Cameron, S. E. Kuharcik, C. S. Reed, S. J. M. O'Connor, *Electrochim. Acta*, **43** (10-11) 1145-50 (1998).
30. K. B. Visscher, H. R. Allcock, *Chem. Mater.*, **6** 2040-50 (1994).
31. P. N. Pintauro, R. Wycisk, *Sulfonated polyphosphazene membranes for direct methanol fuel cells*, this book.
32. H. R. Allcock, M. A. Hofmann, S. N. Lvov, X. Y. Zhou, *Int. Appl.* WO 00/77874 PCT/US00/16006 (2000).
33. H. R. Allcock, M. A. Hofmann, R. M. Wood, *Macromolecules*, **34** 6915-21 (2001).
34. M. V. Fedkin, X. Y. Zhou, M. A. Hofmann, E. Chalkova, J. A. Weston, H. R. Allcock, S. N. Lvov, *Mater. Lett.*, **52** (3) 192-6 (2002).
35. H. R. Allcock, *Chem. Mater.*, **6** 1476-91 (1994).
36. H. R. Allcock, Water-soluble polyphosphazenes and their hydrogels, in E. J. Glass, Ed., "Hydrophilic polymers", *Adv. Chem. Ser.*, **248** 3-29 (1996).
37. H. R. Allcock, A. M. A. Ambrosio, *Biomaterials*, **17** 2295-302 (1996).
38. H. R. Allcock, A. M. A. Ambrosio, *ACS Polym. Preprints*, **41** (1) 732 (2000).
39. J. M. Duval, A. J. B. Kemperman, B. Folkers, M. H. V. Mulder, G. Degrandchamps, C. A. Smolders, *J. Appl. Polym. Sci.,* **54** 409-18 (1994).
40. W. T. Ferrar, B. K. Coltrain, C. J. T. Landry, V. K. Long, T. R. Molaire, D. E. Schildkraut, Polyphosphazene molecular composites – in situ polymerisation of silicon, titanium, zirconium and aluminium alkoxides, in P. Wisian-Neilson, H. R. Allcock, and K. R. Wynne, Eds., "*Inorganic and Organometallic Polymers II*", ACS Symp. Ser., **572** 258-67 (1994).
41. M. K. Harrup, A. K. Wertsching, *ACS Polym. Mater. Sci. Eng.*, **84** 905-6 (2001).

42. M. Guglielmi, G. Brusatin, G. Facchin, M. Gleria, R. De Jaeger, M. Musiani, *J. Inorg. Organom. Polym.,* **6** (4) 221-36 (1996).
43. G. Golemme, G. Facchin, M. Gleria, L. Guarino, E. Drioli, *Phosphorus Res. Bull.*, **10** 736-41 (1999).
44. F. F. Stewart, M. K. Harrup, *J. Appl. Polym. Sci.*, **72** 1085-90 (1999).
45. F. F. Stewart, T. A. Luther, M. K. Harrup, R. P. Lash, *ACS Polym. Preprints,* **41** (1) 577 (2000).
46. I. C. Roman, New directions in membranes for gas separation, Paper presented at the UEF "*Advanced Membrane Technology Conf.*", Barga (I), Oct. 14-19, 2001.
47. J. H. Petropoulos, Mechanisms and theories for sorption and diffusion of gases in polymers, in D. R. Paul, Yu. P. Yampol'skii, Eds., "*Polymeric gas separation membranes*", CRC Press: Boca Raton (FL, USA), 1994; chapt. 2, pp. 17-81.
48. J. G. Wijmans, R. W. Baker, *J. Membrane Sci.*, **107** 1-21 (1995).
49. J. L. Duda, J. M. Zielinski, Free-volume theory, in P. Neogi, Ed., "*Diffusion in polymers*", Marcel Dekker, Inc.: New York, 1996; Chapt. 3, pp. 143-171.
50. L. M. Robeson, *J. Membrane Sci.,* **62** 165 (1991)
51. T. Hirose, K. Mizoguchi, *J. Appl. Polym. Sci.,* 43 891 (1991).
52. C. J. Orme, M. K. Harrup, T. A. Luther, R. P. Lash, K. S. Houston, D. H. Weinkauf, F. F. Stewart, *J. Membrane Sci.*, **186** 249 (2001).
53. H. R. Allcock, C. J. Nelson, W. D. Coggio, I. Manners, W. J. Koros, D. R. B. Walker, L. B. Pessan, *Macromolecules*, **26** 1493-502 (1993).
54. P. Wisian-Neilson, G.-F. Xu, *Macromolecules*, **29** 3457-61 (1996).
55. V. T. Stannet, W. J. Koros, D. R. Paul, H. K. Lonsdale, R. W. Baker, *Adv. Polym. Sci.,* **32** 69 (1979).
56. M. R. Pixton, D. R. Paul, Relationships between structure and transport properties for polymers with aromatic backbones, in D. R. Paul, Yu. P. Yampol'skii, Eds., "*Polymeric gas separation membranes*", CRC Press: Boca Raton (FL, USA), 1994; chapt. 3, pp 83-153.
57. B. D. Bhide, S. A. Stern, *J. Membrane Sci.*, **81** 239-52 (1994).
58. K. Toi, G. Morel, D. R. Paul, *J. Appl. Polym. Sci.,* **27** 2997-3005 (1982).
59. N. A. Platé, Yu. P. Yampol'skii, High free volume materials: structure-transport relationship, in D. R. Paul, Yu. P. Yampol'skii, Eds., "*Polymeric gas separation membranes*", CRC Press: Boca Raton (FL, USA), 1994; chapt. 4, pp. 155-207.
60. M. Kajiwara, *J. Mater. Sci.,* **29** 6268-72 (1994).
61. M. Kajiwara, T. Kimura, *J. Mater. Sci.,* **31** 5457-62 (1996).
62. M. Kajiwara, T. Kimura, *J. Mater. Sci.,* **31** 6659-63 (1996).
63. L. E. Starannikova, D. R. Tur, V. V. Teplyakov, and N. A. Platé; *Polym. Sci.* (Vysokomol. Soedin.), **36** (11) 1610 (1994).
64. A. Ito, S. Duan, Y. Ikenori, A. Ohkawa, *Sep. Purif. Technol.*, **24** 235-42 (2001).
65. S. Kulprathipanja, Mixed matrix membranes development, Paper presented at the UEF *"Advanced Membrane Technology Conf.",* Barga (I), Oct. 14-19, 2001.
66. J. Néel, "*Pervaporation*", Lavoisier Tech. & Doc.: London (UK), 1997.
67. R. W. Baker, J. G. Wijmans, Membrane separation of Organic vapors from gas streams, in D. R. Paul, Yu. P. Yampol'skii, Eds., "*Polymeric gas separation membranes*", CRC Press: Boca Raton (FL, USA), 1994; chapt. 8, pp. 353-97.
68. Y.-M. Sun, C.-L. Lin, Y.-K. Chen, C.-H. Wu, *J. Membrane Sci.*, **134** 117-26 (1997).

69. Y.-M. Sun, C.-H. Wu, C.-L. Lin, *J. Polym. Res. - Taiwan*, **6** (2) 91-8 (1999).
70. T. Uragami, T. Morikawa, *Angew. Makromol. Chem.*, **234** 39-51 (1996).
71. Y.-M. Sun, Y.-K. Chen, C.-H. Wu, A. Lin, *AIChE J.*, **45** (3) 523-34 (1999).
72. M. L. Stone, G. L. Gresham, L. A. Polson, *Anal. Chim. Acta*, **407** 311-7 (2000).
73. F. F. Stewart, M. K. Harrup, R. P. Lash, M. N. Tsang, *Polym. Int.*, **49** (1) 57-62 (2000).
74. F. F. Stewart, M. K. Harrup, E. S. Peterson, M. L. Stone, C. J. Orme, L. A. Polson, *ACS Polym. Mater. Sci. Eng.*, **81** 539-40 (2000).
75. F. F. Stewart, M. K. Harrup, T. A. Luther, C. J. Orme, R. P. Lash, *J. Appl. Polym. Sci.*, **80** 422-31 (2001).
76. C. J. Orme, M. K. Harrup, J. D. McCoy, D. H. Weinkauf, F. F. Stewart, *J. Membrane Sci.*, **197** 89-101 (2002).
77. M. L. Stone, US Pat. 6,093,325 (2000).
78. A. Bac, D. Roizard, P. Lochon, J. Ghanbaja, *Macromol. Symp.*, **102** 225-32 (1996).
79. D. Roizard, R. Clément, P. Lochon, J. Kerres, G. Eigenberger, *J. Membrane Sci.*, **113** 151-60 (1996).
80. W. S. W. Ho, K. K. Sirkar, Eds., "*Membrane Handbook*", Van Nostrand Rheinhold: New York, 1992.
81. M. Rafik, A. Mas, A. Elharfi, F. Schue, *Eur. Polym. J.*, **33** (5) 679-90 (1997).
82. E. S. Peterson, M. L. Stone, C. J. Orme, F. F. Stewart, R. L. Cowan, *Separ. Sci. Technol.*, **32** (1-4) 541-56 (1997).
83. C. Guizard, A. Boyé, A. Larbot, L. Cot, A. Grangeon, *Rec. Progr. en Genie des Procedes*, **6** (22) 27-9 (1992).
84. D. A. Nelson, J. B. Duncan, G. A. Jensen, US Pat. 5,451,322 (1995).
85. D. A. Nelson, J. B. Duncan, G. A. Jensen, S. D. Burton, *J. Membrane Sci.*, **112** 105-13 (1996).
86. J. B. Duncan, D. A. Nelson, *J. Membrane Sci.*, **157** 211-7 (1999).
87. G. H. Koops, "*Nomenclature and symbols in membrane science and technology*", European Society of Membrane Science and Technology: Toulouse (F) and Enschede (NL), 1995.
88. E. L. Cussler, Facilitated and active transport, in D. R. Paul, Yu. P. Yampol'skii, Eds., "*Polymeric gas separation membranes*", CRC Press: Boca Raton (FL, USA), 1994; chapt. 6, pp. 273-300.
89. A. Sarma Kovvali, G. Obuskovic, H. Chen, S. Majumdar, K. K. Sirkar, Novel immobilized liquid membranes for gas/vapor separations, Paper presented at the UEF "*Advanced Membrane Technology Conf.*", Barga (I), Oct. 14-19, 2001.
90. K. D. Kreuer, *J. Membrane Sci.*, **185** 29-39 (2001).
91. G. Alberti, M. Casciola, M. Pica, Preparation of nano-structured polymeric proton conducting membranes and their use in fuel cells, Paper presented at the UEF "*Advanced Membrane Technology Conf.*", Barga (I), Oct. 14-19, 2001.
92. L. Giorno, R. Molinari, E. Drioli, D. Bianchi, P. Cesti, *J. Chem. Technol. Biotechnol.*, **54** 345 (1995).

In: Applicative Aspects of Poly(organophosphazenes) ISBN 1-59454-025-X
Editors: M. Gleria and R. De Jaeger, pp. 207-224

Chapter 10

LINEAR AND CYCLOMATRIX POLYPHOSPHAZENE RESEARCH FOR MEMBRANE APPLICATIONS

Frederick F. Stewart, Thomas A. Luther, Mason K. Harrup, and Christopher J. Orme

Idaho National Engineering and Environmental Laboratory, P.O. Box 1625
Idaho Falls, ID 83415-2208, USA

INTRODUCTION

Polyphosphazenes are endowed with a spectrum of potential chemistries that allow for the tailoring of properties such as solubility. Matching chemical affinities is a valuable ability when dense rubbery polymers are applied to membrane science and technology. Membrane transport of liquids through thin dense film membranes has been described using solution-diffusion theory.[1] Permeability has been defined as a product of the solubility and the diffusivity of a penetrant molecule within the membrane. For liquid transport through rubbery dense film membranes, solubility is the predominate parameter in liquid transport. Several reports have discussed application of polyphosphazene polymers as pervaporation membranes for use in separating organic-organic and organic-water mixtures.[2-6] In these instances, the authors have attempted to enhance separation factors and transmembrane fluxes through matching phosphazene polymers with the solubility characteristics of a particular permeate. Pervaporation is a technique for conducting liquid-liquid separations where the feed is circulated at low pressures and permeation is driven by the application of reduced pressure on the permeate side of the membrane (Figure 1).

Permeation is proposed to occur through solubility interaction between the membrane and the permeate followed by evaporation off the backside of the membrane. Separation factors in pervaporation experiments are controlled by several factors including membrane boundary layer effects, chemical affinity of the membrane, and vapor pressure of the permeate. A complete treatment of the technique is available elsewhere.[7]

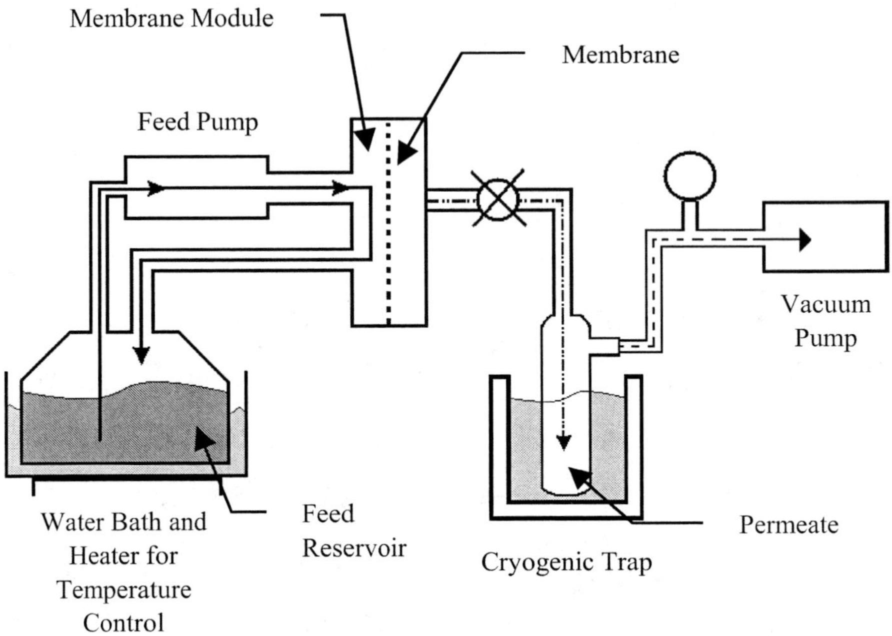

Figure 1. Schematic diagram of a pervaporation apparatus.

A significantly larger body of work addresses gas separations where solubility is less dominant. Mechanistic aspects of gas permeability have been considered for several decades.[1,8,9] A practical application of permeability theory employs two concepts: free volume and solubility. Solubility entails the ability of a penetrant gas to dissolve within a polymer matrix. In general, the small permanent gases have weak solubility interactions with rubbery polymer membranes, thus transport is dominated by free volume considerations which contribute to the diffusivity. Free volume is created through polymer structures that provide voids within the matrix that allow small molecules to permeate. Relationships between the selectivity of a membrane and the flux were addressed in an empirical fashion by Robeson.[10] In this work, the ideal separation factor (α) was defined as a simple ratio of the pure gas permeabilities. Relationships were then drawn from a plot of the ideal separation factor versus the permeability of a particular gas. For example, in general terms, glassy polymers experienced the highest gas selectivity, however, at the expense of flux. Conversely, rubbery polymers exhibited high fluxes, although selectivities were poor. Related reports[11,12] discussed the synthetic modifications required to increase both the flux and the selectivity of polymer membranes. Conclusions drawn included stiffening of the polymer backbone and adding sterically bulky groups within the polymer to increase free volume.

Compilations of gas permeability data using polyphosphazene membranes have been published by Drioli.[13,14] Permeabilities and selectivities of various gases were typical of rubbery polymer membranes. Overall, this is not surprising, however, given the high chemical and physical tolerances of the phosphazene family of materials, these membranes are clearly unique. In a related study, poly[bis(trifluoroethoxy)phosphazene] (PTFEP) was studied for gas permeation where the authors observed a clear solubility interaction between the polar trifluoroethoxy pendant groups and carbon dioxide.[15] PTFEP is a semicrystalline polymer, thus requiring a consideration of the phase upon permeability. Permeation was studied both below and above the mesophase T_1 transition temperature with significant differences in both physical and transport characteristics.[16] It should be noted that above T_1, there was a linear relationship between permeability and temperature, suggesting that the mesophase and the amorphous phase have similar transport characteristics and that the crystalline phase is relatively impermeable.

Several phosphazene homopolymers (one pendant group) and terpolymers (two pendant groups) were studied for gas transport.[17-21] The authors report oxygen gas permeabilities for polyalkoxyphosphazene membranes crosslinked using a variety of techniques including ultraviolet, electron beam, and gamma irradiation. In all cases, mechanical properties were improved. Crosslinking processes using additives also were explored with success. Another report used membranes formed from phosphazene terpolymers with varying proportions of 2,2,2-trifluoroethoxy and adamantyl pendant groups.[22] For these polymers, it was proposed that the small permanent gases were transported using a sieving mechanism with solubility playing a small role. However, light hydrocarbons and CO_2 were observed to have a larger solubility interaction leading to significant contributions of both solubility and diffusivity to the transport mechanism.

Polyphosphazenes are unique materials in that their backbones can have up to three differing configurations (Figure 2).

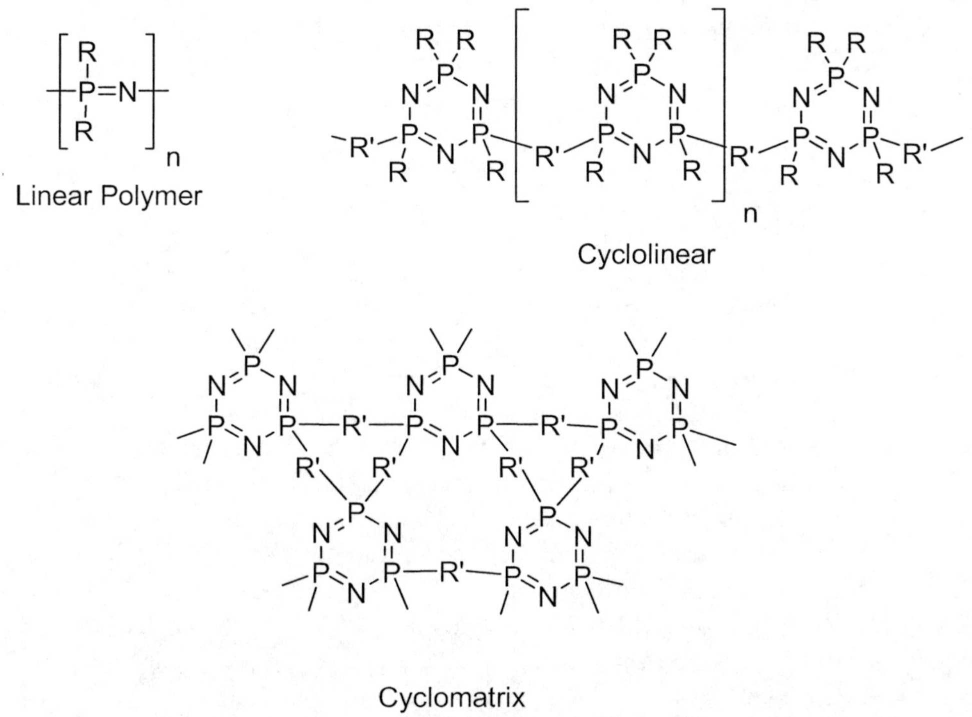

Figure 2. Polyphosphazene Backbone Configurations. R and R′ represent monodentate pendant groups and bidentate crosslinking groups, respectively.

Linear polymers are typically generated through nucleophilic substitution of poly[bis(chloro)phosphazene]. Formation of poly[bis(chloro)phosphazene] is most commonly accomplished through the ring opening polymerization of phosphonitrilic chloride trimer under vacuum at 250 °C, although a recent report may lead to the popular application of a solvent-less "one pot" route.[23] Ring opening, both with and without catalysts, has been shown to initially form long linear chains rapidly (and in low yield) and with molecular weights approaching 10^6.[24] Maximum yields of poly[bis(chloro)phosphazene] approach 40% for reaction times up to 48 hours. Longer heating periods can lead to subsequent reactions yielding branched materials. Mechanistic studies have shown that the polymerization reaction is cationic and is initiated through the loss of chloride ion.[25]

Lability of chloride can be exploited through its use as a leaving group allowing for nucleophilic substitution processes onto phosphazene trimers yielding new compounds. Many of these compounds are fluids and embody unique behaviors, as discussed in a recent contribution.[26] Fluids are commonly synthesized by attaching several differing pendant groups onto the trimer ring, thus inhibiting crystallization.[27] Conversely, solid trimers may be formed using six equivalents of one pendant group to substitute all six sites per ring. A more complete treatment of the chemistry of cyclic phosphazenes has been published recently.[28]

Use of cyclotriphosphazenes as polymer precursors have an advantage over linear polyphosphazenes in that stable monodisperse structures may be formed and characterized. Polymerization of cyclic trimers in a two dimensional manner has been accomplished yielding cyclomatrix polymers. Addition of functional groups onto cyclic trimer rings is conducted in a similar manner as the linear polymers and polymerization is generally effected through the addition of a linker. For example, polymerization of hexakis(4-hydroxyphenoxy)cyclotriphosphazene with hexamethylene tetraamine (HMTA) has been reported to occur at 177 °C yielding insoluble cyclomatrix phosphazene materials.[29,30] A complete review of this chemistry has been recently published.[28]

LINEAR PHOSPHAZENE POLYMERS WITH IMPROVED PROPERTIES

Blending of pendant groups on the phosphazene backbone naturally leads one to the ability to tailor these materials for either specific chemical or physical properties. An initial goal of our synthetic work was to create new polymers for membranes where water and polar organic species would transport selectively. With this goal in mind, 2-(2-methoxyethoxy)ethoxide (MEE) was chosen as the hydrophilic component. Furthermore, 4-methoxyphenoxide (MEOP) was chosen to be the hydrophobic component. Addition of MEOP was required to reduce the water affinity that would be created by incorporation of the polyether onto the polymer and to improve the mechanical properties. If the polymer were too hydrophilic, it would swell excessively leading to structural weakness, and it would be non-selective due to a large fraction of solvent induced free volume. Another possibility would be degradation of the membrane due to its dissolution in the feed stream. Our goal was to increase the solubility of the polymer membrane without creating too much solvent affinity, thus a limited degree of membrane swelling in the feed was sought.

An initial report discussed the synthesis and characterization of MEE/MEOP terpolymers.[31] This synthesis was conducted with sequential addition of polymer and pendant group. The poly[bis(chloro)phosphazene] in toluene was initially added to an anhydrous THF solution of sodium *p*-methoxyphenoxide. The resulting ^{31}P NMR spectrum gave two major signals (Figure 3).

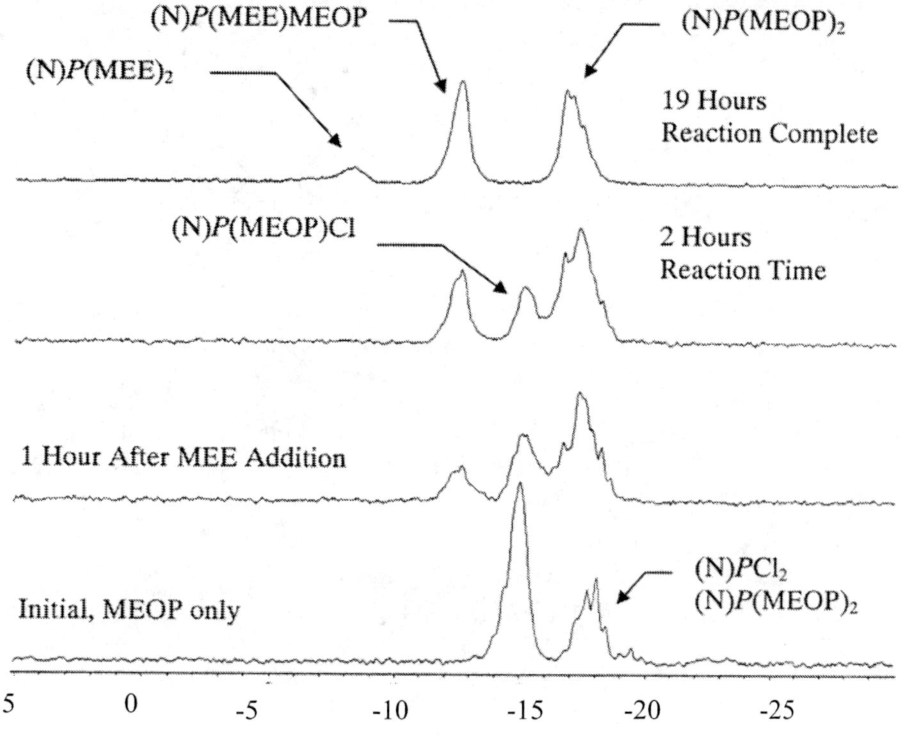

Figure 3. Sequential progression of polymer substitution forming a terpolymer as viewed through ^{31}P NMR spectroscopy.

In the initial spectrum, the broad peak at -15 ppm has been assigned to (N)*P*(MEOP)Cl, while the envelope from -17 to -19 ppm has been assigned to (N)*P*(MEOP)$_2$ and (N)*P*Cl$_2$. Subsequent addition of MEE salt in anhydrous THF followed by heating gave additional peaks at -8 and -13 ppm. The -13 ppm peak has been assigned to (N)*P*(MEE)MEOP. The peak at -8 ppm has been assigned to (N)*P*(MEE)$_2$ in agreement with 100% MEE substituted polyphosphazene (MEEP) as a solution in deuterated chloroform.[32]

A 60% MEE/40% MEOP phosphazene polymer was formed into membranes using solution casting methods.[33] Solution casting entails dissolution of the polymer in an appropriate solvent (2-5 weight % polymer), typically toluene or tetrahydrofuran, followed by application of the solution to a support or substrate. In this instance, the membranes were cast onto a glass plate, allowed to dry in a fume hood, and removed through immersion in a water bath. Once dry, the membranes were placed onto a ceramic support (Whatman Anopore®, 0.2 μm pore size) and mounted into cells modified for pervaporation. Pervaporation experiments, using water/ethylene glycol as the feed, were conducted over seven days and the transmembrane flux was observed to decrease with time. This behavior was attributed to poor mechanical properties. It was proposed that the polymer became highly swollen and entered the pores of the ceramic support, leading to larger effective membrane thicknesses and lower transmembrane flux.

In a previous work, phosphazenes bearing carboxylato-functionality were shown to undergo esterification reactions in neat alcohol with acid catalysis.[34] Using this methodology, poly[bis(4-carboxyphenoxy)phosphazene] was esterified with MEE in the presence of concentrated HCl (Figure 4).

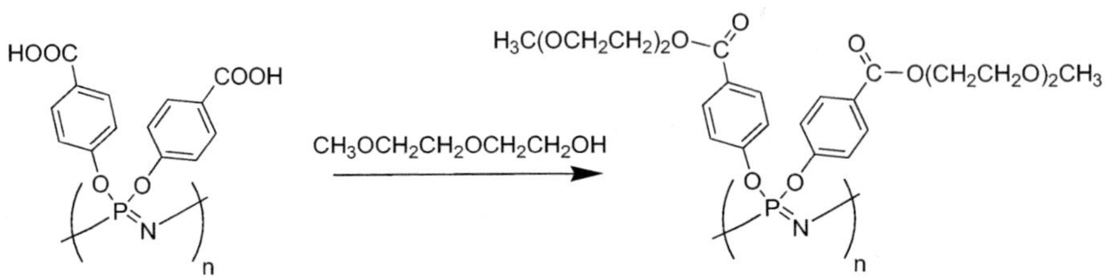

Figure 4. Synthetic scheme for the esterification of poly[bis(4-carboxyphenoxy)phosphazene] with MEE.

The resulting material was a polymer with a significantly higher degree of hydrophilicity. Pervaporation of water-dye solutions with this polymer exhibited similar dimensional instability as the terpolymer discussed above.

Increased dimensional stability of MEE-containing polymers required facile crosslinking pathways to prevent flow of the polymer. A third pendant group, 2-allylphenol (2AP), was added to the mix in small amounts (5-20%) giving a phosphazene heteropolymer (six potential mers) (Figure 5).

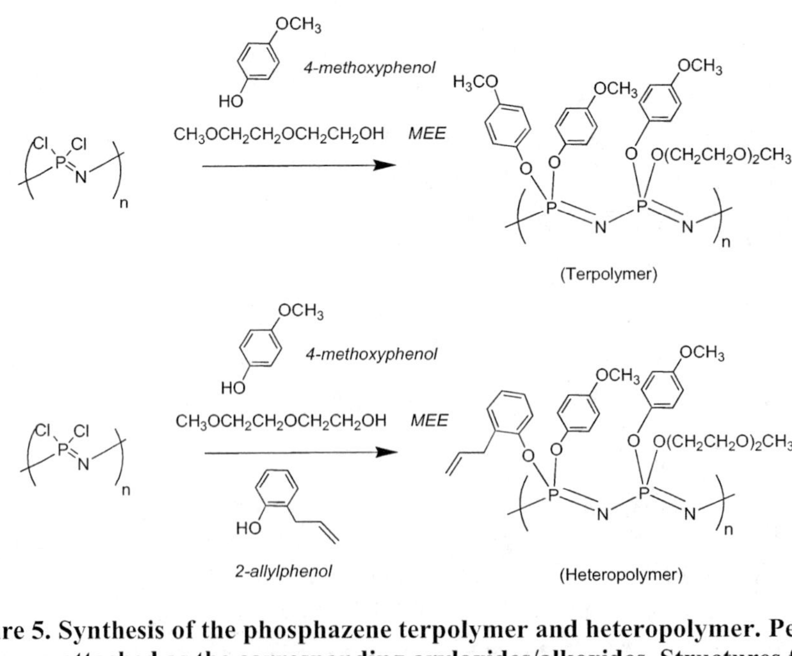

Figure 5. Synthesis of the phosphazene terpolymer and heteropolymer. Pendant groups are attached as the corresponding aryloxides/alkoxides. Structures for each do not represent all possible mers.

Crosslinking using the allylic moiety can readily occur through irradiation (UV, gamma, and electron beam) and can be performed using thermal curing. Thermal curing was found to be effective at reasonable temperatures (130 °C) with an initiator such as benzoyl peroxide or 1,1'-azo-bis(cyclohexanecarbonitrile).[35] Crosslinking is effected through homolytic bond cleavage at the benzylic position where the free radical is stabilized by both the aromatic ring and the allylic group resulting in low bond dissociation energies.

Electron beam irradiation also is highly effective and has one key advantage: incremental dosing. If a required crosslink density has not been achieved, the sample may be exposed to subsequent doses until the desired crosslink level is achieved. However, two significant disadvantages of electron-beam treatment are cost and availability.

NMR spectroscopy was used to determine molar ratios of pendant groups on the polymer. At an operating frequency of 121 MHz, the ^{31}P NMR spectrum revealed three peaks reminiscent of the terpolymer without the 2-allylphenol. Figure 6

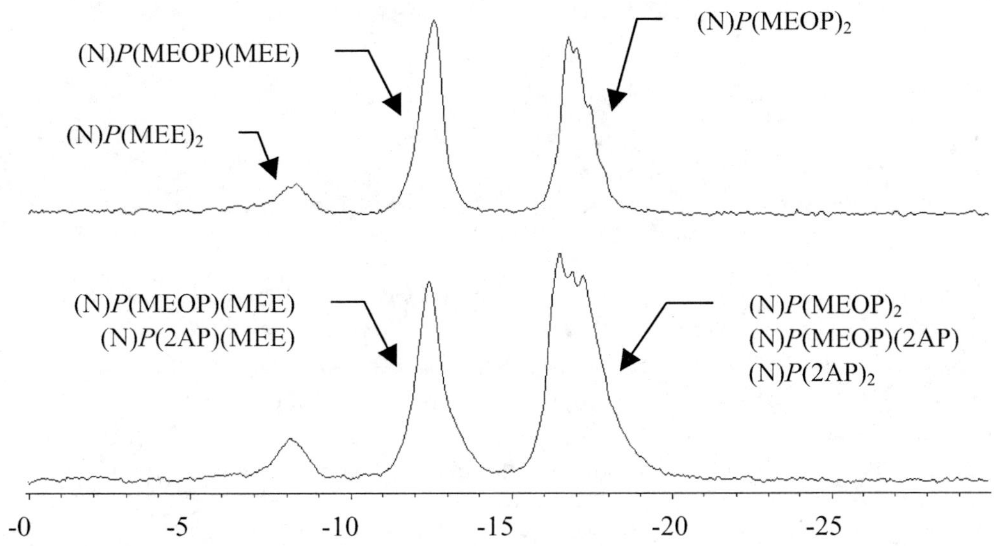

Figure 6. ^{31}P NMR spectra of terpolymer (top) and heteropolymer (below).

shows the ^{31}P spectra for the terpolymer and the heteropolymer where the only significant difference is the peak at ~-17 ppm that is broadened due to three potential aromatic-only mers. To determine the relative amounts of each pendant group on the backbone, integration of ^{1}H NMR spectra was performed (Figure 7).

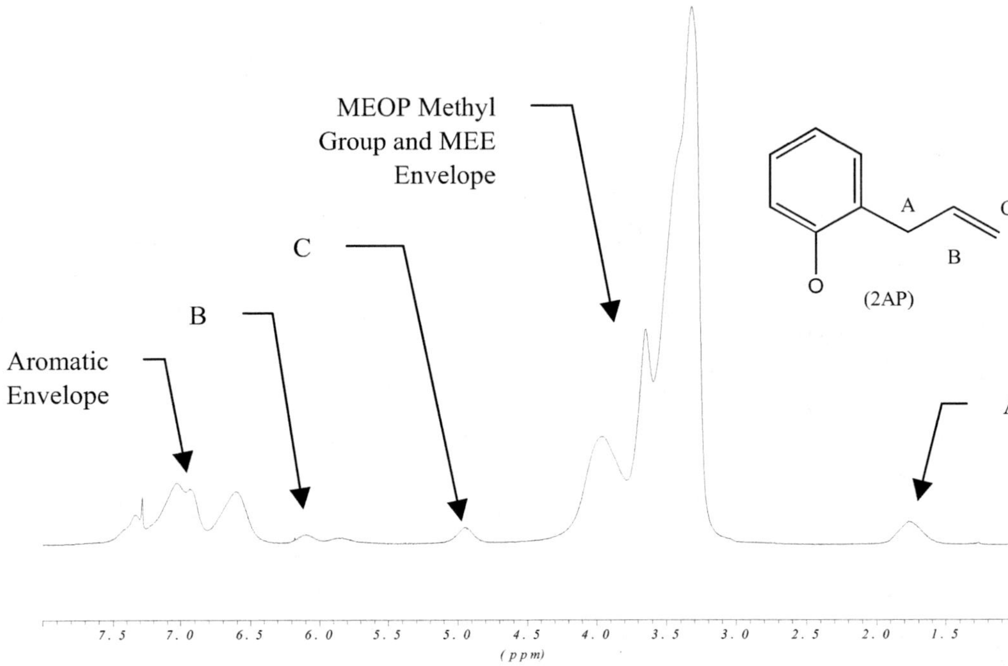

Figure 7. ^{1}H NMR spectrum of a phosphazene heteropolymer. Resonances are indicated for the allyl group of 2AP

From these data, signals for the aromatic protons, etherial protons, and allylic protons were integrated thus allowing for the calculation of molar ratios of each group. Thermal analysis also has been shown to be a valuable characterization tool for these materials. In a recent work[36], it was shown that phosphazene terpolymers follow the Fox equation.[37] This relationship refers to the dependence of the glass transition temperature (T_g) of a copolymer on the relative amounts of each homopolymer and is expressed mathematically as:

$$T_g = M_1T_{g1} + M_2T_{g2} \qquad \text{eq 1.}$$

where M_1 and M_2 are the molar percentages of each homopolymer and T_{g1} and T_{g2} are their respective T_g values. It should be noted that this relationship is valid only for homogeneous copolymers where there is no phase separation. For phosphazenes, although there are three mers present in a binary pendant group mixture, the Fox equation has been found to be accurate in describing the T_g based upon molar ratios derived from ^{1}H NMR data. Extension of this analysis to heteropolymers with six possible mers was found to give good results. Although the T_g of a 2AP-substituted homopolymer is not known, at small amounts this contributor can be ignored yielding a rough correspondence between T_g and composition.[38]

A series of these MEE/MEOP/2AP heteropolymers were studied as membranes for gas transport using an instrument designed to measure pure gas permeabilities (Figure 8).[39] Membranes for these experiments were solution cast (5-10% polymer with 2% 1,1'-azo-bis(cyclohexanecarbonitrile)(polymer weight percent) in tetrahydrofuran) onto a 0.02 μm pore size ceramic disk (Whatman Anopore®) and cured at 130 °C for at least 10 minutes. Using this method, consistent membrane thicknesses of 200 μm were obtained. Permeabilities for several gases were determined using polymers with MEE compositions ranging from 6% to 100% (permeabilities for H_2, N_2, O_2, and CO_2 are shown in Table 1).

In this data, increasing permeability with increasing MEE content was noted as a general trend except for the final data point for MEEP (100% MEE phosphazene). A decrease in permeability was noted for MEEP that was attributed to a decrease in free volume. It should be noted that MEEP required crosslinking via UV irradiation prior to testing to prevent polymer flow.

Table 1. Permeabilities (P, Barrers) and Ideal Separation (α) Data for MEE Containing Phosphazene Polymer Membranes Using Selected Gases.[39] The Ideal Separation Factor for CO_2 and H_2 were Calculated from the Permeability Data for this Work.

% MEE	P_{H2}	P_{N2}	P_{O2}	P_{CO2}	$\alpha_{O2/N2}$	$\alpha_{CO2/H2}$
6	4.2	0.3	1.7	9.3	5.7	2.2
23	13.4	1.8	4.8	38.9	2.7	2.9
25	19.4	2.8	7.7	81.9	2.8	4.2
38	25.6	3.9	10.2	107.7	2.6	4.2
48	23.7	8.3	14.8	115.9	1.8	4.9
74	28.8	10.2	18.0	226.7	1.8	7.9
100	25.0	4.0	7.0	250.0	1.8	10.0

Ideal separation factors for the O_2/N_2 gas pair steadily decreased with increasing MEE suggesting morphological differences in the polymer with increased MEE that could potentially lower free volume. With decreasing free volume, transport of O_2 and N_2 are controlled by their relative weak solubility interactions with the polymer. An opposite trend was observed for the CO_2/H_2 gas pair where α increased with increasing MEE. In this instance, CO_2 is highly condensable and a soluble gas within an MEE containing polymer while H_2 has low solubility within the polymer; thus, transport of CO_2 is favored.

Phosphazene MEE/MEOP/2AP heteropolymers have been used as pervaporation membranes for the selective transport of water and alcohols.[40] Thin films (~15 μm) were formed through solution casting directly onto a porous ceramic support. Studies using membranes directly cast versus membranes formed on a casting plate, lifted, and placed on a support revealed that there was no significant difference between the two methods using these heteropolymers. Molecular weight analyses revealed that the terpolymers, without a crosslinking moiety, have $M_w = 10^5 - 10^6$ while similar heteropolymers have higher M_w values, ranging from $10^6 - 10^7$. It is proposed that the higher M_w values obtained for these heteropolymers are the result of adventitious crosslinking between 2AP groups during polymer synthesis. It is also important to control the amount of 2AP that is used for polymer synthesis since amounts greater that 20% can lead to an intractable product.

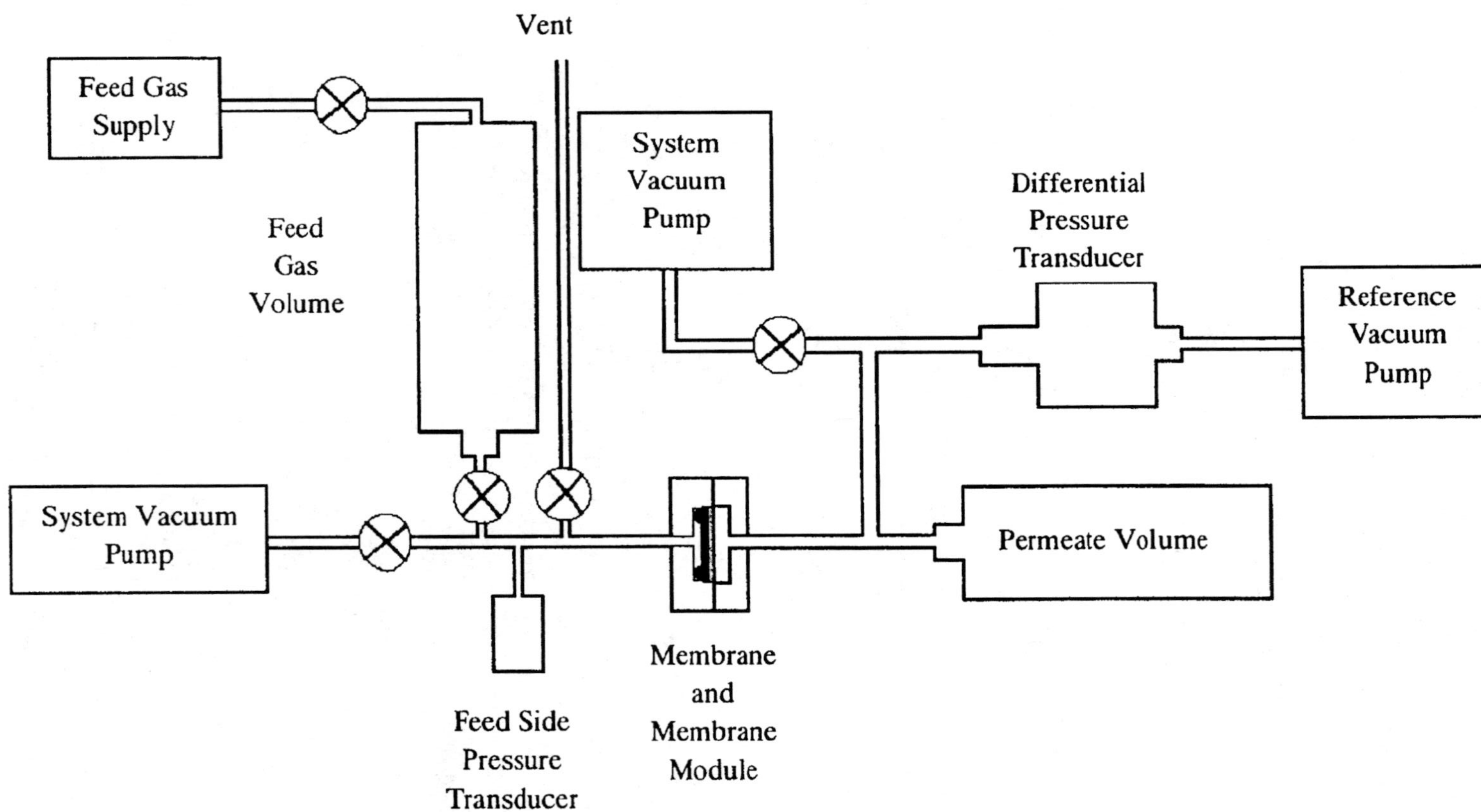

Figure 8. Apparatus for measuring pure gas permeabilities through phosphazene membranes.

An important advantage in the use of these heteropolymers for membranes is their inherent chemical and thermal stability. For example, these heteropolymers have decomposition temperatures (T_d) in excess of 250 °C as measured by thermogravimetric analysis suggesting that these materials may be usable in high temperature membrane applications.[38] In these experiments, T_d is noted as the temperature at which mass loss in the sample is observed. It is proposed that the majority of mass loss at these temperatures is due to pendant group degradation. This conclusion is supported by the lower T_d values noted for polymers with increased percentages of MEE. MEEP decomposes at 260 °C and the 100% MEOP polymer does not lose mass until 320 °C is exceeded. In general, T_d values for the MEE/MEOP/2AP heteropolymers were measured in between 260 °C and 320 °C with a trend of lower values with increasing MEE content.

Chemical stability in MEE/MEOP/2AP heteropolymers has been demonstrated using highly basic feeds in a membrane separation.[38] A pervaporation experiment using an ethylene diamine/water feed showed good stability of the membrane and consistent water fluxes indicating no membrane degradation. In fact, magnitudes of the fluxes were consistent with other experiments using only water as the feed. In another experiment, a terpolymer with a similar distribution of MEE and MEOP pendant groups degraded rapidly in the feed suggesting that the crosslinked structure of the heteropolymer provided the materials with increased resistance to chemically aggressive feeds.

A heteropolymer with 48% MEE, 48% MEOP, and 4% 2AP was studied for the pervaporation of alcohol/water mixtures, where the alcohols were methanol, ethanol, and 2-propanol.[40] In all cases, the alcohol was the favored permeate, although the separation factors were modest (α = 2-7). Also observed was a typical inverse relationship between flux and separation factor as the concentration of alcohol was increased in the feed. With increasing concentrations of alcohol, fluxes increased significantly, however the separation factors decreased.

Additional studies to characterize the effect of blending pendant groups used solubility parameters.[40] Single point Hildebrand parameters commonly are employed to describe effectively the polymer-solvent interactions for non-polar systems. However, Hildebrand parameters often are not sufficient to describe interactions in polar systems. To describe polar systems better, Hansen parameters have been devised where three parameters are used to describe solubility. Using this scheme, solubility behavior is separated into dispersive, polar, and hydrogen bonding interactions. Dispersive interactions are the non-polar attractive forces such as van der Waals and London forces, which generally are well represented using the single point Hildebrand parameter. The polar parameter describes dipole interactions while the hydrogen-bonding parameter gives a relative measure of the hydrogen-bonding capability of the substrate in question.

Application of Hansen parameters to phosphazenes was conducted through swelling experiments where the interaction between the polymer and each solvent was determined through immersion, followed by gravimetric determination of solvent uptake into the polymer.[40] Classes of solvents analyzed included ethers (diethyl ether, tetrahydrofuran, 1,4-dioxane), polyethers (ethoxyethoxyethoxyethanol, tetraethylene glycol), alcohols (methanol, isopropanol), aromatics (toluene), and hydrocarbons (cyclohexane, pentane, octane). Hansen parameters for the solvents were obtained from literature sources[41] while parameters for the polymers were estimated using group contribution theory.[42] Literature sources supplied all of the necessary constants for the pendant groups, however, estimations were required to account for the contributions derived from the polymer backbone.[40] Using this method, the solubility behaviors of a MEE/MEOP/2AP heteropolymer, MEEP, and 100% MEOP polymer were presented and the behavior was qualitatively described. Relative degrees of solubility interaction were described in terms of swelling. For example, "good" solvents were ones that were observed to have a high amount of solvent uptake into the polymer matrix as measured gravimetrically. Likewise, "poor" solvents caused little swelling. For the 100% MEOP polymer, "good" solvent behavior was limited to a narrow region represented by solvents such as tetrahydrofuran. Alcohols, that are more polar and have greater hydrogen-bonding interactions, were found to be "poor" solvents for the 100% MEOP polymer. On the other hand, MEEP exhibited "good" solubility behavior with the solvents that showed "good" behavior with the 100% MEOP and the alcohols, polyethers, ethers (diethyl ether) and aromatics. Instead of clear and distinct shift in solubility behavior, the addition of MEE expands this behavior to more classes of solvents that have a degree of polarity. In fact, the magnitude of the solvent interaction between most of the solvent classes and the phosphazene heteropolymers appeared to be proportional to the molar ratio of MEE. One clear exception was the hydrocarbons where little interaction was observed regardless of the MEE content. The small degree of phosphorus-nitrogen backbone polarity prevented extensive solubility of the polymers in non-polar hydrocarbons.

SOLUBLE CYCLOMATRIX POLYPHOSPHAZENES

Polymerization of functionalized phosphazene trimers forming cyclomatrix structures offer a route to new materials that could potentially be used as membrane materials. Hexakis-(4-hydroxyphenoxy)cyclotriphosphazene (Figure 9)

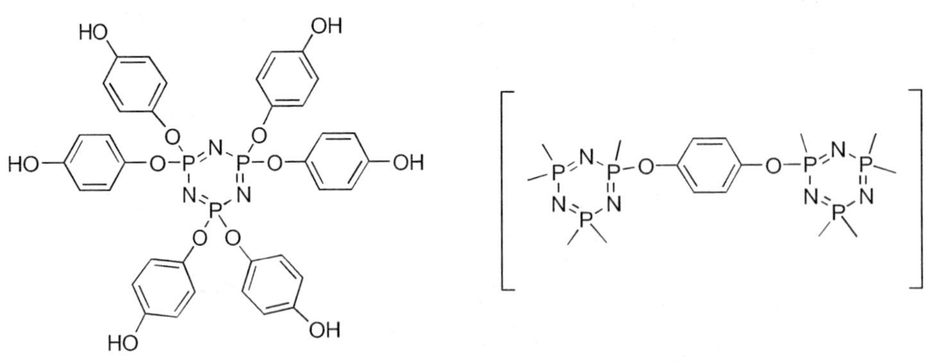

Figure 9. Structure of hexakis-(4-hydroxyphenoxy)cyclotriphosphazene (left) and proposed structure of a hydroquinone bridging crosslink (right).

was of interest due to the hydroxyl functionality that allows attachments through simple organic chemistry. Several syntheses of this material have been published. A simple one-pot route was first reported in 1964[29] where hexachlorocyclotriphosphazene was reacted with hydroquinone in the presence of pyridine using carbon tetrachloride as the solvent. Isolation and characterization of the product yielded a polydisperse polymeric product where, in some instances, reaction of both hydroxyls with the trimer ring yielded linked trimer rings (Figure 9). Further supporting the polymeric nature of this product was a complex ^{31}P NMR spectrum.[43] A discrete cyclic trimer species would be expected to show only one peak as the phosphorus atoms are magnetically equivalent.

A two-step route to hexakis-(4-hydroxyphenoxy)cyclotriphosphazene involved attachment of a pendant group with a masked hydroxyl functionality, 4-benzoxyphenol or 4-methoxyphenol, followed by hydrolysis.[44] The advantage of this longer route was that discrete monomeric cyclic trimers were obtained without crosslinking and the resulting ^{31}P NMR spectrum showed the product to be a sharp singlet.

An alternate synthesis using the methodology from the "one-pot" method and obtaining monomeric products was demonstrated using *tert*-butylhydroquinone as the pendant group forming hexakis-[(3-*tert*-butyl)(4-hydroxy)phenoxy]cyclotriphosphazene.[45] In this work, *tert*-butylhydroquinone in the presence of 4-picoline in cyclohexane was found to substitute for chlorine on the cyclotriphosphazene ring with regiospecificity where the *tert*-butyl group was positioned away from the trimer ring (Figure 10).

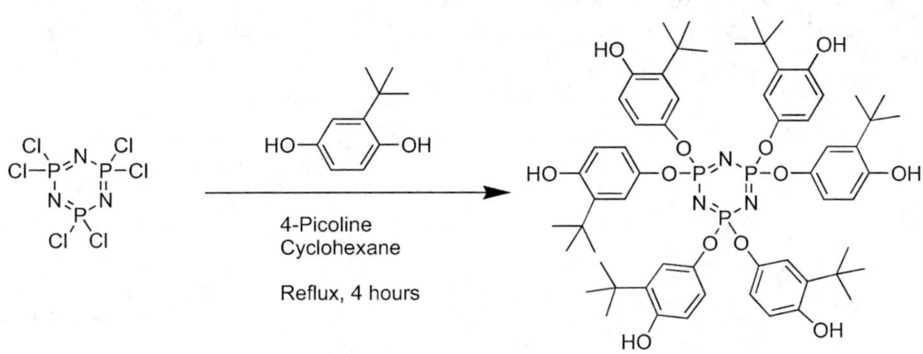

Figure 10. Synthesis of hexakis[(3-*tert*-butyl)(4-hydroxy)phenoxy]cyclotriphosphazene.

It was proposed that the *tert*-butyl group shielded the ortho-hydroxyl (with respect to the alkyl group) preventing attachment under the reaction conditions. Support for this structural determination was provided by ^{1}H and ^{31}P NMR spectroscopy. Using ^{1}H NMR nuclear Overhauser experiments, the position of the *tert*-butyl was ascertained. Further, only one peak was observed in the ^{31}P NMR spectrum suggesting a homogenous magnetic environment consistent with the proposed structure.

Steric influences that force regiospecific attachment of *tert*-butyl hydroquinone was investigated through the reaction of methylhydroquinone with hexachlorocyclotriphosphazene.[46] In this instance, non-regiospecific attachment of methylhydroquinone was observed where the methyl group was found to be ortho to the phosphorus-bearing oxygen in approximately 17% of the pendant groups as measured by integration of the ^{13}C NMR signals. A loss of regiospecificity was attributed to the substantially lower steric bulk of a methyl group as compared to a *tert*-butyl group.

With both hydroxyls available for chemical attachment to the phosphazene ring, polymerization becomes a possibility. In fact, the ^{31}P NMR signal attributable to the products was a complex multiplet suggesting that oligomerization had occurred. To probe the degree of polymerization, gel permeation chromatography was employed where detection was performed using laser light scattering. Molecular weights (M_w) for the *tert*-butylhydroquinone, methylhydroquinone, and hydroquinone (one-pot synthesis method) adducts were determined (Table 2).

Table 2. Calculated and Measured Molecular Weights (M_w) for Three Cyclic Trimers as Indicated by Pendant Group. Calculated M_w Values are for Discrete Cyclic Trimers.

Pendant Group	Calculated M_w (g/mol)	Measured M_w (g/mol)
Hydroquinone	790	452,000 ± 9000
Methylhydroquinone	874	4970 ± 60
tert-Butylhydroquinone	1126	1070 ± 100

The *tert*-butylhydroquinone adduct was found to have an M_w nearly identical to the calculated value suggesting no oligomerization. Methylhydroquinone gave an oligomerized structure with M_w representing a chain with approximately six mers. The hydroquinone adduct was significantly polymerized with M_w nearly one-half million indicating that, in the absence of steric encumbrances, crosslinking of individual trimers rings occur to a significant degree. However, a hindering group, such as *tert*-butyl, can block attachment to the trimer ring.

The free hydroxyl of hexakis-[(3-*tert*-butyl)(4-hydroxy)phenoxy]cyclotriphosphazene was found to be reactive to alkyl halides in the presence of weak base (Figure 11).[47]

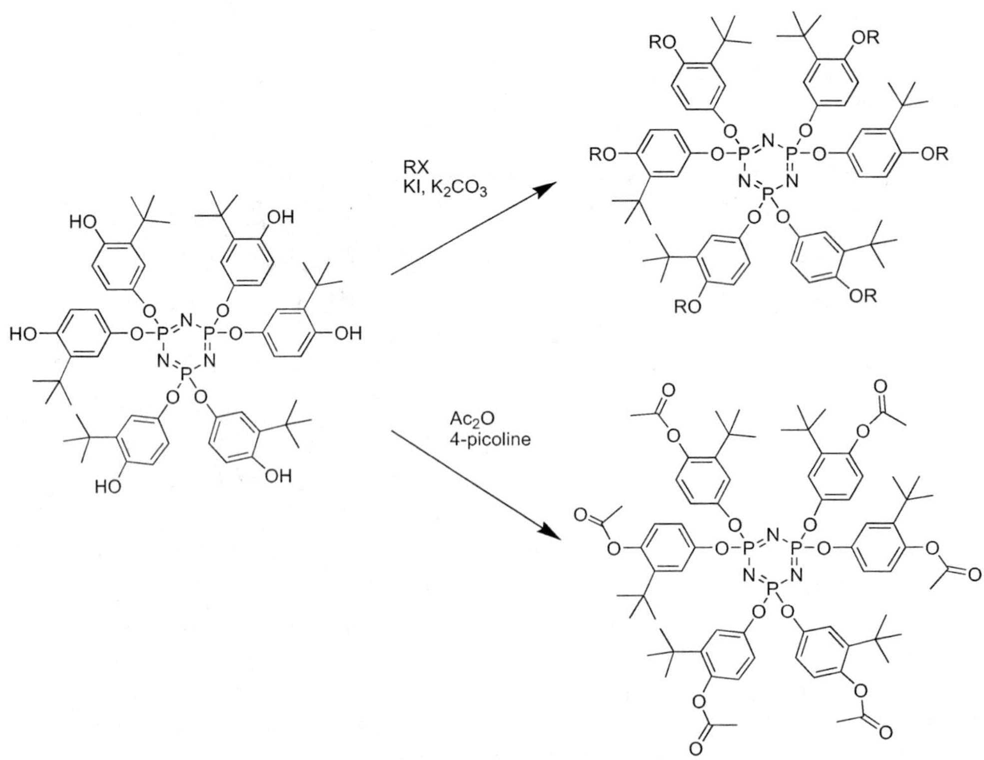

Figure 11. Reactions of hexakis-[(3-*tert*-butyl)(4-hydroxy)phenoxy]cyclotriphosphazene. R = n-pentyl and n-hexyl; X = Cl, Br, I.

Reaction times for these substitutions were dependent on the halide, where the alkyl iodide required 5 days at reflux and the corresponding chloride took up to three weeks. This reaction can be more accurately thought of as a substitution process on the alkyl group, where the halide was substituted for the oxo-nucleophile of the hexakis-[(3-*tert*-butyl)(4-hydroxy)phenoxy]cyclotriphosphazene. The oxo-nucleophile was created by deprotonation of the free hydroxyl upon exposure to base. In this case, the data supported an S_N2 mechanism for the substitution reaction on terminal carbon of the alkyl group. In another example of reactivity, acetylation was observed through reaction of hexakis-[(3-*tert*-butyl)(4-hydroxy)phenoxy]cyclotriphosphazene with acetic anhydride in the presence of 4-picoline.[47] ^{13}C NMR characterization of the acetylated product was unambiguous with the expected six aromatic resonances and the appearance of a signal at 169.4 ppm that was assigned to the acetyl carbonyl carbon. Thus, the reactivity of the free hydroxyl was demonstrated suggesting that the steric bulk imposed by the *tert*-butyl group may not be the only directing factor in the chemistry of the free hydroxyl. Formation of a phosphorus-oxygen bond at the hydroxyl *ortho* to the *tert*-butyl group also may be prevented by steric concerns imposed by the trimer itself.

Demonstration of the reactivity of the free hydroxyls towards electrophiles suggested that polymerization of trimer rings could be accomplished using commercially available bidentate acid chlorides. In this instance, acid chlorides connected by alkyl chains could serve as linkers between trimer rings. Hexakis-[(3-*tert*-butyl)(4-hydroxy)phenoxy]cyclotriphosphazene was found to readily react in 1,4-dioxane solvent with adipoyl chloride (C_6), succinyl chloride (C_4), and dodecanedioyl dichloride (C_{12}) forming soluble polymeric materials.[47] In general terms, the ^{13}C NMR spectra of the polymeric products for all three dichlorides were identical for the aromatic region suggesting ester formation consistent with the previously discussed acetylation chemistry. Additionally, two peaks were noted to be carbonyls. For example, the chemical shifts for the product formed using adipoyl chloride were 171.3 ppm and 174.0 ppm. The peak at 171.3 ppm was assigned to the ester carbonyl analogous to the acetylated product. Carboxylic acids typically fall between 170 ppm and 180 ppm, thus the peak at 174 ppm was assigned to a free carboxylic acid resulting from monodentate binding of the adipoyl chloride. With one end of the acid chloride attached through ester formation, the unattached end was converted to the acid form by hydrolysis during purification.

Molecular weights of the polymers were found to vary according to the stoichiometry of the two components. At a ratio of 1:1 (trimer:dodecanedioyl dichloride), the polymerized product was intractable and assumed to be high molecular weight material. At 1:2, soluble polymer materials were formed with $M_w = 1.2 \times 10^6$ and a polydispersity index (PDI) of 3.03. A PDI as high as 3.03 indicated that the product had a broad distribution of molecular weights, as expected from a non-linear polymerization that must yield branched materials. Polymerization was proposed to occur where there was a range of attachments to each of the trimer rings. A proposed structure is shown with a distribution of bridging linkers, terminal linkers (monodentate attachment) and unreacted hydroxyl sites (Figure 12).

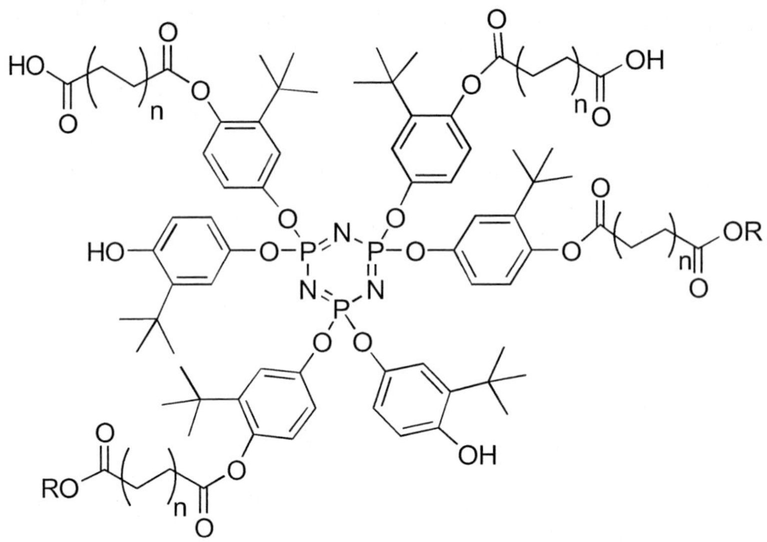

Figure 12. Proposed structure for the polymeric product of the reaction of hexakis-[(3-*tert*-butyl)(4-hydroxyphenoxy)]cyclotriphosphazene with dodecanedioyl dichloride. R = another cyclotriphosphazene ring and n = 1, 2, 5.

In some cases, trimer rings were terminal where only one bridging attachment was formed. In other cases, more attachments are envisioned. Thus, the structure of these polymers was described as a cyclomatrix where the properties were a function of both the cyclic trimer and the linker group. Differences in the physical characteristics could then be attributed to the linkers. The two polymers formed from the shorter linkers, adipoyl chloride and succinyl chloride, were glassy materials. Conversely, the polymers formed from the twelve-carbon dodecanedioyl dichloride were waxy, reflecting the aliphatic content of the polymer.

ACKNOWLEDGEMENTS

The work described in this paper was supported by the United States Department of Energy through Contract DE-AC07-99ID13727.

REFERENCES

1. Wijmans, J. G.; Baker, R. W. *J. Membr. Sci.* **1995**, *107*, 1-21.
2. Sun, Y. M.; Lin, C. L.; Chen, Y. K.; Wu, C. H. *J. Membr. Sci.* **1997**, *134*, 117-126.
3. Peterson, E. S.; Stone, M. L.; Cummings, D. G.; McCaffrey, R. R. *Sep. Sci. Technol.* **1993**, *28*, 271-281.
4. Peterson, E. S.; Stone, M. L.; Orme, C. J.; Stewart, F. F.; Cowan, R. L. *Sep. Sci. Technol.* **1997**, *32*, 541-556.
5. McCaffrey, R. R.; McAtee, R. E.; Grey, A. E.; Allen, C. A.; Cummings, D. G.; Appelhans, A. D. *J. Membr. Sci.* **1986**, *28*, 47-67.
6. Sun, Y. M.; Chen, Y. K.; Wu, C. H.; Lin, A. *AICHE J.* **1999**, *45*, 523-534.

7. Aptel, P.; Neel, J. In *Synthetic Membranes: Science, Engineering and Applications*; Series C: Mathematical and Physical Sciences Vol. 181 ed.; Bungay, P. M., Lonsdale, H. K., de Pinho, M. N., Eds.; D. Reidel Publishing Company: Dordrecht, 1986; pp 403-436.
8. Barrer, R. M. *Transactions of the Faraday Society* **1939**, *35*, 628.
9. Amerongen, G. J. v. *Journal of Applied Physics* **1946**, *17*, 972.
10. Robeson, L. M. *J. Membr. Sci.* **1991**, *62*, 165-185.
11. Freeman, B. D. *Macromolecules* **1999**, *32*, 375-380.
12. Alentiev, A.; Yampolskii, Y. P. *J. Membr. Sci.* **2000**, *165*, 201.
13. Drioli, E.; Zhang, S.-M.; Basile, A.; Golemme, G.; Gaeta, S. N.; Zhang, H.-C. *Gas Separation and Purification* **1991**, *5*, 252.
14. Golemme, G.; Drioli, E. *J. Inorg. Organomet. Polym.* **1996**, *6*, 341-365.
15. Hirose, T.; Kamiya, Y.; Mizoguchi, K. *J. Appl. Polym. Sci.* **1989**, *38*, 809-820.
16. Mizoguchi, K.; Kamiya, Y.; Hirose, T. *J. Polym. Sci. Pt. B-Polym. Phys.* **1991**, *29*, 695-703.
17. Kajiwara, M.; Kimura, T. *J. Mater. Sci.* **1996**, *31*, 5457-5462.
18. Kajiwara, M. *J. Mater. Sci.* **1994**, *29*, 6268-6272.
19. Kajiwara, M.; Kimura, T. *Inorganic and Organometallic Polymers II* **1994**, *572*, 268-278.
20. Kajiwara, M. *J. Mater. Sci.* **1994**, *29*, 4339-4341.
21. Kajiwara, M. *Polymer* **1989**, *30*, 1536-1539.
22. Nagai, K.; Freeman, B. D.; Cannon, A.; Allcock, H. R. *J. Membr. Sci.* **2000**, *172*, 167-176.
23. Allen, C. W.; Hneihen, A. S.; Peterson, E. S. U.S. Patent # 6,309,619, October 30, 2001.
24. Harrup, M. K. *Personal Communication* **2001**.
25. Allcock, H. R.; Best, R. J. *Can.J. Chem.* **1964**, *42*, 447-455.
26. Singler, R. E.; Gomba, F. J. In *Synthetic Lubricants and High-Performance Fluids*; Rudnick, L. R., Shubkin, R. L., Eds.; Marcel Dekker: New York, 1999; pp 297-311.
27. Singler, R. E.; Bieberich, M. J. In *Synthetic Lubricants and High-Performance Functional Fluids*; Shubkin, R. L., Ed.; Marcel Dekker: New York, 1992; pp 215-228.
28. De Jaeger, R.; Gleria, M. *Prog. Polym. Sci.* **1998**, *23*, 179-276.
29. Rice, R. G.; Geib, B. H.; Kaplan, L. A. U.S. Patent # 3,121,704, February 18, 1964.
30. Stone, M. L. *Material Research Society Symposium Proceedings* **1993**, *305*, 85-91.
31. Stewart, F. F.; Peterson, E. S.; Stone, M. L.; Singler, R. E. *Polymer Preprints* **1997**, *38*, 836.
32. Harrup, M. K.; Stewart, F. F. *J. Appl. Polym. Sci.* **2000**, *78*, 1092-1099.
33. Stewart, F. F.; Harrup, M. K.; Peterson, E. S.; Stone, M. L.; Orme, C. J.; Polson, L. A. In *Abstr. Pap. Am. Chem. Soc.*, 1999; Vol. 81, p 539.
34. Stewart, F. F.; Lash, R. P.; Singler, R. E. *Macromolecules* **1997**, *30*, 3229-3233.
35. Stewart, F. F.; Singler, R. E.; Harrup, M. K.; Peterson, E. S.; Lash, R. P. *J. Appl. Polym. Sci.* **2000**, *76*, 55-66.
36. Stewart, F. F.; Harrup, M. K.; Lash, R. P.; Tsang, M. N. *Polym. Int.* **2000**, *49*, 57-62.
37. Fox, T. G. *Bulletin of the American Physics Society* **1956**, *1*, 123.
38. Stewart, F. F.; Harrup, M. K.; Luther, T. A.; Orme, C. J.; Lash, R. P. *J. Appl. Polym. Sci.* **2001**, *80*, 422-431.
39. Orme, C. J.; Harrup, M. K.; Luther, T. A.; Lash, R. P.; Houston, K. S.; Weinkauf, D. H.; Stewart, F. F. *J. Membr. Sci.* **2001**, *186*, 249-256.

40. Orme, C. J.; Harrup, M. K.; McCoy, J. D.; Weinkauf, D. H.; Stewart, F. F. *J. Membr. Sci.* **2002**, In press.
41. Hansen, C. M.; Beerbower, A. In *Kirk-Othmer Encyclopedia of Chemical Technology, Supplemental Volume*; Standen, A., Ed.; Interscience: New York, 1971; p 889.
42. Barton, A. F. M. *Handbook of solubility parameters and other cohesion parameters*; 2nd ed.; CRC Press: Boca Raton, FL, 1991.
43. Femec, D. A.; McCaffrey, R. R. *J. Appl. Polym. Sci.* **1994**, *52*, 501-505.
44. Medici, A.; Fantin, G.; Pedrini, P.; Gleria, M.; Minto, F. *Macromolecules* **1992**, *25*, 2569-2574.
45. Stewart, F. F.; Harrup, M. K. *J. Appl. Polym. Sci.* **1999**, *72*, 1085-1090.
46. Luther, T. A.; Stewart, F. F.; Lash, R. P.; Wey, J. E.; Harrup, M. K. *J. Appl. Polym. Sci.* **2001**, *82*, 3439-3446.
47. Stewart, F. F.; Luther, T. A.; Harrup, M. K.; Lash, R. P. *J. Appl. Polym. Sci.* **2001**, *80*, 242-251.

In: Applicative Aspects of Poly(organophosphazenes)
Editors: M. Gleria and R. De Jaeger, pp. 225-254
ISBN 1-59454-025-X

Chapter 11

SULFONATED POLYPHOSPHAZENE MEMBRANES FOR DIRECT METHANOL FUEL CELLS

Peter N. Pintauro** and Ryszard Wycisk
Department of Chemical Engineering
Case Western Reserve University
Cleveland, OH 44106
*VOICE: 216-368-4150
FAX: 216-368-3016
E-mail: pnp3@po.cwru.edu

1. INTRODUCTION

Fuel cells are attracting increased attention as an alternative power source for electric vehicles, portable electronic devices, household equipment, and various military applications. A fuel cell, like a battery, provides dc electricity from spontaneous electrochemical reactions that occur at two electrodes. Whereas the fuel and oxidant in a battery are wholly contained within the confines of the device, fuel cells utilize outside air and a fuel that is stored separately and pumped into the device as needed. The capacity of a fuel cell to deliver power for extended periods of time is only restricted by the size of the fuel tank. Thus, it can be stated that, while a battery (either primary or secondary) is simply an energy storage device, a fuel cell should be treated as an energy conversion device.

Fuel cells are classified according to their electrolyte: proton-exchange membrane, alkaline, phosphoric acid, molten carbonate and solid oxide. Proton-exchange membrane (PEM) fuel cells operate at relatively low temperatures (at or below about $100^{o}C$), exhibit a high power density, and are well suited to a variety of consumer-oriented applications. At the present time, there are two common types of PEM fuel cells, which operate on either H_2 and air (oxygen) or methanol and air. The key component in both systems is a "membrane-electrode-assembly" (abbreviated as MEA) that consists of a thin proton conducting polymeric ion-exchange membrane coated on both sides with a sparse layer of precious metal catalyst powder. In the hydrogen/air fuel cell, humidified hydrogen gas is pumped past a platinum-on-carbon anode where the hydrogen is oxidized to H^+ and electrons. The protons migrate through the ion-exchange membrane and contact a Pt/C cathode where they combine with electrons and oxygen (from air) to form water. Useful work is extracted from the electrons that flow through the external circuit. Humidified hydrogen is needed to maintain the ion-exchange membrane in a hydrated, proton conducting state. Additional background information on this and other fuel cells can be obtained from a variety of literature references (see, for example, references 1-3).

In a direct methanol fuel cell (DMFC), methanol is oxidized in the presence of water at the anode to produce carbon dioxide, protons, and electrons. Protons migrate across the cation-exchange membrane to the cathode where they react with oxygen from air to produce water. The individual electrode reactions and electric potentials are:

Anode:	$CH_3OH + H_2O \rightarrow CO_2 + 6H^+ + 6e^-$	$E_o = 0.046$ V
Cathode:	$3/2O_2 + 6H^+ + 6e^- \rightarrow 3H_2O$	$E_o = 1.23$ V

The overall reaction and cell potential are:

$$CH_3OH + 3/2O_2 + H_2O \rightarrow CO_2 + 3H_2O \qquad E_o = 1.18 \text{ V}$$

A schematic diagram of direct methanol fuel cell is shown in Figure 1. A Pt/Ru catalyst anode is typically employed (the presence of Ru reduces poisoning by CO during CH_3OH oxidation). The anode feed is usually a 0.5 or 1.0 M aqueous methanol solution. Direct methanol fuel cells have not yet been commercialized because of two, as yet unsolved, technological problems: (1) the need for a more active anode catalysts that is not poisoned by CO and (2) the need for an ion-exchange membrane that possesses a high proton conductivity, yet minimizes methanol cross-over (leakage) from the anode to the cathode. Methanol cross-over is particularly troublesome because it leads to: (1) cathode depolarization (i.e., a loss of fuel cell power) when methanol contacts the cathode catalyst and is oxidized chemically by O_2, (2) excessive water production and possible electrode flooding when methanol is chemically oxidized at the air cathode, and (3) a reduction in the efficiency of methanol fuel utilization due to the loss of methanol from the anode feed stream.

For a proton-exchange membrane to be used in a PEM fuel cell, it must possess the following properties: (i) a high protonic conductivity (> 0.01 S/cm) with zero electronic conductivity, (ii) low oxygen and fuel (either H_2 or methanol) permeability, (iii) dimensional stability (low swelling), (iv) high mechanical strength, (v) minimal water transport, and (vi) a high resistance to oxidation, reduction, and hydrolysis (4-6). Studies of fuel cell membranes have focused, for the most part, on perfluorosulfonic acid membranes, in particular Dupont's Nafion®, because it is highly stable to chemical degradation and possess nearly all of the required properties. In fact, Nafion is the membrane of choice for H_2/air PEM fuel cells. Unfortunately, there is unacceptably high methanol cross-over when Nafion is used in methanol fuel cells.

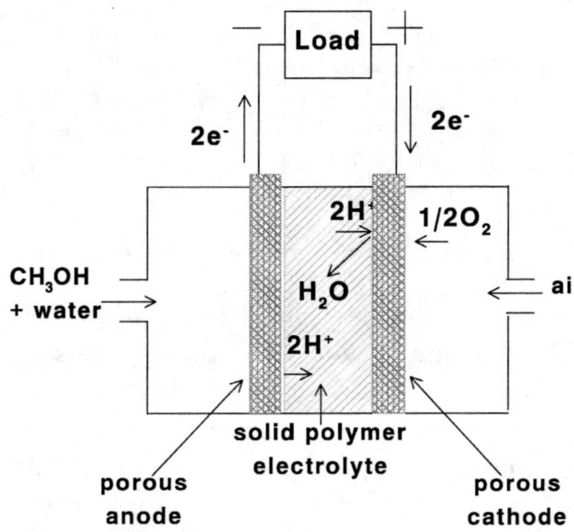

Figure 1 – A Schematic Diagram of a Direct Methanol PEM Fuel Cell

Polymer scientists and electrochemists have recognized the need to investigate new polymers to replace Nafion in direct methanol PEM fuel cells. New membranes materials that have been or are currently under investigation include sulfonated polyetherketone (7,8), sulfonated polysulfones and polyethersulfones (9), sulfonated polyimides (10), and phosphoric acid-doped polybenzimidazole (PBI) (11). Of these membranes, PBI appears to be the most promising, with a high electrical conductivity at elevated temperatures (0.0025-0.04 S/cm at 130-190°C) and methanol crossover rates ten or more times less than that in Nafion. Acid-doped PBI membranes, however, can only be used with methanol vapor feeds; if liquid water or methanol were to contact the membrane, phosphoric acid will leach from the film with a subsequent loss in proton conductivity.

1.1 Improving the Performance of Direct Liquid Methanol PEM Fuel Cells with Polyphosphazenes

Polyphosphazenes and especially polyorganophosphazenes offer new possibilities as potential membrane materials for fuel cells. These polymers should be regarded as hybrid polymers because their backbone is inorganic while the side groups are of organic nature.

From a synthetic viewpoint, polyphosphazenes (whose basic structure is shown in Figure 2) are the most highly developed of all the inorganic-backbone polymer systems (12,13). With appropriate functionalization of the basic poly(dichlorophosphazene) backbone (R_1 and R_2 in Figure 2), an unlimited number of specialty polymers can be synthesized. Unlike hydrocarbon polymers, polyphosphazenes are generally resistant to both thermal and chemical degradation (13), which makes them a viable alternative to fluoropolymer materials.

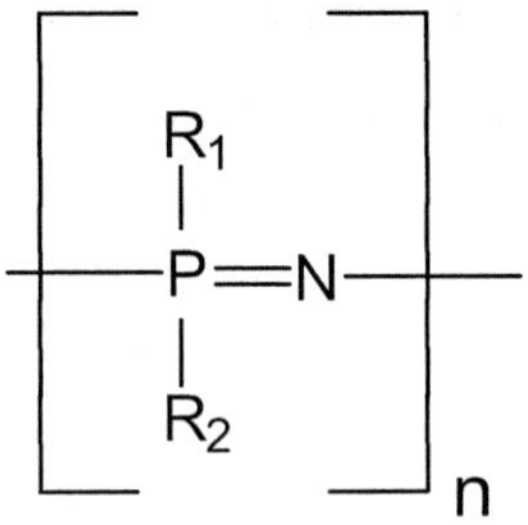

Figure 2– The repeating monomer unit of polyphosphazene

The use of polyphosphazene membranes in electrochemical devices is not entirely new. Shriver and co-workers, in collaboration with H. Allcock, carried out studies of polyphosphazenes doped with lithium salts as solvent-free polymer electrolytes for solid-state lithium batteries (14-16). Here the phosphazene polymer itself acts as the solvent through which anions and lithium cations migrate when subjected to an electric field or concentration gradient (as compared to PEM films where water or a concentrated acid electrolyte is contained in the membrane). Other examples of the use of polyphosphazenes as membrane materials include radiation crosslinked films containing hydrophilic $NHCH_3$ and hydrophobic OCH_2CF_3 or OC_6H_5 side chains for biomedical applications (17) and uncrosslinked poly[bis(phenoxy)phosphazene] membranes for the separation of mixed gases (e.g., H_2S and CH_4) and the selective removal of low concentrations of organic pollutants (e.g., methylene chloride) from water by pervaporation (18,19).

In order to create a polyphosphazene-based proton-exchange membrane, acidic functional groups (e.g., sulfonic acid groups) must be incorporated into the polymer. There are limitations, however, to the proton-exchange group concentration in the film, imposed by both the polymer chemistry and osmotic stability of the membrane (i.e., if the concentration of fixed charges were too high, the polymer would solubilize in water). To control swelling and possible dissolution in water, the polymer can be either crosslinked or blended with an uncharged, hydrophobic polymer that will not swell in polar solvents.

In this chapter, methods for preparing and characterizing, sulfonated polyphosphazene-based membranes for direct liquid methanol PEM fuel cells are described. The goal of this work was to fabricate films with: (i) A high ion-exchange capacity and (ii) an optimized degree of crosslinking for high ionic conductivity, low swelling, and low methanol cross-over. The preparation and properties of films composed of a crosslinked and sulfonated polyphosphazene that is blended with either poly(vinylidene fluoride) or polyacrylonitrile are also described. Here, the presence of the inert, hydrophobic component contributes to the modification of both the mechanical and transport properties of the membrane.

2. Membrane Preparation

There are two synthesis routes leading to water insoluble sulfonic acid membranes from polyphosphazene polymers. The first method involves crosslinking a polyphosphazene film, followed by heterogeneous sulfonation. In the second technique, homogeneous or heterogeneous polymer sulfonation is carried out first, followed by film casting and crosslinking. Both methods require a polymer with sidegroups that can be used for crosslinking. The idea of sulfonating polyaryloxyphosphazenes is not new; three research groups have synthesized successfully sulfonic acid polyphosphazenes, although none of the polymers was examined as a membrane material. Montoneri *et al.* (20,21) sulfonated poly[bis(phenoxy)phosphazene] in dichloroethane with SO_3. Allcock and coworkers (13,22,23) sulfonated various aryloxy- and arylaminophosphazenes with sulfuric acid (either concentrated or fuming). Depending on the type of polyphosphazene polymer and the reaction conditions, various sulfonation degrees were obtained, with the reaction taking place primarily at the *para*-position. Kurachi and Kajiwara (24) prepared poly[(anilinosulfamic)phosphazenes] by the reaction of poly(anilinophosphazenes) and chlorosulfonic acid in tetrachloroethane.

2.1 Polyphosphazene Sulfonation

Pintauro and co-workers (25,26) investigated the sulfonation of five alkylphenoxy-substituted polyphosphazenes with sulfur trioxide. The starting polymers were poly[(3-methylphenoxy)(phenoxy)\phosphazene], poly[(4-methylphenoxy)(phenoxy)phosphazene], poly[(3-ethylphenoxy)(phenoxy)phosphazene], poly[(4-ethylphenoxy)(phenoxy) phosphazene] and poly[bis(3-methylphenoxy)phosphazene]. The chemical structure of these compounds is shown in Figure 3. The molecular weight of these polymers was in the range of 1.6-5.0 x 10^6 and their glass transition temperature was between –24.2°C and –1.5°C.

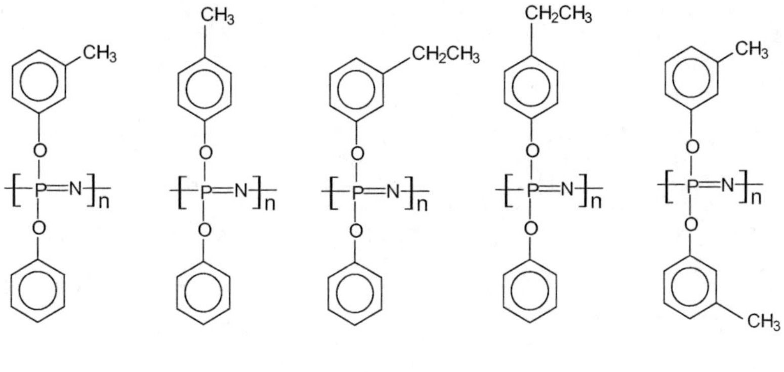

Figure 3 – The chemical structure of: A - poly[(3-methylphenoxy)(phenoxy)phosphazene], B - poly[(4-methylphenoxy)(phenoxy)phosphazene], C - poly[(3-ethylphenoxy)(phenoxy) phosphazene], D - poly[(4-ethylphenoxy)(phenoxy)phosphazene], and E - poly[bis(3-methylphenoxy)phosphazene]

To sulfonate the polyphosphazenes, a known weight of polymer (1-2 grams) was dissolved in 40-80 ml of 1,2-dichloroethane (DCE). A given amount of SO_3 in DCE was then added to the polymer solution at 0°C in a dry nitrogen atmosphere. The resulting precipitate was stirred for ca. 3 hours, followed by the addition of NaOH (in a water/methanol solvent) to terminate the reaction. During the sulfonation of poly[bis(3-methylphenoxy)phosphazene] with SO_3, aromatic carbon sulfonation and the formation of a $\equiv N \rightarrow SO_3$ backbone complex occurred, as shown by the reaction scheme in Figure 4.

After solvent evaporation, the sulfonated polymer was preconditioned by soaking sequentially in water (to wash out salt and hydrolyze $\equiv N \rightarrow SO_3$ complex), 0.1 M NaOH (to neutralize $\equiv N^{+}H \cdots HSO_4^{-}$ which is produced by the hydrolysis of $\equiv N \rightarrow SO_3$), distilled water, 0.1 M HCl (to insure that the sulfonate sites are in the H^+ form), and finally distilled water. The polymer product was then dried thoroughly and dissolved in N,N-dimethylacetamide (DMAc). Membranes were cast from this solution onto a polypropylene or Teflon™ plate and dried at 70°C for 24 hours.

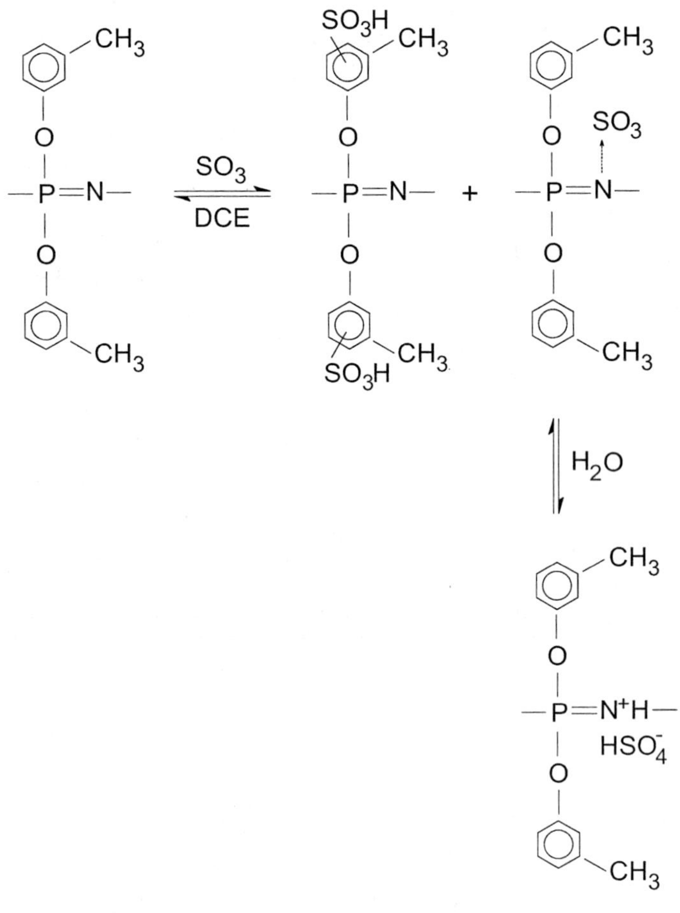

Figure 4 – The reaction scheme for sulfonation of poly[bis(3-methylphenoxy)phosphazene] with SO_3.

The polyphosphazene most studied for sulfonation and membrane fabrication was poly[bis(3-methylphenoxy)phosphazene], henceforth denoted as POP. The measured ion-exchange capacity (IEC) of sulfonated POP is plotted vs. the SO_3/monomer molar ratio in Figure 5. A gradual increase in IEC with sulfur trioxide addition was observed for SO_3/POP ratios > 0.64 (during the initial stages of SO_3 addition, there was no polymer sulfonation, but rather the formation of SO_3-nitrogen complex). Water insoluble polymers were prepared with ion-exchange capacities as high as 2.1 mmol/g (for comparison purposes, the IEC of a Nafion 117 membrane is 0.909 mmol/g). No polymer degradation was observed during sulfonation and the sulfonation yield was as high as 44% at an IEC of 2.1 mmol/g (26).

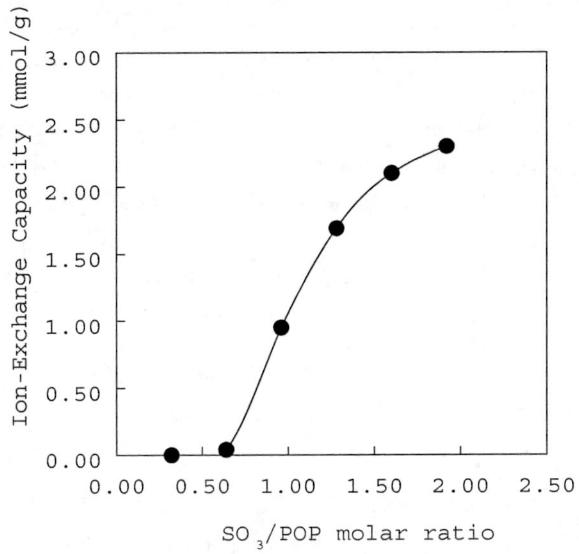

Figure 5 – Ion-exchange capacity of sulfonated poly[bis(3-methylphenoxy)phosphazene] as a function of the SO_3 monomer molar ratio.

The location of SO_3 attack on the methylphenoxy rings of sulfonated poly[bis(3-methylphenoxy)phosphazene], as determined by solid-state NMR, was found to be in the *para*-position when the sulfonation degree was low, and was multi-positioned when the sulfonation degree was high (26). The sulfonated bis-methylphenoxy polymer was thermally stable up to 350°C in dry air; above this temperature the polymer's main chains and aromatic rings decomposed. The non-sulfonated base polymer was thermally stable up to 420°C in dry air.

Sulfonation results similar to those in Figure 5 were obtained with poly[(3-methylphenoxy)(phenoxy)phosphazene and poly[(4-methylphenoxy)(phenoxy)phosphazene (25). The functional dependence of IEC on the SO_3/monomer ratio, however, was found to change somewhat with polymer molecular weight (the data in Figure 5 was collected with a 2.0 x 10^6 MW polymer). Ethylphenoxy-substituted polyphosphazenes could not be sulfonated with SO_3, but rather underwent severe degradation (25). The larger ethyl groups on the phenoxy sidechains (as compared to methyl moieties) may have sterically blocked sites for aromatic ring sulfonation, thus allowing SO_3 to degrade the SPOP backbone, but the exact cause of our inability to sulfonate ethylphenoxy polymers with sulfur trioxide is not well understood at this time.

The glass transition temperature of poly[bis(3-methylphenoxy)phosphazene] varied from –28°C for the base polymer to about 90°C when the IEC was 2.1 mmol/g. The change in T_g was attributed to electrostatic interactions between fixed-charge sulfonic acid sites and stiffening of the polymer backbone. Wide-angle X-ray diffractograms of poly[bis(3-methylphenoxy)phosphazene] prior to and after sulfonation (with an IEC of 0.4 and 0.6 mmol/g) are shown in Figure 6. The base polymer was found to be semi-crystalline, with two main reflection peaks at 12.72 Å and 11.34 Å and several other reflections centered between 8.08 Å and 3.58 Å. For the sulfonated polymer with an IEC 0.4 mmol/g, all of the base-polymer diffraction peaks were preserved, although the intensity of the peaks between 8.08 and 3.58 was reduced, indicating that the main crystal structure of the phosphazene polymer was retained. At a higher IEC (0.6 mmol/g), all diffraction peaks between 8.08 Å and 3.58 Å smeared out into a amorphous halo centered at 4.77 Å, but the main peak centered at 11.85 Å was still present (a similar result was found for sulfonated polymers with an IEC > 0.6 mmol/g). This data indicates that the polymer lost its three-dimensional crystal structure when the IEC was ≥ 0.6, but a two dimensional (2-D) ordered phase was retained, as evidenced by the X-ray peak at 11.85 Å (which has been identified as an intermolecular reflection peak).

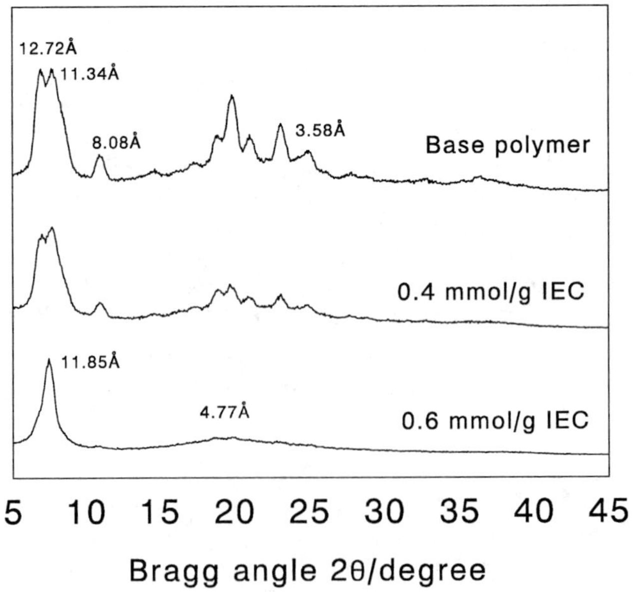

Figure 6 – Wide-angle x-ray diffractograms of non-sulfonated and sulfonated poly[bis(3-methylphenoxy)phosphazene]

The equilibrium swelling of sulfonated poly[bis(3-methylphenoxy)phosphazene] films in water at 25°C was found to be dependent on the polymer's ion-exchange capacity, as illustrated by the data in Table 1. Water swelling was quantified in terms of the percent increase in dry membrane weight, according to the following formula,

$$\%\ swelling = \frac{m_{wet} - m_{dry}}{m_{dry}} x100 \qquad (1)$$

where m_{wet} and m_{dry} are the weights of the water-swollen and dry films, respectively (all films in the H^+ form). The base polymer prior to exposure to SO_3 was highly hydrophobic and did not absorb water. Once the polyphosphazene contacted SO_3 (even at a low SO_3/POP molar ratio), water uptake was observed, due to the presence of $\equiv N^+H\cdots HSO_4^-$ species and a small number of sulfonic acid sites. As expected, the extent of polymer swelling increased with increasing IEC. Membranes prepared from the sulfonated polymers with a SO_3/POP ratio of 1.92 (an IEC ≥ 2.1 mmol/g) were found to be water soluble. For comparison purposes, a sulfonated poly[bis(3-methylphenoxy)phosphazene] membrane with an ion-exchange capacity of 0.95 mmol/g swelled 42% at room temperature, whereas a Nafion® 117 perfluorosulfonic acid membrane (with an IEC of 0.909 mmol/g) swelled 34% (27). Thus, the use of high IEC polyphosphazene membranes will require crosslinking and/or blending to limit swelling and prevent the loss of mechanical properties, as will be discussed in the following section.

Table 1 – Equilibrium water swelling of sulfonated poly[bis(3-methylphenoxy)phosphazene] membranes at 25°C.

SO_3/POP Molar Ratio	Membrane Ion-Exchange Capacity (mmol/g)	Membrane Swelling (% dry membrane weight)
0.64	0.04	5
0.80	0.80	25
0.96	0.95	42
1.28	1.69	230
1.6	2.1	900

The microstructure of sulfonated poly[bis(3-methyphenoxy)phosphazene] (abbreviated as SPOP) was studied using small-angle X-ray scattering (SAXS) (28). A reflection peak, attributed to the presence of ionic clusters, was observed in the small-angle X-ray diffraction patterns from hydrated and dry polymers when the IEC was ≥ 0.6mmol/g. The ionic cluster Bragg spacing was about 30 Å for non-hydrated (dry) polymers in the H^+ form with ion-exchange capacities in the range of 0.8 - 1.6 mmol/g (see Figure 7). When the ion-exchange capacity was < 0.6 mmol/g, no reflection peak was observed. SAXS scans of fully hydrated SPOP films with different IECs in the H^+ form are shown in Figure 8. Unlike the results from dry polymer samples, the X-ray scattering maximum of hydrated films shifted towards smaller angles and the scattering intensity increased with increasing IEC. As can be seen by the data in Table 1, water uptake in sulfonated poly[bis(3-methylphenoxy)phosphazene] membranes increased with IEC and the Bragg spacing data in Figure 8 suggest that this water absorbs into and expands the ionic clusters. At the same time, the increase in water content causes the electron density of the clusters to decrease to a greater extent than that of the surrounding uncharged polymer medium, which resulted in the observed increase in reflection intensity with IEC.

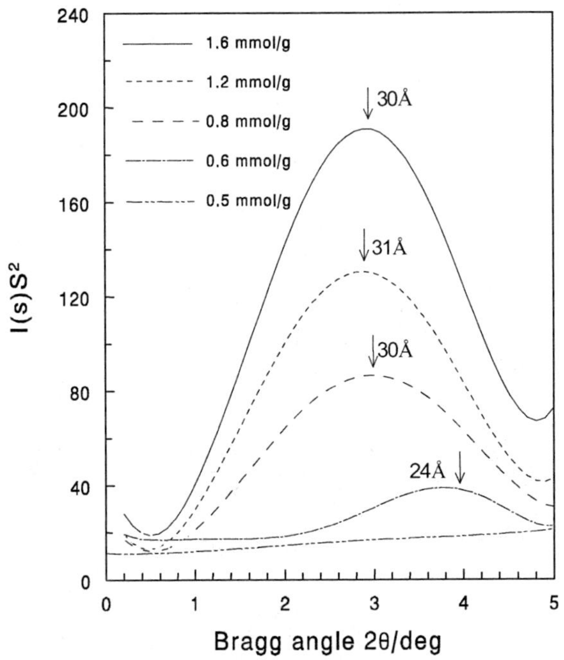

Figure 7 – Small-angle invariant X-ray scans of sulfonated poly[bis(3-methylphenoxy)phosphazene] in dry (non-hydrated) film form at different polymer ion-exchange capacities.

2.2 Polyphosphazene Crosslinking

Although the non-sulfonated form of poly[bis(3-methylphenoxy)phosphazene] is highly hydrophobic, there can be considerable water sorption into the polymer after addition of sulfonic acid fixed-charge groups, as evidenced by the data in Table 1. Some water is needed in the polymer, in order to insure dissociation and migration of protons during current passage when a SPOP membrane is used in a fuel cell. A high degree of polymer swelling, however, is undesirable because it weakens the mechanical properties of the membrane and increases the wet membrane porosity (which allows for higher methanol diffusion rates). Without polymer crosslinking, membranes can only be prepared with a sulfonated polymer that is water insoluble. With crosslinking, one can utilize higher ion-exchange capacity sulfonated polyphosphazenes that are water soluble in their non-crosslinked form, but do not dissolve when crosslinked.

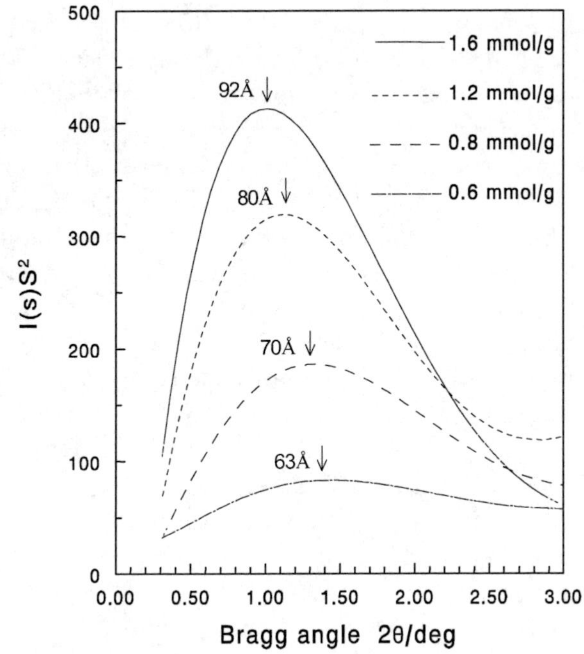

Figure 8 – Effect of ion-exchange capacity on the small-angle invariant X-ray scans for hydrated poly[bis(3-methylphenoxy)phosphazene] films in the H^+ form.

2.2.1 Crosslinking of Non-Sulfonated Polymers

For non-sulfonated alkylphenoxy-substituted polyphosphazenes, polymer crosslinking was accomplished using UV-radiation in the presence of a photo-initiator or electron-beam radiation. Extensive and fundamental UV crosslinking studies were performed on non-sulfonated poly[(methylphenoxy)(phenoxy)phosphazene], poly[(ethylphenoxy)(phenoxy) phosphazene], and poly[(isopropylphenoxy)(phenoxy)phosphazene] (29,30). Most of the work focused on the use of benzophenone as a UV light photoinitiator, which was soluble in the polyphosphazene membrane casting solution, remained dissolved in a polymer film after solvent evaporation, and exhibited a high rate of hydrogen abstraction during UV light exposure. The UV crosslinking mechanism of poly[(3-methylphenoxy)(phenoxy) phosphazene] with benzophenone (henceforth abbreviated as BP) is shown in Figure 9. When BP is irradiated with UV light of wavelength 340 nm, it is excited to a singlet state, which by intersystem crossing reverts to a reactive triplet state. This state can be considered as a 1,2-diradical, which can abstract a hydrogen from the benzylic carbon of POP, creating a macro-radical and a ketyl radical. Recombination of the macro-radicals leads to the formation of a C-C crosslink, while the ketyl radicals recombine to give pinacol. A similar mechanism holds for electron-beam crosslinking, but polymer macro-radicals are created solely by ionizing radiation, without the need for a photoinitiator.

In preliminary experiments, poly[(4-ethylphenoxy)(phenoxy)phosphazene] was crosslinked with different concentrations of benzophenone (29). The glass transition temperature of the polymer increased from –8.8°C to 53.5°C and the equilibrium swelling of polymer films in dimethylacetamide (the original membrane casting solvent) decreased from infinity to 31% as the BP/polyphosphazene molar ratio was increased from 0 to 0.5 (29). Both of these observations indicate the formation of polymer crosslinks. In a later study (30), a series of alkylphenoxy/phenoxy-substituted polymers were crosslinked with BP and UV light, where the alkyl chain was either methyl, ethyl, or isopropyl. The purpose of this study was to assess the relative ease of hydrogen abstraction from the alkylphenoxy side groups, which should be in the order isopropyl > ethyl > methyl, vs. the ease for crosslink formation, which was expected to follow the opposite order due to steric effects (31). The ordering of benzophenone disappearance during exposure of dry polymer films to UV light was methylphenoxy > ethylphenoxy > isopropylphenoxy, indicating that the rate controlling step for benzophenone disappearance was the consumption of the ketyl radical, either by benzopinacole formation (with the creation of a polymer crosslink) or by the reaction of the ketyl radical with the polymer macro-radical to form a ketyl adduct. The presence of ketyl adducts in ethylphenoxy/phenoxy and isopropylphenoxy/phenoxy phosphazene films was verified by solid-state NMR. The adduct was not observed in NMR spectra of crosslinked methylphenoxy-substituted polymers. The ordering of polymer swelling (for a given initial benzophenone concentration) when crosslinked polyphosphazene films were equilibrated in DMAc was isopropylphenoxy/phenoxy > ethylphenoxy/phenoxy > methylphenoxy/phenoxy. Within experimental accuracy, swelling of the methylphenoxy polymers was independent of both the irradiation temperature (25°C or 70°C) and the position of the methyl substituent on the phenoxy ring (*meta* vs *para*). Polymers containing 4-ethylphenoxy or 4-isopropylphenoxy side chains underwent some degradation during exposure to UV light and swelled more in DMAc than the corresponding polyphosphazenes with the alkyl group in the *meta* position. Additional evidence of crosslink formation was provided by DSC measurements, where the glass transition temperature of all crosslinked phosphazene polymers increased with increasing benzophenone concentration.

From the sum total of all crosslinking data collected, it was concluded that the 3-methylphenoxy/phenoxy or 4-methylphenoxy/phenoxy phosphazene was the best polymer for UV photo-crosslinking with benzophenone. The smaller and less sterically hindered methyl substituent was superior to the bulkier ethyl and isopropyl moieties on the polymer's phenoxy side chains, in terms of facilitating the carbon-carbon macro-radical crosslinking reaction, with minimal ketyl adduct formation. While an increase in temperature during UV irradiation was advantageous from the standpoint of increasing the reactivity of the photo-initiator and polymer radicals, it did not appear to produce a higher degree of crosslinking nor did it result in increased amounts of ketyl adduct in the methylphenoxy-substituted phosphazenes.

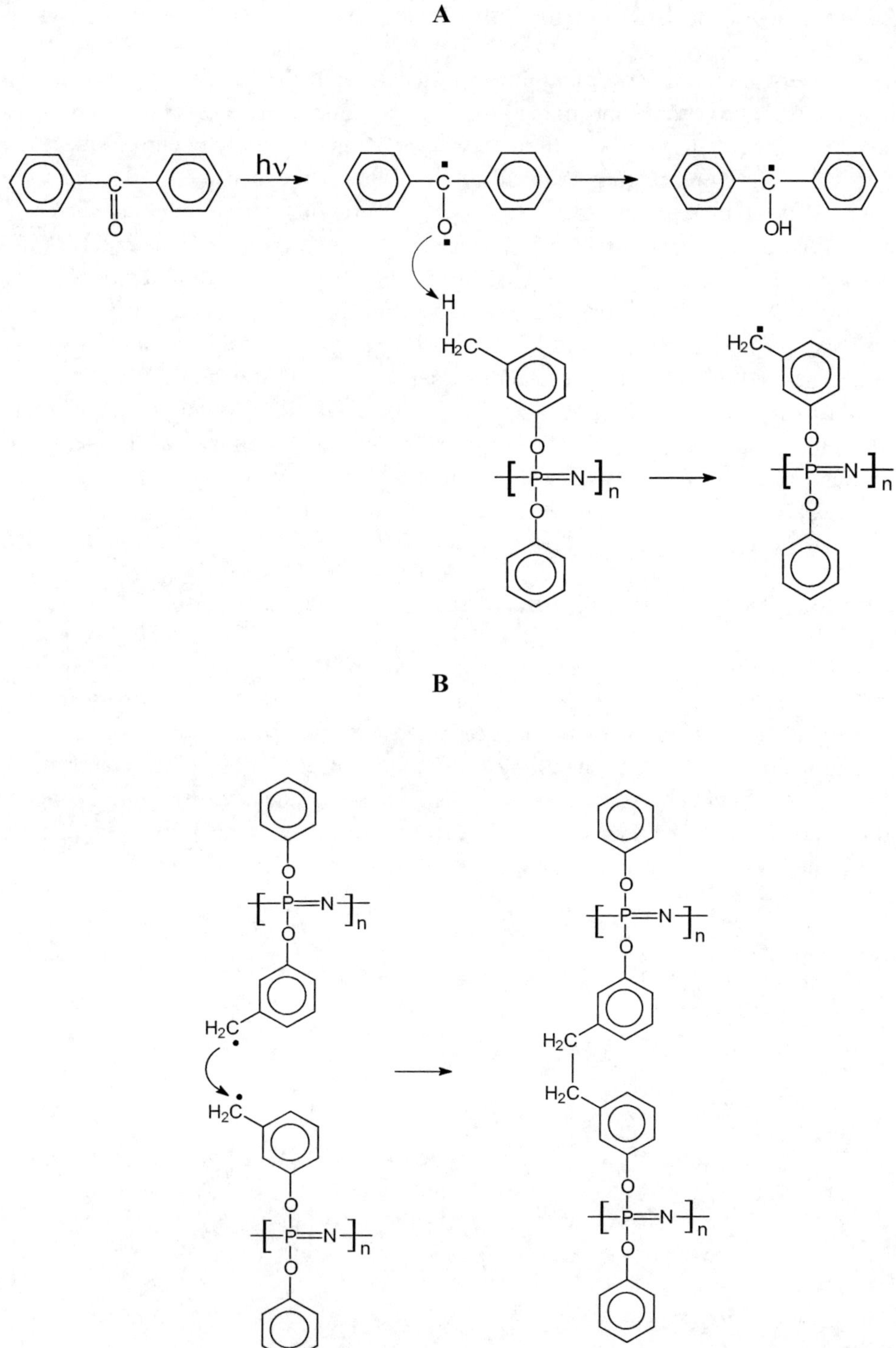

Figure 9 – UV Photo-crosslinking mechanism for sulfonated poly[bis(3-methylphenoxy)phosphazene]: (A) Generation of macro-radicals and (B) Recombination of macro-radicals.

2.2.2 Crosslinking of Sulfonated Polyphosphazenes

The successful crosslinking of an unsulfonated polyphosphazene with benzophenone and UV light does not necessarily mean that a similar procedure can be used to form crosslinks in a sulfonated polymer. The bulky sulfonic acid groups on the methylphenoxy sidechains may interfere sterically with the formation of C-C crosslinks between methyl groups. Also, the reduction in polymer chain flexibility with increasing IEC may adversely affect recombination of the macro-radicals during UV irradiation. For moderately sulfonated polyphosphazenes, such as a 1.4 mmol/g IEC SPOP, the glass transition temperature is close to room temperature and there are numerous unsulfonated methylphenoxy sidechains. Such a polymer was crosslinked in dry film form with 365 nm UV light and benzophenone concentrations of 1-10 wt%. The results of these experiments are shown in Figure 10, where the equilibrium water swelling at 25°C of crosslinked films is plotted against the BP photoinitiator concentration. The reduction in water swelling with increasing BP concentration is evidence of crosslink formation. Unfortunately, even at 10% benzophenone, polymer swelling may be too high (≈80%) for many membrane applications.

2.3 Blending and Crosslinking Sulfonated Polyphosphazenes

The data in Figure 10 suggest that crosslinking alone may not be sufficient to control (i.e., lower) the swelling of a 1.4 IEC SPOP membrane in water. We also observed that highly sulfonated and crosslinked polyphosphazenes were brittle in the dry state, which could cause serious problems with film handling, membrane-electrode-assembly fabrication, and installation into a fuel cell. These results led us to consider another method to decrease water absorption and improve the mechanical properties of dry membranes, that being the blending of a sulfonated polyphosphazene with a non-sulfonated semi-crystalline polymer followed by UV crosslinking.

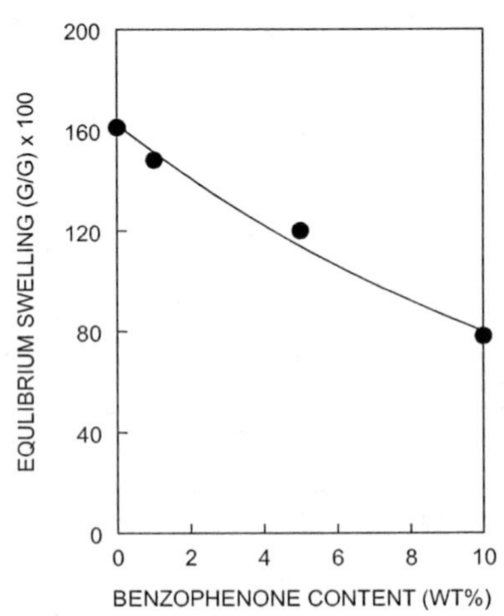

Figure 10 – Effect of benzophenone content on the equilibrium water swelling (at 25°C) of a UV-crosslinked sulfonated poly[bis(3-methylphenoxy)phosphazene] membrane (1.4 mmol/g IEC). Irradiation conditions: 3h at 15 mW/cm^2.

Blending of polymers is always associated with the problem of compatibility of the components (32). In most cases it is not possible to achieve a true, molecular level miscibility, because the entropic contribution to the free energy of mixing is small and dispersive or Van der Waals interactions lead to positive values of the enthalpy of mixing. The resultant free energy of mixing is positive and from a thermodynamic standpoint immiscibility should be expected. Thus, in general, solution casting results in blends of domain-type morphology. If the domain size of the minor component is in the nanometer range, then the polymers are regarded to be compatible and the resultant blend usually possesses the expected properties. This kind of blend is obtained when there are some specific interactions between the polymers (33). If there are no specific interactions, the domain size is usually on the order of several to tens of micrometers. The mechanical properties of such incompatible blends may be worse than those predicted by a combination of the pure components' properties.

When blending ionic and non-ionic polymers, the situation is even more complex. The ionic groups tend to form multiplets and clusters that make mixing with the uncharged component difficult. Thus, in order to obtain a blend with very fine domains, either the ionic fixed-charge sites must be effectively masked or the polymeric blending component should exhibit some level of mutual ionic interactions. For fuel cell applications, it is also required that the neutral polymer component of the blend be chemically (oxidatively) stable. There is, however, a potential complication with the blending approach for fuel cell ion-exchange membrane fabrication, that being the dilution of the electroactive polymer. In order to achieve an acceptable electrochemical performance (e.g., protonic conductivity) with a blended membrane, it may be necessary to use a sulfonated polymer with a very high ion-exchange capacity. For example, if the desired effective IEC of a blended membrane were to be 1.2 mmol/g and the inert polymer content were 50%, then the required IEC of the sulfonated component would need to be 2.4 mmol/g. Such a high IEC polymer may be water soluble and the blended membrane might degrade in aqueous solutions. Therefore, special care must be taken to either crosslink the sulfonated component of the blend or graft both polymers in order to ensure membrane stability.

Literature reports related to blended proton-exchange membranes for fuel cells can be divided into three groups: (1) interpenetrating polymer networks (IPN) of sulfonated polystyrene and various fluoropolymers (34-37), (2) acid-base type blends of selected sulfonated polyarylenes with basic polymers, like polybenzimidazole, polyethylene imine or aminated polysulfone (8,38) and, (3) blends containing Nafion (39,40).

Our approach to polyphosphazene membrane preparation was to blend a sulfonated and crosslinkable polyphosphazene with a mechanically tough and chemically resistant polymer, and then crosslink the polyphosphazene component with UV radiation. The most promising results have been obtained when sulfonated poly[bis(3-methylphenoxy)phosphazene] (SPOP) was blended with either a PVDF copolymer (Kynar® Flex) or polyacrylonitrile (PAN). To prepare a blended and UV-crosslinked SPOP membrane with Kynar Flex or PAN, a two-step procedure was used. First, all three components: SPOP, Kynar Flex (or PAN) and benzophenone were dissolved in dimethylacetamide (DMAc) and cast into a membrane. In the second step, the dry blend was irradiated under an argon atmosphere with a UV lamp of wavelength 365 nm. The membrane was then soaked with 1M HCl and finally washed with distilled water several times over 24 hours.

From preliminary experiments (water swelling and proton conductivity) it was found that the useful effective ion-exchange capacity of a blended SPOP membrane should be about

0.90 - 1.2 mmol/g. This capacity was achieved by: (1) blending SPOP of moderate IEC (1.4-2.2 mmol/g) with Kynar Flex or PAN as the minor component or (2) blending SPOP of high IEC (2.5-4.0 mmol/g) with Kynar Flex or PAN as the major component.

Initial studies were directed towards blending SPOP with Kynar Flex. It was found that Kynar Flex was generally immiscible on a molecular level with the SPOP, but there was some degree of compatibility and blended films (especially those containing high IEC SPOP) showed no signs of macroscopic phase separation for a relatively wide composition range. Three groups of SPOP/Flex blended membranes were synthesized, with an effective IEC in the range of 0.9-1.2 mmol/g: (1) ≈70% 1.6 mmol/g SPOP + Flex, (2) ≈50% 2.1 mmol/g SPOP + Flex and (3) ≈30% 3.5 mmol/g SPOP + Flex. All of the membranes contained 5% benzophenone and each film was crosslinked by exposure to UV light. The mechanical properties of membranes belonging to group 1 were not acceptable and macroscopic phase separation was frequently observed. Membranes belonging to group 2 had good mechanical properties, with no phase separation. Proton conductivities of films in water at 25°C ranged from 0.01 S/cm to 0.08 S/cm. There was no macroscopic phase separation with membranes belonging to group 3, but the membranes lost SPOP immediately after immersing in water. This was an indication of insufficient UV-crosslinking of the high-IEC SPOP component.

Most of our research was centered on membranes belonging to group 2. Figure 11 shows SEM micrographs of the cross-sections of three SPOP/Flex membranes, where the counterion of the sulfonic acid groups in the SPOP component was either tetrabutylammonium (TBA), Na^+, or Cs^+ (counterion substitution was carried out prior to blending and film casting). To obtain SEMS, films were manually fractured after reaching thermal equilibrium in liquid nitrogen. Specimens were sputter coated with gold to a final thickness of approximately 2 nm and imaged on a JEOL JSM-820 scanning electron microscope at 15 kV. It can be seen that significant phase separation was observed in blends of Flex and SPOP in the Cs^+ form. The best compatibility was achieved when SPOP was in the TBA form. Unfortunately, it was impossible to crosslink these blends, presumably due to steric interference by the bulky TBA cation. Consequently, all of our follow-on studies of blended membranes were performed with SPOP in the Na-form (which could be crosslinked and which exhibited minimal phase separation).

The results of tensile tests performed on UV-crosslinked membranes composed of SPOP (Na-form) and Flex are presented in Table 2. These tensile measurements were performed on an Instron 5567. Specimens were tested in a water-swollen state at 25°C. The extension rate was 50 mm/s. Based on the stress-strain curve, the ultimate strength and elongation at break were determined. As expected, the addition of Kynar Flex into a sulfonated polyphosphazene film led to a significant improvement in mechanical properties. The tensile strength increased from 3.4 to 6.4 MPa and the extension at break increases from 150 % to 186 % when the Kynar Flex content was increased from 40 wt% to 60 wt.%.

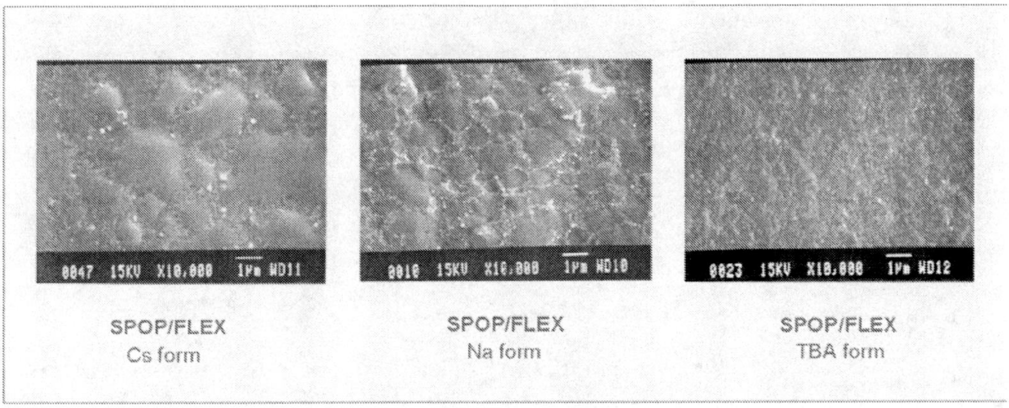

Figure 11 - SEM micrographs of the cross-sections of three samples of 2.1 mmol/g IEC SPOP with 40% Flex membranes that were UV crosslinked with 5% benzophenone

Table 2 – Results of Tensile Tests on Blends of 2.1 Mmol/G IEC Poly[Bis(3-Methylphenoxy)Phosphazene] and Kynar Flex.

KYNAR FLEX CONTENT (%)	EFFECTIVE IEC (mmol/g)	ULTIMATE STRENGTH (MPa)	ELONGATION AT BREAK (%)
40	1.25	3.39	150
45	1.15	4.68	152
50	1.05	5.42	166
55	0.95	6.03	178
60	0.85	6.38	186

The influence of blending on equilibrium water swelling and proton conductivity (of water-equilibrated samples) is shown in Figure 12. These membranes contained 2.1 mmol/g SPOP in the Na form and they were UV crosslinked with 5% fluorobenzophenone photoinitator (both benzophenone and fluorobenzophenone were found to work equally well as UV photoinitiators in dry SPOP membranes). By appropriate blending with different amounts of Flex (between 40% and 60%), the effective IEC of the membranes was varied from 0.85 to 1.25 mmol/g. As can be seen, there was a decrease in water swelling from 66% to 35% and a decrease in conductivity from 0.054 to 0.013 S/cm when the Kynar Flex content in the membranes was increased from 40% (1.25 mmol/g effective IEC) to 60% (0.85 mmol/g effective IEC).

3. Membrane Characterization and Testing

In this section, the transport and mechanical properties of polyphosphazene-based ion-exchange membranes are presented. Experimental methods and the resulting data are presented for membranes composed of pure SPOP (i.e., sulfonated poly[bis(3-

methylphenoxy)phosphazene]) with and without UV crosslinking, UV-crosslinked SPOP/polyacrylonitrile blended membranes, and SPOP/Kynar Flex membranes that were crosslinked using either UV light or electron beam radiation.

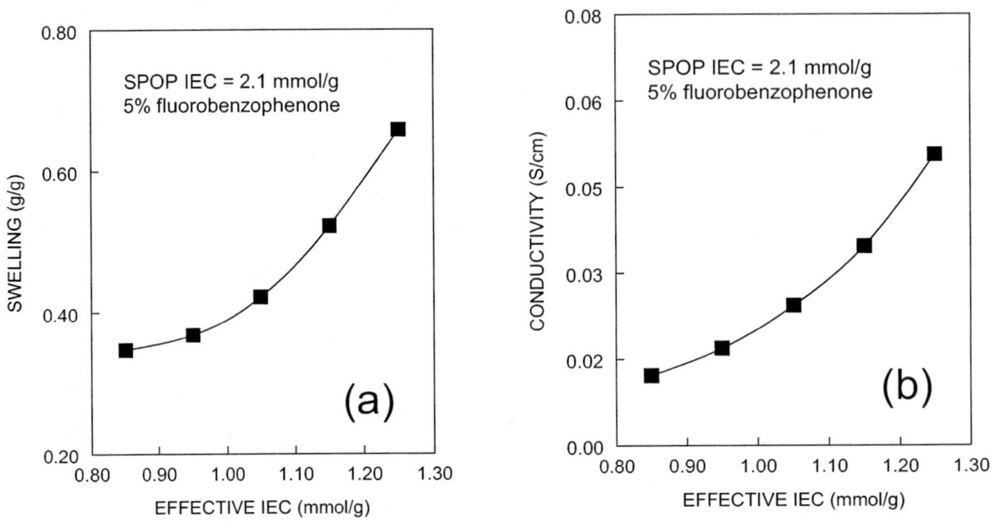

Figure 12 – The effect of blending SPOP and Kynar Flex on: (a) equilibrium membrane water swelling and (b) proton conductivity of water-equilibrated membranes. All measurements were performed at 25°C.

3.1 Proton Conductivity

The electrical conductivity of water-equilibrated polyphosphazene membranes in the H^+ form (where only protons associated with fixed charges can carry the current) was determined using an AC impedance technique. Membrane samples were first soaked in deionized and distilled water for 24 hours. The longitudinal (x-y) conductivity was measured using a pair of pressure-attached, high surface area platinum electrodes, as described elsewhere (41). The mounted membrane sample was immersed in deionized and distilled water at a given temperature and measurements were made from 1Hz to 10^5Hz using a PAR Model 5210 amplifier and a PAR Model 273 potentiostat/galvanostat. Both real and imaginary components of the impedance were measured and the real Z-axis intercept determined. The cell constant was calculated from the spacing of the electrodes, the thickness of the membrane, and the area of the platinum electrodes.

Proton conductivities in crosslinked films composed of sulfonated poly[bis(3-methylphenoxy)phosphazene] and Kynar Flex at 25°C are plotted against the equilibrium membrane water swelling in Figure 13. Crosslinking was achieved by either electron beam radiation or by UV light with benzophenone (or fluorobenzophenone) photoinitiator. Each data point in this figure represents a different membrane. The variation in membrane swelling was achieved by changing the amount of sulfonated polyphosphazene (10-50%) and its IEC

(1.4-2.2 mmol/g) in the blended films. The following conclusions can be drawn from the data in Figure 13: (1) membranes that swelled > 20% in water possessed a proton conductivity that was sufficiently high (> 0.01 S/cm) for direct methanol fuel cell applications, (2) there was no significant difference in the swelling vs. conductivity behavior of membranes crosslinked with benzophenone and UV light or electron-beam radiation, and (3) the conductivity of blended/crosslinked polyphosphazene films was comparable to that of Nafion (42) only when the sulfonated POP membranes exhibited a swelling > 80%.

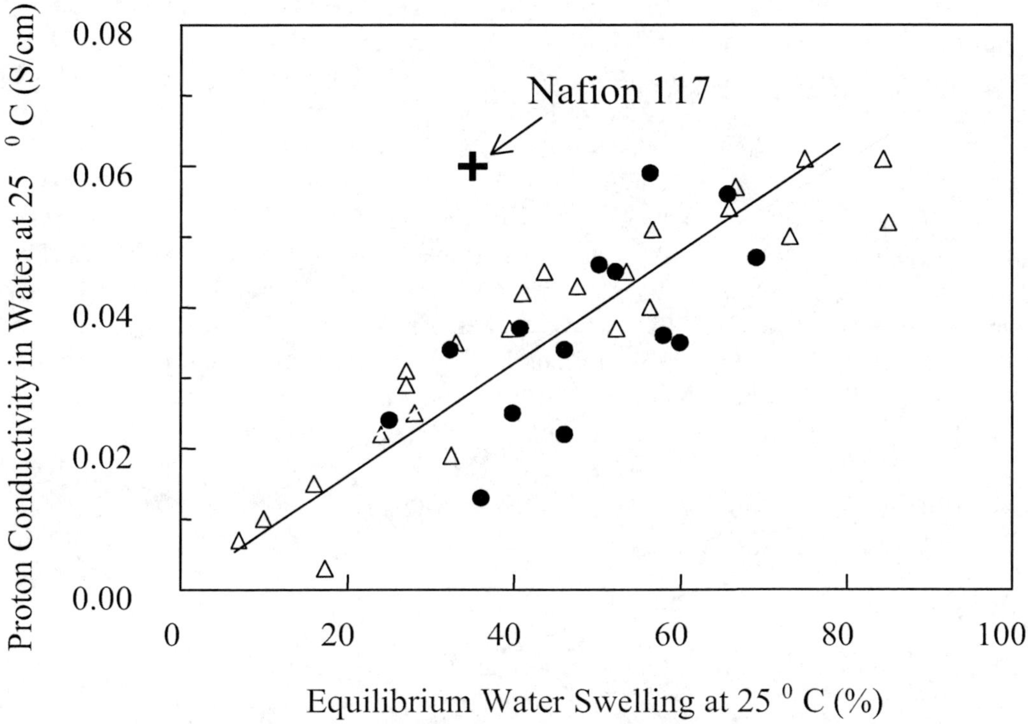

Figure 13 - Proton conductivity (in water) vs. equilibrium water swelling for blended and crosslinked membranes composed of sulfonated poly[bis(3-methylphenoxy)phosphazene] and Kynar Flex. Δ- films crosslinked with UV light and a photoinitiator (either benzophenone or fluorobenzophenone); ● films crosslinked with electron beam radiation.

3.2 Water Diffusion Coefficient Measurements

Water diffusion coefficients in pure SPOP films were determined under controlled water vapor activity conditions using a McBain sorption apparatus (43,44). Diffusion coefficients were determined by a weight loss method, using a thick membrane (400-600 μm) to minimize surface drying effects. After a membrane sample (in the H^+ form) was fully equilibrated with water vapor at an activity of 0.98, the vapor activity was lowered by 5% and the membrane weight loss was recorded with time. The diffusion coefficient was computed from the initial straight-line slope of a weight loss vs. square-root of time plot (time measured in seconds) and the following equation (45),

$$D = \frac{\pi\delta^2 (slope)^2}{16} \qquad (2)$$

where δ is the dry membrane thickness.

The temperature dependence of measured water diffusion coefficients in crosslinked and non-crosslinked polyphosphazene membranes (1.4 mmol/g IEC) is presented in Figure 14, along with literature data for the self diffusion coefficient of water in Nafion 117 (41,42,46). The water diffusion coefficients in both SPOP membranes (e.g., 1.2×10^{-7} cm^2/s for a crosslinked membrane at 25°C) were significantly smaller than that in Nafion 117 due to a combination of one or more of the following: (i) The presence of water molecules that are bonded to the P-N polymer backbone of SPOP polymers (due to the lone-pair electrons on backbone nitrogens), as compared to that in Nafion with a hydrophobic PTFE backbone, (ii) a higher percentage of hydrogen-bonded water in the SPOP membranes, as compared to Nafion, (iii) a low H_2O/SO_3^- ratio in crosslinked SPOP membranes at high water activities, indicating that there are fewer loosely bound water molecules available for diffusive transport, and (iv) less solvent swelling in crosslinked POP films, which increased the tortuous pathway for species diffusion. Items (i) and (ii) were confirmed by preliminary Attenuated Total Reflectance FTIR spectroscopic analyses of water in hydrated, acid-form SPOP membranes. These two effects are most clearly demonstrated by the fact that a non-crosslinked 1.4 IEC SPOP membrane, which swells more than Nafion in water (see Table 1), was more effective than Nafion in restricting water diffusion. As expected, the water diffusion coefficient in an uncrosslinked 1.4 IEC SPOP membrane is greater than that in a crosslinked film due to greater water swelling when the polymer is not crosslinked.

3.3 Methanol Crossover Measurements

Proton-exchange membranes for direct methanol fuel cells must exhibit a low methanol crossover (permeability). Different experimental techniques have been used to measure the methanol barrier properties of fuel cell membranes, including the determination of a methanol diffusion limited current density (47,48), the measurement of methanol transport rates in a standard diffusion cell (49), and the analysis of methanol vapor adsorption/desorption using a McBain sorption balance (27).

For SPOP membranes exposed to aqueous methanol solutions, two techniques were used to assess methanol permeability: (1) ^{1}H NMR analysis of methanol desorption from a film sample (50) and (2) measurement of methanol fluxes across a membrane by monitoring the downstream rate of methanol oxidation. A novel tracer-diffusion H^1-NMR method was developed and used to measure the mutual diffusion coefficient of methanol in UV-crosslinked and non-crosslinked membranes composed of sulfonated poly[bis(3-methylphenoxy)phosphazene]. The technique measures the growth of a methanol NMR signal in the bulk (external) water solution as methanol diffuses out of a thin film membrane. The transient increase in methanol peak height during methanol desorption was fitted to a simple theoretical diffusion model using the methanol diffusion coefficient as an adjustable parameter. This method was found to be fast, reproducible and accurate to within about ±20% and can be used when the more common pulsed field gradient NMR method is not suitable (e.g., when the internal solute signal is either too small or too broad to be utilized or when the diffusion rate is too slow to yield observable attenuation of the NMR signal with commonly available gradient strengths or possible gradient application times). Successful application of the tracer-desorption NMR approach requires that the external solute signal be observable in a single pulse and that the membrane desorption rate not be too fast (the diffusion coefficient should be $< 10^{-6}$ cm^2/s).

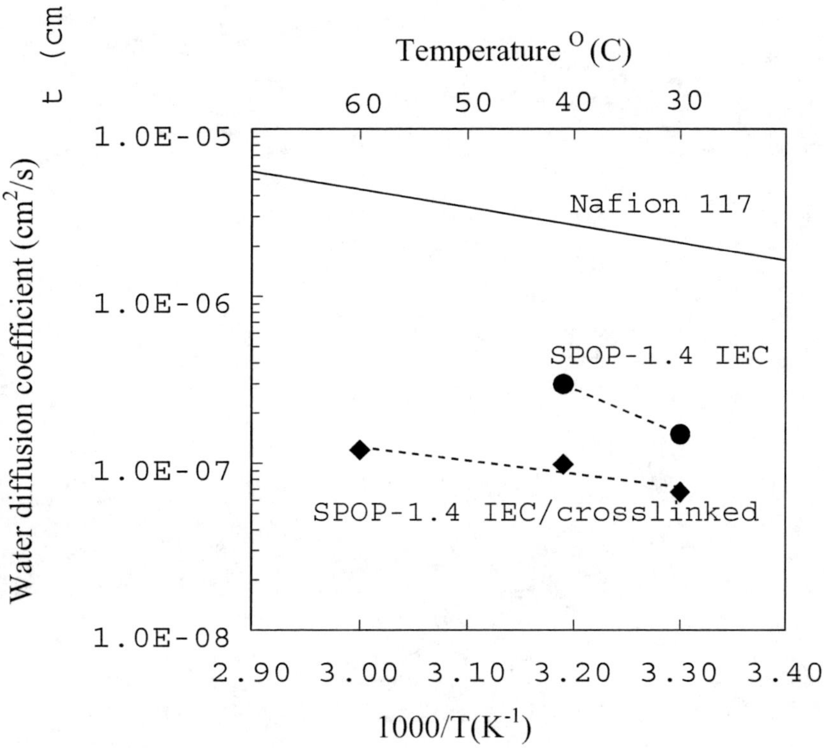

Figure 14 – Temperature dependence of water diffusion coefficients in crosslinked and uncrosslinked membranes composed of sulfonated poly[bis(3-methylphenoxy)phosphazene] (data from Reference 27).

Measured mutual diffusion coefficients of methanol in crosslinked and non-crosslinked 1.4 mmol/g IEC sulfonated polyphosphazene cation-exchange membranes are shown in Figure 15, along with literature data for a Nafion 117 membrane (48). Diffusion coefficients at 25°C were in the range of 8.0 x 10^{-8} cm^2/s to 4.0 x 10^{-7} cm^2/s for methanol concentrations of 1.0-5.0 M and were significantly smaller than those reported for Nafion.

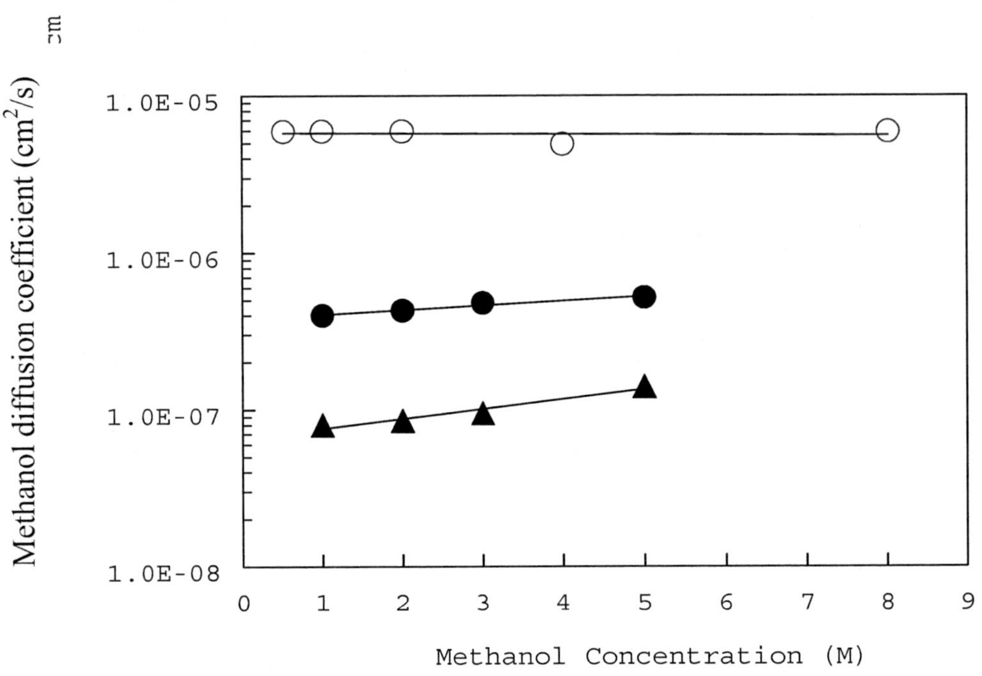

Figure 15 – Concentration dependence of methanol mutual diffusion coefficients in UV-crosslinked and non-crosslinked membranes composed of 1.4 mmol/g IEC sulfonated poly[bis(3-methylphenoxy)phosphazene]; ● non-crosslinked polyphosphazene; ▲ UV-crosslinked polyphosphazene (with 15% benzophenone); ○ diffusion coefficients in Nafion 117 (41).

In a second technique, a polyphosphazene-based ion-exchange membrane was placed in a 5 cm^2 fuel cell test apparatus. An aqueous methanol solution was pumped past one side of the membrane and the opposite face was in contact with flowing air. On the air side, a commercial Pt/C fuel cell cathode was pressed against the membrane surface. In such an arrangement, any methanol that permeated through the membrane would be chemically oxidized to CO_2 and water at the platinum catalyst. An electronic sensor (Vaisala GMM12B) placed in the downstream air line monitored the concentration of the CO_2 generated from methanol oxidation. A schematic diagram of this approach is shown in Figure 16. From the air flow rate, the increase in CO_2 concentration over background air, and the membrane area exposed to the methanol solution, one can calculate the trans-membrane methanol flux.

Typical steady-sate methanol flux data at 60°C and 70°C for two blended and UV-crosslinked polyphosphazene membranes (polyphosphazene blended with either Kynar Flex or polyacrylonitrile) and for Nafion 117 are plotted against the methanol solution concentration in Figure 17.

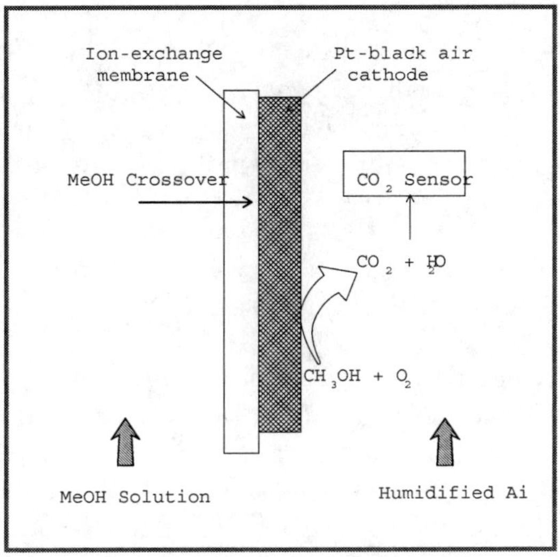

Figure 16 – Schematic diagram of the apparatus used to measure methanol permeation through a polyphosphazene-based proton-exchange membrane.

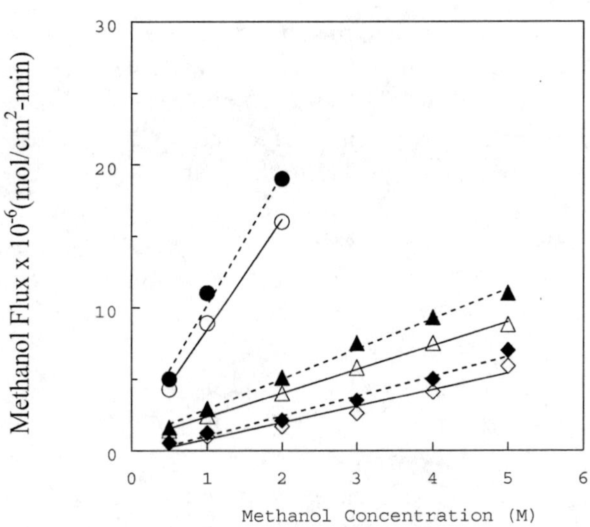

Figure 17 – Trans-membrane methanol flux for simple diffusion across Nafion 117 and blended SPOP membranes.

▲ and Δ: SPOP/Kynar Flex blend (52% 1.9 mmol/g IEC SPOP + 48% Flex, crosslinked with 60 megarads of electron beam radiation; 130 μm wet thickness; κ=0.015 S/cm at 25°C);

and : SPOP/Kynar Flex blend (50% 2.0 mmol/g IEC SPOP + 40% Flex + 10% difluorobenzophenone, crosslinked with UV radiation; 220 μm wet thickness; κ=0.037 S/cm at 25°C);

● and ○: Nafion 117 (220 μm wet thickness). Solid symbols are data at 70°C, open symbols at 60°C.

Taking into account the differences in film thickness (130-140 μm for the polyphosphazenes vs. 220 μm for Nafion), thickness-corrected methanol fluxes for the SPOP membranes were 6-11 times lower than those obtained for Nafion. The data in Figure 17 do not show as dramatic a drop in methanol permeability as the diffusion coefficient data in Figure 15. This may be due to: (1) different methanol barrier properties of pure SPOP vs. blended SPOP membranes and/or (2) differences in the use of transient vs. steady-state data to estimate methanol permeability.

3.4 Chemical Stability and Mechanical Properties

The chemical stability and thermomechanical properties of UV-crosslinked pure SPOP cation-exchange membranes were evaluated (27). During direct methanol fuel cell operation, the reduction of O_2 gas at the cathode may proceed through a peroxide intermediate. H_2O_2 can react with trace metal ions in the membrane to form HO and HO_2 radicals, which can chemically attack the polymer. To evaluate whether a SPOP membrane can withstand a strong oxidizing environment without degradation, membrane samples were soaked in an aqueous 3% H_2O_2 solution containing 4 ppm Fe^{2+} at 68°C. Periodically, over a 24 hour period, the membrane was removed from the peroxide solution, wiped with filter paper to remove excess liquid, and weighed.

As shown in Figure 18, a minimal membrane weight loss (< 5%) was observed, confirming our hypothesis and literature references regarding the chemical stability of phosphazene polymers. A comparison of FTIR spectra before and after the 24 hour peroxide soak showed: (i) No new IR peaks and (ii) no change in the P-N (1243 cm^{-1}), S=O (1085 cm^{-1}), and P-O- (1140 cm^{-1}) stretching bands of the original membrane. For comparison purposes, the chemical stability data for a Nafion 117 membrane and a Tokuyama Soda CMX cation-exchange membrane (composed of sulfonated polystyrene, crosslinked with divinyl benzene) are also presented in Figure 18. As expected, Nafion exhibited excellent chemical resistance with a weight loss of < 1%, whereas the CMX membrane underwent rapid and nearly complete degradation.

The mechanical properties of UV-crosslinked and non-crosslinked films of pure SPOP were measured as a function of temperature using a TMA 2940 Thermomechanical analyzer, operating in the probe penetration mode. Membrane samples (≈150 μm in thickness) were heated in air at a rate of 10°C/min. The load for the penetration probe ranged from 0.1-0.5 N (pressures of 160-800 kPa).

Typical TMA data showing the effect of temperature on polymer softening for crosslinked (10% benzophenone) and non-crosslinked SPOP films with an IEC = 1.4 mmol/g are shown in Figure 19. Crosslinking greatly improved the polymer's mechanical properties at elevated temperatures. The non-crosslinked SPOP membrane began to soften and deform at 76°C for a small penetration probe pressure of 160 kPa (24 psig), whereas the crosslinked POP film was mechanically stable up to 173°C for a probe pressure as high as 800 kPa (118 psig). Direct liquid methanol fuel cells operate at a temperature of 60-80°C and a pressure of about 3 atm. Under these conditions, softening and deformation of the non-crosslinked POP membrane is possible.

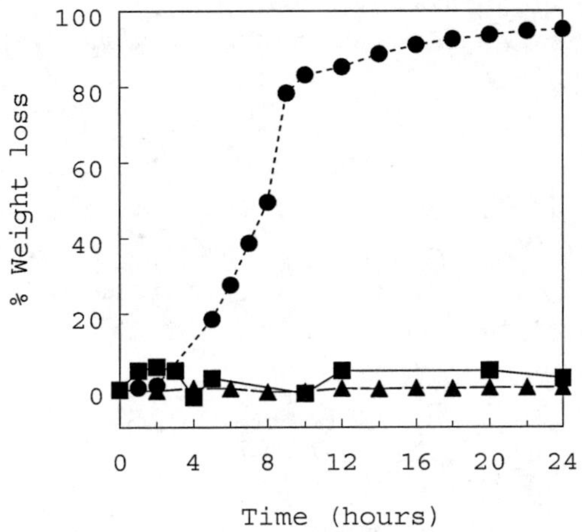

Figure 18 – Degradation (measured weight loss vs. time) of membranes immersed in a 3% H_2O_2/4 ppm Fe^{2+} solution at 68°C. ■ 1.4 IEC/UV-crosslinked poly[bis(3-methylphenoxy)phosphazene] membrane with 15% benzophenone photoinitiator; ▲ Nafion 117; • Tokuyama Soda CMX cation-exchange membrane (sulfonated polystyrene, crosslinked with divinyl benzene).

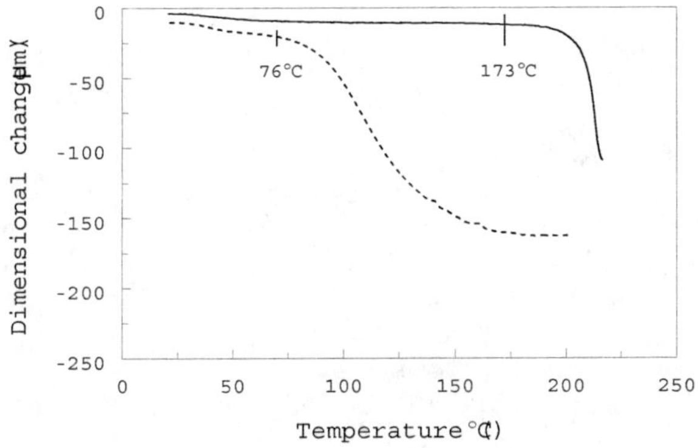

Figure 19 – Probe penetration depth vs. temperature from TMA measurements. (– – – – –) 1.4 IEC poly[bis(3-methylphenoxy)phosphazene] membrane with no crosslinking; (————) 1.4 IEC poly[bis(3-methylphenoxy)phosphazene] membrane with UV-crosslinking (15% benzophenone photoinitiator). The penetration probe load was 0.5 N and the film thickness was 150 μm.

3.5 Preliminary Direct Methanol Fuel Cell Experiments

Direct methanol fuel cell tests were carried out with UV-croslinked membranes composed of sulfonated poly[bis(3-methylphenoxy)polyphosphazene] and polyacrylonitrile (PAN). Electrodes (5 cm^2 in area) were fabricated by painting catalyst ink containing Nafion binder onto E-Tek (Andover, MA) carbon paper electrodes and evaporating the solvent. The anode catalyst was 4.4 mg/cm^2 Pt-Ru powder with 15% (dry catalyst basis) Nafion binder. The cathode was composed of 4.7 mg/cm^2 Pt-black powder with 10% Nafion. Membrane-electrode-assemblies (MEAs) were prepared by hot pressing the electrodes onto a 130 μm thick (dry thickness) SPOP/PAN membrane at 120°C and 125 psi for 5 minutes

A representative cell voltage (V) vs. current density (i) plot with a SPOP/PAN membrane is shown in Figure 20. This membrane was composed of 52% 2.1 mmol/g SPOP, 43% polyacrylonitrile, and 5% benzophenone. Direct methanol fuel cell tests were performed with a 1.0 M aqueous methanol feed to the anode and air fed to the cathode at a flow rate of 600 standard cm^3/min. (SCCM). Experiments were performed at 60°C using a Scribner Fuel Cell Test Station. Methanol crossover was determined by monitoring the carbon dioxide concentration in the cathode air exhaust using a Vaisala GMM12B CO_2 detector. As compared to a Nafion 117 MEA, whose performance is also shown in Figure 20, there was a higher open circuit potential with the polyphosphazene membrane (indicative of lower methanol crossover). The performance of the SPOP-based MEA was good, but the V-i curve fell below that of Nafion. These results are acceptable and promising, considering the fact that there was no optimization of electrode ink composition and hot-pressing conditions. The measured methanol crossover with the SPOP/polyacrylonitrile membrane at a current density of 0.1 A/cm^2 was found to be 2.5 times lower than that in Nafion 117.

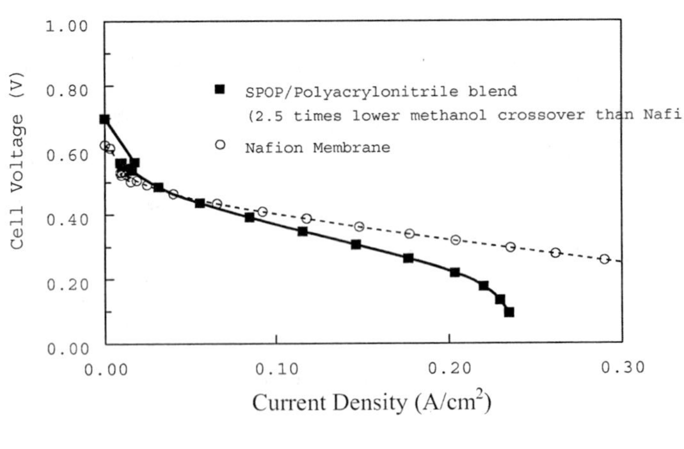

Figure 20 - Direct Methanol Fuel Cell Performance of a UV-Crosslinked SPOP/polyacrylonitrile Membrane. Cell Temp. = 60°C; 1 M Methanol; Humidified air at 60°C and 600 SCCM; Anode: Pt-Ru at 4.4 mg/cm^2 w/15% Nafion binder; Cathode: Pt-Black at 4.7 mg/cm^2 w/10% Nafion binder; electrode area = 5 cm^2. Membrane composition: 52% 2.1 IEC SPOP/43% PAN/5% benzophenone; Dry membrane thickness ≈130 μm;

ACKNOWLEDGEMENTS

Some of the work contained in this chapter was funded by the U.S. Army Research Office, Grant No. DAAD19-00-1-0517. We also wish to acknowledge the assistance of Science Research Laboratories, Inc., Sommerville, MA who performed the electron beam crosslinking of polyphosphazene membranes.

REFERENCES

1. Prater, K., The renaissance of the solid polymer fuel cell, *J. Power Sources*, 1990, **29**, 239-250.
2. Larminie, J. and Dicks, A., *Fuel Cell Systems Explained*, John Wiley, New York, 2000.
3. Heinzel, A., Nolte, R., Ledjeff-Hev, K., and Zedda, M., Membrane fuel cells – concepts and system design, *Electrochim. Acta*, 1998, **43**, 3817-3820.
4. Appleby, A. J. and Foulkes, F. R., *Fuel Cell Handbook*, Van Nostrand Reinhold, New York, 1989.
5. Wakizoe, M., Velev, O. A., and Srinivasan, S., Analysis of Proton Exchange Membrane Fuel Cell Performance with Alternate Membranes, *Electrochim. Acta*, 1995, **40**, 335-344.
6. Appleby, A. J., The Electrochemical Engine for Vehicles, *Scientific American*, 1999, July, 74-79.
7. J. Kerres, J., Urllich, A., Haring, Th., Preidel, W., Baldauf, M., and Gebhardt, U., Preparation, Characterization, and Fuel Cell Application of New Acid-Base Blend Membranes, *J. New Mater. Electrochem. Sys*., 2000, **3** 229-239.
8. Kerres, J., A., Development of ionomer membranes for fuel cells, *J. Membr. Sci.*, 2001, **185**, 3-27.
9. Nolte, R., Ledjeff, K., Bauer, M., and Mulhaupt, R., Partially sulfonated poly(arylene ether sulfone) – a versatile proton conducting membrane material for modern energy conversion technologies, *J. Membr. Sci.*, 1993, **83**, 221-220.
10. Faure, S., Pineri, M., Aldebert, P., Mercier, R., and Sillion, B., Sulfonated polyimides, membranes and fuel cell, European Patent 897407 A1, 1999.
11. Wainright, J.S., Wang, J.-T., Weng, D., Savinell, R. F., and Litt, M. H., Acid-doped polybenzimidazoles: a new polymer electrolyte, *J. Electrochem. Soc.*, 1995, **142**, L121-L123.
12. Potin, Ph. and DeJaeger, R., Polyphosphazenes: Synthesis, Structures, Properties, Applications, *Eur. Polym. J.*, 1991, **27**, 341-348.
13. Allcock, H. R., Fitzpatrick, R. J., and Salvati, L., Sulfonation of (Aryloxy)- and (Arylamino)phosphazenes: Small-Molecule compounds, Polymers, and Surfaces, *Chem. Mater.*, 1991, **3**, 1120-32.
14. Blonsky, P. M., Shriver, D. F., Austin, P., and Allcock, H. R., Polyphosphazene Solid Electrolytes, *J. Am. Chem. Soc.*, 1984, **106**, 6854-6855.
15. Tonge, J. S., Blonsky, P. N., Shriver, D. F., Allcock, H. F., Austin, P. E., Neenan, T. X., and Sisko, J. T., Effect of Sidechain and Crosslinking on Ionic Conductivity in Polyphosphazene Solid Electrolytes, *Proc.-Electrochem. Soc., 87-1 (Proc. Symp. Lithium Batteries)*, 1987, 533-44.

16. Ganapathiappan, S., Chen, K., and Shriver, D. F., A New Class of Cation Conductors: Polyphosphazene Sulfonates, *Macromolecules*, 1988, **21**, 2299-2301.
17. Allcock, H. R.,Gebura, M., Kwon, S., and Neenan, T. X., Amphiphilic Polyphosphazenes as Membrane Materials: Influence of Side Group on Radiation Cross-Linking, *Biomaterials*, 1988, **9**, 500-508.
18. Peterson, E. S., Stone, M. L., McCaffrey, R. R., and Cummings, D. G., Separation of Hazardous Organics from Gas and Liquid Feedstreams Using Phosphazene Polymer Membranes, *Sep. Sci. and Techn.*, 1993, **28**, 271-281.
19. Peterson, E. S., Stone, M. L., McCaffrey, R. R., and Cummings, D. G., Mixed-Gas Separation Properties of Phosphazene Polymer Membranes, *Sep. Sci. and Techn.*, 1993, **28**, 423-440.
20. Montoneri, E., Gleria, M., Ricca, G., and Pappalardo, G. C., On the reaction of catenapoly(diphenoxy-λ^5-phosphazene) with sulfur trioxide, *Makromol. Chem.*, 1989, **190**, 191-202.
21. Montoneri, E., Gleria, M., Ricca, G., and Pappalardo, G. C., New acid-polyfunctional water-soluble phosphazenes: synthesis and spectroscopic characterization," *J. Macromol. Sci. Chem.*,1989, **A26**, 645-661.
22. Mark, J. E., Allcock, H. R., and West, R., *Inorganic Polymers*, Prentice Hall, Englewood Cliffs, NJ, 1992.
23. Allcock, H. R., Klingenberg, E. H., and Welker, M. F., Alkane-sulfonation of cyclic and high polymeric polyphosphazenes, *Macromolecules*, 1993, **26**, 5512-5519.
24. Kurachi, Y., and Kajiwara, M., Synthesis and properties of poly(organophosphazenes) ionomers $[NP(HNC_6\text{-}H_5)_{2-(x+y)}(HNC_6H_4SO_3H)\,(HNC_6H_4SO_3Li)_y]_n$, *J. Mater. Sci.*, 1991, **26**, 1799-1802.
25. Wycisk, R., and Pintauro, P. N., Sulfonated polyphosphazene ion-exchange membranes, *J. Membr. Sci.*, 1996, **119**, 155-160.
26. Tang, H., Pintauro, P. N., Guo, Q., and O'Connor, S., Polyphosphazene Membranes. III. Solid-state characterization and properties of sulfonated poly[bis(3-methylphenoxy) phosphazene], *J. Appl. Polym. Sci.*, 1999, **71**, 387-399.
27. Guo, Q., Pintauro, P. N., Tang, H., and O'Connor, S., Sulfonated and Crosslinked Polyphosphazene-Based Proton-Exchange Membranes, *J. Memb. Sci.*,1999, **154**, 175-181.
28. Tang, H. and Pintauro, P. N. Polyphosphazene Membranes IV: Polymer Morphology and Proton Conductivity in Sulfonated Poly[bis(3-methylphenoxy)phosphazene], *J. Appl. Polym. Sci.*, 2001, **79**, 49-59.
29. Wycisk, R., Pintauro, P. N., Wang, W., and O'Connor, S., Polyphosphaze Membranes. I. Solid-State Photocrosslinking of Poly[(3-ethylphenoxy) (phenoxy)phosphazene], *J. Appl. Polym. Sci.*, 1996, **59**, 1607-1617.
30. Graves, R. and Pintauro, P. N., Polyphosphazene Membranes. II. Photocrosslinking of Poly[(alkylphenoxy)(phenoxy)phosphazene] Films, *J. App. Polym. Sci.*, 1998, **68**, 827-836.
31. Rabek, J. F., *Mechanisms of Photophysical Processes and Photochemical Reactions in Polymers,* John Wiley, Chichester, 1987.
32. Utracki L. A. and Weiss, R. A., *Multiphase Polymers: Blends and Ionomers*, American Chemical Society, Washington, D.C., 1989.

33. Smith, P., Hara, M., Eisenberg, A., in *Current Topics in Polymer Science*, vol. II, Ottenbrite/Utracki/Inoue (Eds.), Hauser Publisher, N.Y., p.255, 1987.
34. Büchi, F N., Gupta, B., Haas, O., and Scherer, G. G., Study of radiation-grafted FEP-g-polystyrene membranes as polymer electrolytes in fuel cells, *Electrochim. Acta*, **40**, 1995, 345-353.
35. Lehtinen, T., Sundholm, G., Holmberg, S., Sundholm, F., Bjornbom, P., and Bursell, M., Electrochemical characterization of PVDF-based proton conducting membranes for fuel cells, *Electrochim. Acta*, **43**, 1998,1881-1890.
36. Holmberg, S., Lehtinen, T., Näsman, J., Ostrovskii, D., Paronen, M., Serimaa, R., Sundholm, F., Sundholm, G., Torell, L. and Torkkeli, M., Structure and Properties of Sulfonated Poly[(vinylidene fluoride)-g-styrene] Porous Membranes, *J. Mater. Chem.* **6**, 1996,1309-1317.
37. Liebhavsky, H. A. and Cairns, E. J., in *Fuel Cells and Batteries*, Wiley, N.Y, 1968.
38. Kerres, J., Ullrich, A., Meier, F., and Haring, Th., Synthesis and Characterization of Novel Acid-Base Polymer Blends for the Application in Membrane Fuel Cells, *Solid State Ionics*, 1999, **125**, 243-249.
39. Lin, J-C., Ouyang, M., Fenton, J. M., Kunz, H. R., Roberstein, J. T., and Cutlip, M. B., Study of Blend Membranes Consisting of Nafion and Vinylidene Fluoride-Hexafluoropropylene Copolymer, *J. Appl. Polym. Sci*., 1998, **70,** 121-127.
40. Jia, N., Lefebvre, M. C., Halfyard, J., Qi, Z., Pickup, P. G., Modification of Nafion Proton Exchange Membranes to Reduce Methanol Crossover in PEM Fuel Cells, *Electrochem. Solid-State Letters,* 2000, **3**, 529-531.
41. Zawodzinski, T., Neeman, M., Sillerud, L. O., and Gottesfeld, S., Determination of Water Diffusion Coefficient in Perfluorosulfonate Ionomeric Membranes, *J. Phys. Chem*., 1991, **95**, 6040-6044.
42. Zawodzinski. T., Springer, T. E., Davey, J., Jestel, R., Lopez, C., Valerio, J., and Gottesfeld. S., A Comparative Study of Water Uptake by and Transport Through Ionomeric Fuel Cell Membranes, *J. Electrochem. Soc*., 1993, **140**, 1981-1985.
43. Felder, R. M. and Huvard, G. S., Diffusion and sorption of gases and vapors, in R. A. Fava (Ed.), *Methods of Experimental Physics*, vol. 16, part C, Academic Press, New York, pp. 315-377, 1980.
44. Ponangi, R., Pintauro, P. N., and De Kee, D., Free volume analysis of organic vapor diffusion in polyurethane membranes, *J. Membr. Sci*., 2000, **178**, 151-164.
45. Crank, J., *The Mathematics of Diffusion*, Oxford University Press, London, p.228, 1956.
46. Zawodzinski. T., Derouin, C., Radzinski, S., Sherman, R., Smith, V. T., Springer, T. E., and Gottesfeld. S., Water Uptake by and Transport Through Nafion 117 Membranes, *J. Electrochem. Soc*., 1993, **140**, 1041-1047.
47. Ren, X., Zawodzinski, T., Uribe, F., Dai, H., and Gottesfeld, S., Methanol Cross-over in Direct Methanol Fuel Cells, *in Proton Conducting Membrane Fuel Cells I,* Electrochemical Society Proceedings, **Vol. 95-23**, Pennington, NJ, pp. 284-298, 1995.
48. Ren, X., Springer, T. E., Zawodzinski, T. A., and Gottesfeld, S., Water and Methanol Uptakes in Nafion Membranes and Membrane Effects on Direct Methanol Cell Performance, *J. Electrochem. Soc*., 2000, **147**, 92-98.
49. Pivovar, B. S., Wang, Y., and Cussler, E. L., Pervaporation membranes in direct methanol fuel cells, *J. Membr. Sci*., 1999, **154**, 155-162.

50. Carter, R., Evilia, R., and Pintauro, P.N., Tracer-Desorption H^1-NMR Method for Measuring Methanol Diffusion Coefficients in Polyphosphazene Ion-Exchange Membranes, *J. Phys. Chem. B*, 2001, **105**, 2351-2355.

In: Applicative Aspects of Poly(organophosphazenes) ISBN 1-59454-025-X
Editors: M. Gleria and R. De Jaeger, pp. 255-265

Chapter 12

POLYORGANOPHOSPHAZENE METAL CATALYSTS

Paolo Pertici[1], Giovanni Vitulli[1], Piero Salvadori[1], Emanuela Pitzalis[2] and Mario Gleria[3]

1- Istituto per la Chimica dei Composti Organometallici del C.N.R., Sezione di Pisa, Via Risorgimento 35, 56126 Pisa, Italy

2-Istituto per i Processi Chimico-Fisici del CNR, Area della Ricerca di Pisa, via G. Moruzzi 1, 56124 Pisa, Italy.

3-Istituto di Scienze e Tecnologie Molecolari del C.N.R., c/o Dipartimento C.I.M.A. dell'Università di Padova, Via Marzolo 1, 35131 Padova, Italy, and Istituto per la Sintesi Organica e la Fotoreattività, Consiglio Nazionale delle Ricerche, Viale dell'Università, 2 – 35020 Legnaro, Padova, Italy.

INTRODUCTION

Macromolecular systems containing transition metal nanoparticles trapped into macromolecules are of great interest in catalysis[1]. A valuable approach in their preparation is the gentle decomposition of organometallics in the presence of an appropriate polymer[2].

Suitable organometallics for this goal are transition metal mononuclear or cluster compounds containing organic ligands which can be easily removed from the metal under mild conditions, such as zerovalent olefin transition metal complexes[3] or solvated metal atoms obtained by reactions of metal vapor with weakly stabilizing organic ligands[4].

As far as polymeric substrates are concerned, a variety of different macromolecules (e.g. polystyrene[5], or polyvinylpirrolidone[6]) have been exploited.

In this area poly(organophosphazenes) are very attractive supports[7,8] for at least two reasons:

(1) they allow to combine the versatility and processability of organic macromolecules with the better thermal stability of inorganic polymers[9];
(2) the nitrogens of the phosphazene chain are very basic and are able to be protonated or to form coordination complexes with metal salts, as evidenced by investigations carried out by Wisian-Neilson[10,11] and Gallazzi[12].

In this paper, attention will be focused on the catalytic activity of Ruthenium, Rhodium and Palladium systems supported onto poly[bis(aryloxy)]phosphazenes, i.e. poly[bis(phenoxy)phosphazene] (PBPP); poly[bis(*p*-methoxyphenoxy)phosphazene] (PB*p*MPP); poly[bis(*m*-methoxyphenoxy)phosphazene] (PB*m*MPP) and poly[bis(*p*-benzoylphenoxy)phosphazene] (PB*p*BPP), whose structure is reported below:

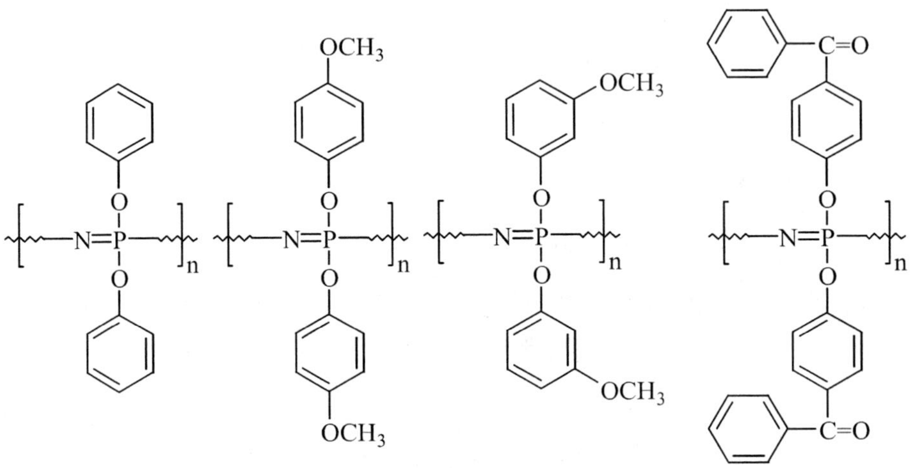

or poly(alkyl/arylphosphazenes), such as poly(dimethyl)phosphazene

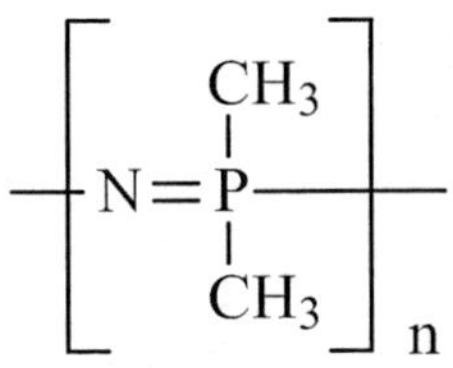

in a wide range of hydrogenation (Ru and Rh) and Heck-type reactions (Pd).

POLYORGANOPHOSPHAZENE RUTHENIUM CATALYSTS

Polyorganophosphazene ruthenium catalysts have been prepared by decomposition under hydrogen (1 atm, 20°C) of the complex Ru(η^6-cycloocta-1,3,5-triene)(η^4-cycloocta-1,5-diene), Ru(COT)(COD), **1**, in the presence of the polymer (Scheme 1).

Complex **1**, which is easily obtained in high yield by reaction of ruthenium trichloride hydrated and 1,5-cyclooctadiene with zinc dust (Scheme 1a)[13], reacts with hydrogen giving cycloctane and ultrafine metallic ruthenium[3,14]. Working in the presence of the polyorganophosphazene, the metal nanoparticles deposit on the polymer (Scheme 1b).

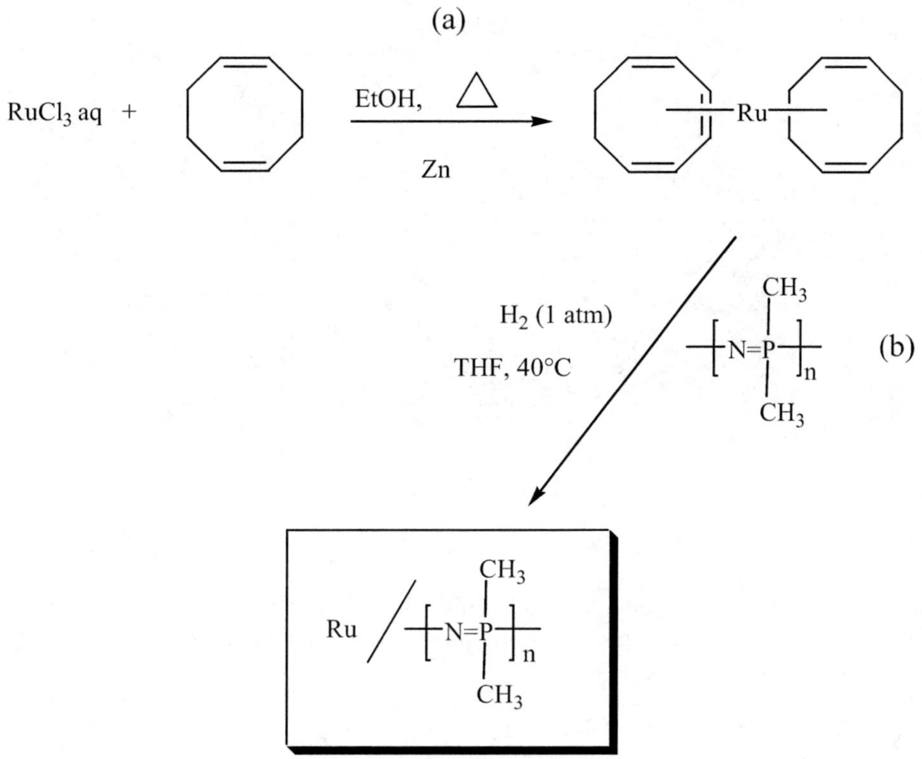

Scheme 1

Information on the structural peculiarities of the ruthenium nanoparticles has been obtained by High Resolution Transmission Electron Microscopy (HRTEM) analysis. The HRTEM micrograph of the ruthenium on polydimethylphosphazene sample (5 wt. % Ru), reported in figure 1 together with the histogram of the particle size dispersion, shows that the metal particles are homogeneously dispersed on the support and the statistical analysis of their dimension indicates a narrow distribution in the range 1-3 nm, with mean diameter d_m = 1.55 nm.

(a)

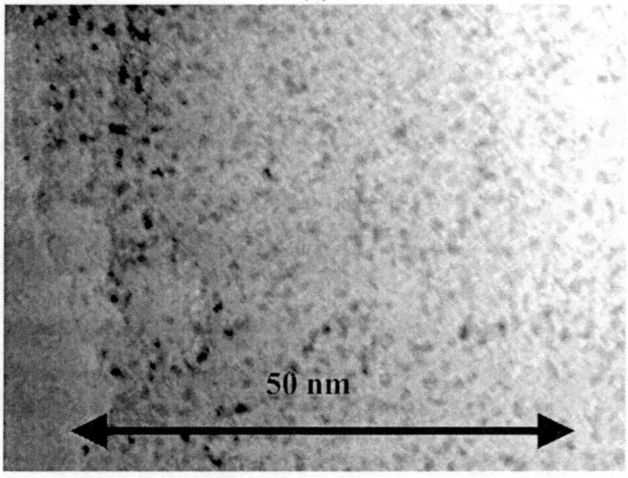

(b)

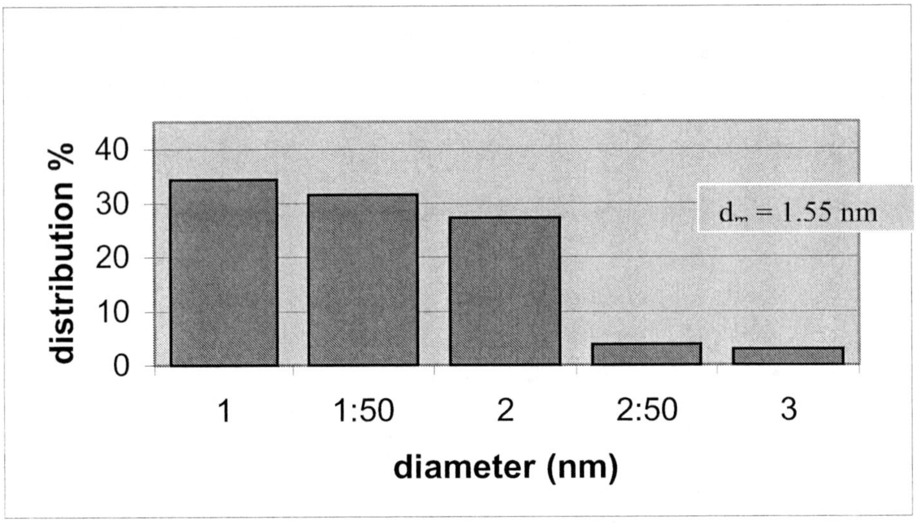

Figure 1. (a) Electron micrograph of the ruthenium on polydimethylphosphazene 5 wt% Ru catalyst. (b) Metal particle size distribution.

The polyorganophosphazene/ruthenium systems show a valuable catalytic activity in the hydrogenation of a wide range of unsaturated compounds. Some significative examples of the catalytic activity of Ru deposited on polydimethylphosphazene (PDMP), the first term of the polyalkylphosphazene series, in the hydrogenation of olefins, carbonyls and aromatic compounds are reported in Table 1.

Table 1: Hydrogenation of Olefins, Carbonyls and Aromatic Compounds with Ruthenium Nanoparticles Deposited on Polydimethylphosphazene[a].

Run	Substrate	Solvent	Phase[b]	Time (h)	Product[c] (%)
1[d]	1-Octene	Ethanol	Hom	12	Octane (100)
2[d,e]	"	"	Hom	"	Octane (100)
3[d]	"	THF	Het	"	Octane (30)
4[d]	Cyclohexanone	Ethanol	Hom	10	Cyclohexanol (100)
5[d]	Ethyl acetoacetate	"	Hom	24	Ethyl 3-hydroxybutanoate (100)
6[d]	Ethyl pyruvate	"	Hom	6	Ethyl lactate (100)
7[d]	"	THF	Het	"	Ethyl lactate (20)
8[f]	Toluene	Ethanol	Hom	48	Methylcyclohexane (60)
9[g]	Methyl benzoate	"	Hom	90	Methyl cyclohexanoate (100)
10[g]	"	THF	Het	"	Methyl cyclohexanoate (30)

[a]Substrate (13 mmol); catalyst Ru on PDMP (5 wt % Ru), 0.26 g ($1.3x10^{-3}$ gat); solvent (15 ml).
[b]Hom = homogeneous phase; Het = heterogeneous phase.
[c]Determined by GC analysis.
[d]Performed at 25°C and at 1 atm hydrogen pressure.
[e]Catalyst recovered from the previous run by precipitation with pentane.
[f]Performed at 80°C and at 50 atm hydrogen pressure.
[g]Performed at 25°C and at 50 atm hydrogen pressure.

The Ru/PDMP system has been employed in the hydrogenation of the aromatic ring of the *p*-aminomethylbenzoic acid to give the 4-aminomethylcyclohexancarboxylic acid, a product of considerable pharmaceutical interest[15] (Scheme 2).

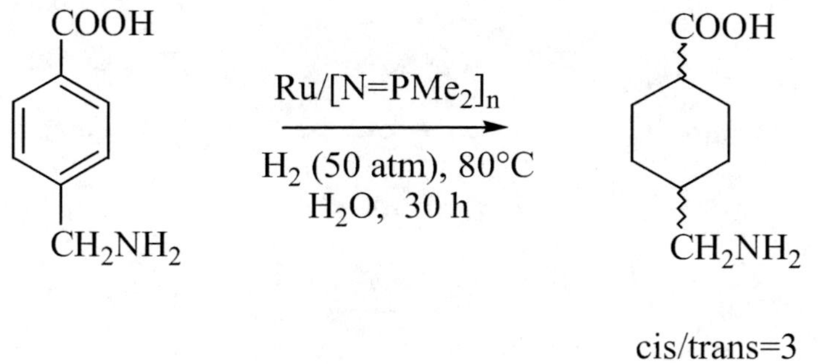

T(°C)	Conv. %
20	25
80	100

Scheme 2. Substrate/ruthenium molar ratio = 100.

At room temperature and 50 atm of hydrogen a conversion of 25% is obtained after 30 hours while at 80°C, in the same time, the reaction is complete. It is worth noting that the reaction is performed in water, an enviromental friendly solvent, working in homogeneous phase and the catalyst, recovered by precipitation with THF, mantains unchanged its activity.

The influence of the structure of the polyorganophosphazene on the catalytic activity of the ruthenium has been also investigated. The polymers employed as supports are poly[bis(aryloxy)phosphazenes], $[N{=}P(OAr)_2]_n$, a class of polymers easily available by simple synthetic procedures[16]. In particular, the used polymers have been the poly[(bis(phenoxy)phosphazene] (PBPP), the most simple term of the series, the poly[bis(p-methoxyphenoxy)phosphazene] (PB*p*MPP) and the poly[bis(*m*-methoxyphenoxy) phosphazene] (PB*m*MPP), in which the aromatic ring contains an activating group in different position, and the poly[bis(*p*-benzoylphenoxy)phosphazene] (PB*p*BPP), in which the aromatic ring is deactivated.

The Ru/-$[N{=}P(OAr)_2]_n$- catalysts have been employed in the hydrogenation of the aromatic compounds methyl benzoate and *p*-aminomethylbenzoic acid, being these reactions the slowest between the reaction catalyzed by Ru/PDMP. The results, reported in scheme 3, show that the Ru/-$[N{=}P(OAr)_2]_n$- catalysts are active in these reactions and that their activity decreases in the order Ru/PB*m*MPP ≅ Ru/PB*p*MPP > Ru/PBPP > Ru/PB*p*BPP, being always lower than that shown by Ru/PDMP in the same reaction conditions. It indicates a strong influence of the structure of polymeric support on the catalytic activity of the incorporated nanoparticles.

POLYORGANOPHOSPHAZENE RHODIUM CATALYSTS

Rhodium nanoparticles have been deposited on poly(alkyl)phosphazenes utilizing the Metal Vapour Synthesis (MVS) technique[4]. Metal atoms generated in a high vacuum apparatus (10^{-4}-10^{-5} mbar) by heating the metal in a high temperature source are co-

condensed with a weakly coordinating ligand in large excess (solvent) on the cold wall of a reactor immersed in liquid nitrogen (T = -196°C). The solid co-condensed material, by warming to the full melting, is recovered and manipuled following the usual technique of the organometallic chemistry.

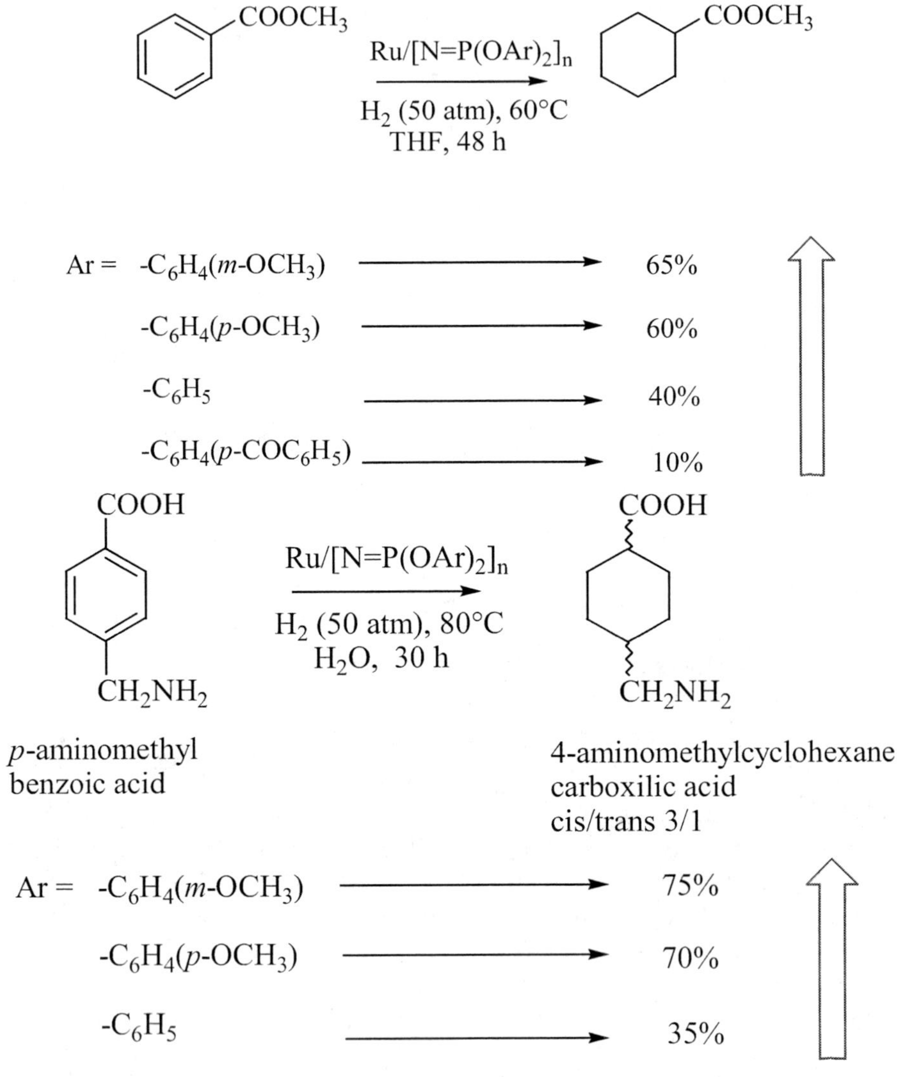

Scheme 3. Substrate/ruthenium molar ratio = 100.

A solution containing the metal atoms stabilized by the weak coordinating ligand is obtained (Solvated Metal Atoms, SMA) (Scheme 4a). Such SMA solutions are thermally unstable and they can be used as precursors of metal nanoparticles. By warming at room temperature in presence of support a complete deposition of the metal nanoparticles on the support is observed. Interestingly with this technique the deposition is performed under mild conditions avoiding the formation of large particles, generally exhibiting less catalytic activity, and no further activation process is necessary[2]. Following this method nanoparticles of rhodium have been deposited on polyorganophosphazenes using mesitylene as solvating organic ligand (Scheme 4b).

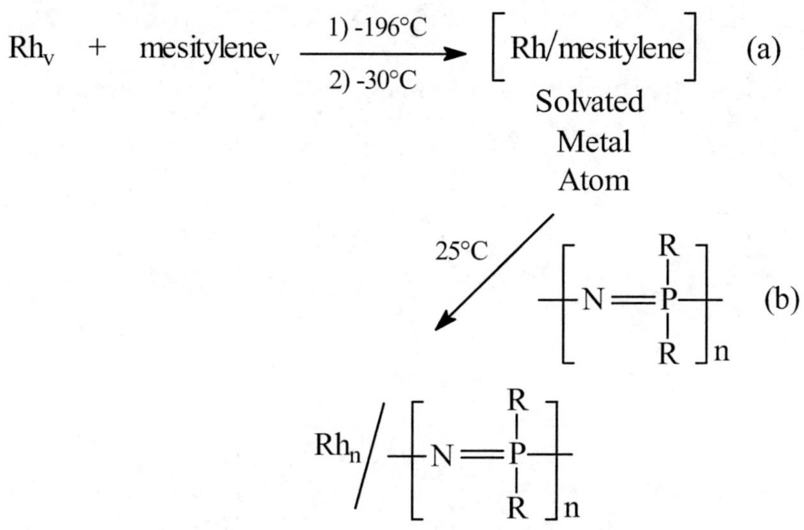

Scheme 4

Generally, in the catalytic systems obtained by MVS, the metal particles are homogeneously distributed with size ranging 2-4 nm in diameter[17].

The polyorganophosphazenes Rh systems are valuable catalysts for the selective hydrogenation of α,β-unsaturated aldehydes and ketones to the corresponding saturated carbonyl derivatives. Here some data are reported on the hydrogenation of cinnamaldehyde, **1**, a widely investigated reference reaction to evaluate the activity and the selectivity of catalysts[18]. The products, which can be formed in this reaction, are the 3-phenylpropanal, **2**, and the cinnamyl alcohol, **3**, deriving from the hydrogenation of the double bond and of the carbonyl group, respectively, and the 3-phenyl-1-propanol, **4**, as final product of the reduction (Scheme 5).

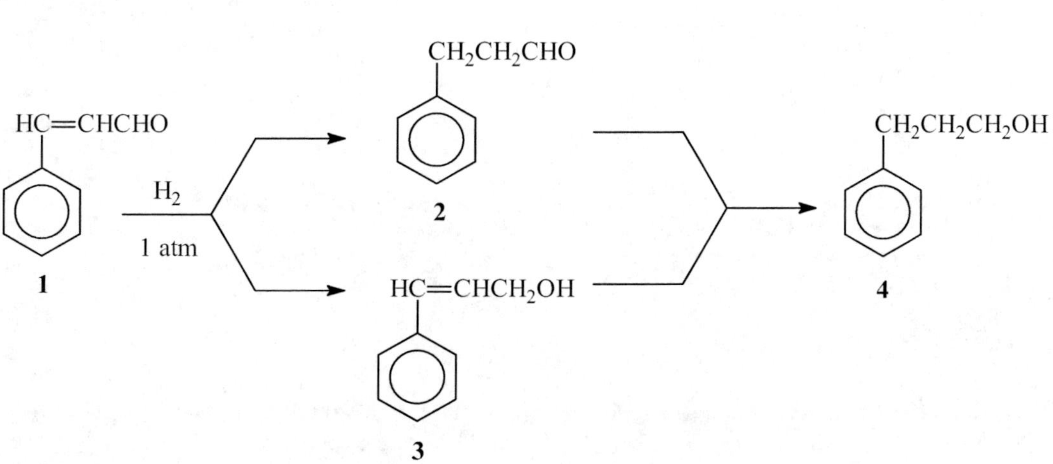

Scheme 5

The results obtained using Rh on PDMP as catalyst are summarized in Table 2, together with those obtained utilizing a commercially available rhodium on γ-alumina catalyst. After 24 h the conversion is ca. 50% with Rh on PDMP, while, at the same time, the yield is only 5% with the commercial Rh on γ-alumina catalyst.

Table 2. Hydrogenation of Cinnamaldehyde, 1, in the Presence of Rhodium on PDMP and Rhodium on γ-Al_2O_3[a].

Catalyst	time (h)	conv (%)	Products (%)[b]		
			2	3	4
Rh/PDMP	12	38	93	6	1
	24	51	92	6	2
	36	68	92	6	2
	96	98	91	6	3
Rh/γ-Al_2O_3	24	5	80	10	10
comm.	96	30	80	10	10

[a] Reaction conditions: cinnamaldehyde (2 ml, 16 mmol); cinnamaldehyde/rhodium molar ratio = 100; solvent = THF, 10 ml; hydrogen pressure = 1 atm; temperature = 25 °C.
[b] Determined by GC analysis.

With Rh on PDMP the conversion is complete after 96 hours (Table 2) and the chemoselectivity toward the formation of **2** is very high (91 %). In figure 2 the progress of the hydrogenation of **1** with the Rh/PDMP catalyst is shown.

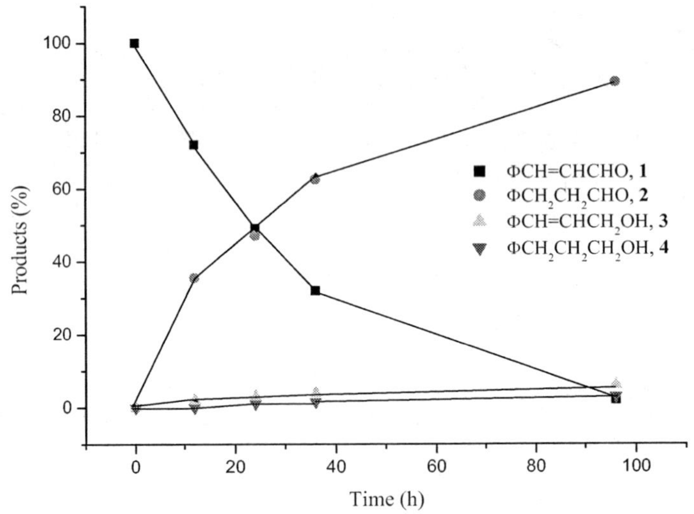

Figure 2. Hydrogenation of cinnamaldehyde catalyzed by rhodium on polydimethylphosphazene (5 wt % Rh) at 25°C and atmospheric pressure of hydrogen (Substrate/Rh molar ratio = 100)

POLYORGANOPHOSPHAZENE PALLADIUM CATALYSTS: PREPARATION AND CATALYTIC ACTIVITY IN THE HECK REACTION

Palladium nanoparticles have been deposited on poly(alkyl)phosphazenes utilizing the Metal Vapour Synthesis (MVS) technique as reported for rhodium (Scheme 4) and the palladium on poly(dimethyl)phosphazene system has been tested as catalyst in the Heck reaction. The Heck reaction, that is the reaction of vinylation of aryl halydes, is widely studied for its application in organic synthesis and fine chemicals[19]. As far as the kind of catalyst employed in this reaction, palladium-based catalysts are resulted particularly active in this reaction[20]. With this in mind the reaction of iodobenzene and methyl acrylate to methyl cinnamate in the presence of palladium on PDMP was a very attractive test for the evaluation of the activity of this polymeric catalyst (Scheme 6). The results obtained are reported in Table 3.

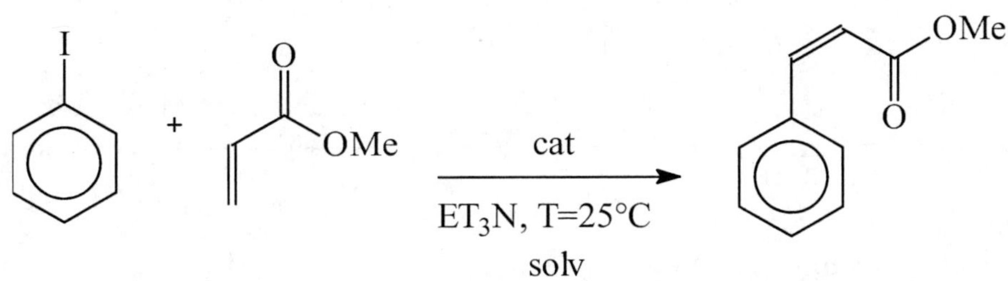

Scheme 6

Table 3. Reaction between Iodobenzene and Methylacrylate to Methyl Cinnamate in the Presence of Palladium Catalysts[a]

Run	Catalyst	Time (h)	Conversion[b] (%)	Specific activity[c]
1	Pd/PDMP	1	80	780
2	Pd/carbon	1	20	235
3	$Pd(OAc)_2$	1	50	500

[a]Reaction conditions: iodobenzene (5 mmol); methylacrylate (10 mmol); triethylamine (5 mmol); palladium catalyst ($5x10^{-3}$ mgatoms); solvent=N-methylpirrolidone (9 ml); T=75°C.
[b]Determined by GC analysis. [c]Moles of methyl cinnamate/g.atom of Pd × hour.

After one hour the conversion is very large (ca. 80%) becoming almost quantitative after three hours (Table 3, run 1). The poly(dimethyl)phosphazene dissolves in the reaction medium allowing to the palladium nanoparticles to operate substantially as an homogeneous catalyst. Interestingly, by addition of tetrahydrofuran, the catalyst precipitates and it can be reused in the same reaction showing an activity very similar to that presented by the starting material, pointing out the high stability of the catalytic system. The Pd on PDMP resulted considerably more active than homogeneous and heterogeneous commercially available palladium catalysts such as the palladium acetate and palladium on carbon, widely employed in the Heck reaction[20]. After one hour the conversion is 20% and 50% with Pd on carbon and $Pd(OAc)_2$, respectively (Table 3, runs 2 and 3).

The examples reported here show that the Ru, Rh and Pd on polyorganophosphazene systems are very attractive catalysts. They can be indifferently used as homogeneous or heterogeneous systems, selecting the appropriate solvent, and can be easily isolated from the reaction mixtures and reused without valuable loss of activity.

Such peculiarities should greatly improve studies on these very unusual catalytic systems.

ACKNOWLEDGEMENTS

Authors wish to thank Prof. Maria C. Gallazzi, Polytechnic of Milan (Italy), for the assistance in the preparation of poly(dimethyl)phosphazene and Drs. Manh Hoang and Terence W. Turney, CSIRO Division of Materials Science and Technology, Clayton, Victoria (Australia) for the HRTEM measurements.

REFERENCES

1. Ciardelli, F., Catalysis in Macromolecular Metal Complexes, In *Macromolecule-Metal Complexes*; Ciardelli, F., Tsuchida, E., Woerle, D., Eds.; Springer-Verlag: Berlin, 1996, p 212-233.
2. a) Klabunde, K. J.; Li, Y. X.; Tan, B. J. *Chem. Mater.* **1991**, *3*, 30.
   b) Vitulli, G.; Verrazzani, A.; Caporusso, A.M.; Pitzalis, E.; Pertici, P.; Salvadori, P. In *Syntheses and Methodologies in Inorganic Chemistry, New Compounds and Materials*; Daolio, S.; Tondello, E.; Vigato, P.F. Eds.; 1997, Vol.7, pp 52-58.
3. Pertici, P.; Vitulli, G. *Comments Inorg. Chem.*, **1991**, *11*, 175.
4. a) Klabunde, K.J. *Free Atoms, Clusters and Nanoscale Particles*; Academic Press, New York, 1994.
   b) Vitulli, G.; Pitzalis, E.; Aronica, L.; Pertici, P.; Bertozzi, S.; Caporusso, A.M.; Salvadori, P.; Coluccia S.; Martra, G. In *Syntheses and Methodologies in Inorganic Chemistry, New Compounds and Materials*; Armelao, L.; Barreca, D.; Daolio, S.; Tondello, E.; Vigato, P.F. Eds.; 2000, Vol. 9, pp 43-50.
5. a) Ciardelli, F.; Pertici, P. *Z. Naturforsch.* **1985**, *404*, 133.
   b) Cardenas-Trivino, G.; Retamal, C.C.; Klabunde, K.J. *Bol. Soc. Chil. Quim.* **1990**, *35*, 223.
6. a)Yu, W.; Liu, H.; Liu, M.; Tao, Q. *J. Mol. Cat. A, Chemicals* **1999**, *138*, 273.
   b) Toshima, V.; Shiraishi, Y.; Terrarishi, T.; Miyate, M; Tominaga, T.; Watanabe, H.; Brijoux, W; Bönneman, H.; Schmid, G. *Appl. Organomet. Chem.* **2001**, *15*, 178.
7. De Jaeger, R.; Gleria, M. *Prog. Polym. Sci.* **1998**, *23*, 179.
8. Gleria, M.; De Jaeger, R. *J. Inorg. Organomet. Polym.* **2001**, *11*, 1.
9. Mark, J. E.; Allcock, H. R.; West, R. *Inorganic Polymers*; Prentice Hall: Englewood Cliffs, New Jersey, USA, 1992, Chapt. 3, p 61.
10. Wisian-Neilson, P. In *Inorganic and Organometallic Polymers II. Advanced Materials and Intermediates*; Wisian-Neilson, P., Allcock, H. R., Wynne, K. J., Eds.; ACS Symposium Series: Washington, DC, USA, 1994; Chapt. 19, Vol. 572, p 246.
11. Wisian-Neilson, P. *this book*, Chapter 5.
12. Gallazzi, M. C. *this book*, Chapter 6.

13. Pertici, P.; Vitulli, G. *Inorg. Synth.* **1983**, *22*, 1976.
14. Pertici, P.; Vitulli, G.; Carlini, C.; Ciardelli, F. *J. Mol. Catal.* **1981**, *11*, 353.
15. Runti, C. *Fondamenti di Chimica Farmaceutica*; Lint: Trieste, 1989; Vol. III, p 465.
16. Singler, R. E.; Hagnauer, G. L.; Schneider, N. S.; LaLiberte, B. R.; Sacher, R. E.; Matton, R. W. *J. Polym. Sci., Polym. Chem. Ed.* **1974**, *12*, 433.
17. Vitulli, G.; Pitzalis, E.; Pertici, P.; Salvadori, P.; Coluccia, S.; Martra, G.; Lampugnani, L.; Mascherpa M. *Material Science and Engineering C* **2001**, *15*, 207.
18. Galezot, P.; Giroir-Fendler, A.; Richard, D. in *Catalysis of Organic reactions*; Pascoe, W. Ed.; Marcel Dekker, New York, 1991, pg. 1. (b) Galezot, P.; Richard, D. *Catal. Rev. Sci. Eng.* **1998**, *40*, 81.
19. Beletskaya, I. P.; Cheprakov, A. V. *Chem. Rev.* 2000, *100*, 3009 and references therein.
20. Bräse, S.; De Meijere, A. in *Metal-Catalyzed Cross-Coupling Reactions*; Dieterich, F., Stang, P. J. Eds.; VCH-Wiley, Weinheim, 1998, p. 99.

In: Applicative Aspects of Poly(organophosphazenes)
Editors: M. Gleria and R. De Jaeger, pp. 267-300
ISBN 1-59454-025-X

Chapter 13

SYNTHESIS AND APPLICATIONS OF PHOSPHAZENE COMPOUNDS

Meisetsu Kajiwara[1] and Keiichi Moriya[2]

1-Nagoya University, School of Engineering, Department of Applied Chemistry, Chikusa-ku, Furo-cho, Nagoya 464-8601, Japan and School of Dentistry, Aichi-Gakuin University, Kusumoto-cho, 1-100, Chikusa-ku 464-0074, Japan

2-Gifu University, Faculty of Engineering, Department of Chemistry, Yanagido, Gifu 501-1193, Japan

1. INTRODUCTION

Polyorganophosphazenes have been synthesized by means of thermal, solution, and radiation ring-opening polymerization of cyclic trimers $(NPCl_2)_3$, followed by macromolecular substitution of chlorines with nucleophiles. This macromolecular substitution reaction has been employed as a standard synthetic route for many phosphazene derivatives. During the last 30 years, over 700 such polymers have been produced. Among them, the PNF elastomer was the first product commercialized by Firestone Tire and Rubber Co. An alternative approach for preparing polyphosphazenes involved polymerization of dichlorophosphinyl phosphorimidic trichloride $Cl_3P{=}NP(O)Cl_2$ as described by DeJaeger. Although, Atochem Co. attempted the aforementioned method for commercialization, the project was now discontinued. Another way of preparing the polymer is the condensation reaction as depicted in Scheme(1):

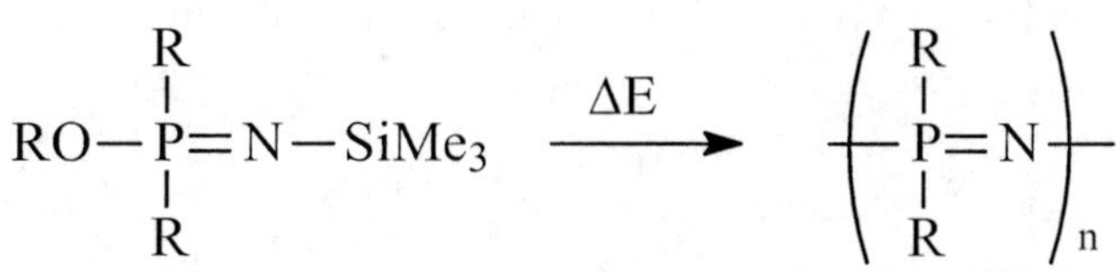

Scheme (1)

However, the variety of side groups that can be introduced in this synthetic approach was limited. Recently, Allcock reported the living cationic polymerization of phosphoramines via an ambient temperature route to afford phosphazenes with controlled molecular weights as shown in Scheme(2). Concurrently, a new type of high molecular weight phosphazene polymers containing 2,2'-dioxyphenyl groups was reported by Carriedo et al, who showed that the direct reaction of $(NPCl_2)_n$ with the bifunctional reagent and K_2CO_3 in THF gave soluble, linear polyorganophosphazenes.

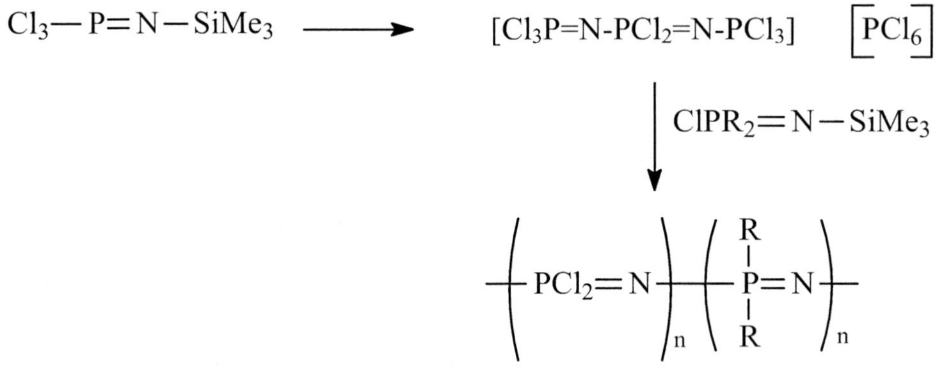

Scheme (2)

While replacement of chlorine in $(NPCl_2)_3$ by the group R is certainly possible, it is difficult to achieve high polymeric phosphazenes from the cyclic organophosphazene compounds. Manners demonstrated that the trimer having a ferrocenyl group can be polymerized to afford high polymers even if halogen atoms are not present. Furthermore, over the last 25 years, six-membered cyclic thionyl phosphazenes such as $(NPCl_2)_2$-NS(O)X(X=Cl, F) have been thoroughly investigated by Manners who reported that cyclic thionyl phosphazene undergoes thermal polymerization to yield novel substituted labile high molecular weight sulfur-nitrogen phosphazene polymers.

In this article, we describe synthesis and applications of some polyorganophosphazene films, for use as new contact lens materials by virtue of enhanced oxygen permeability and applications in electro-optical devices such as photolithography and/or flat panel liquid crystal TV displays. Using dichlorocyclotriphosphazene $(NPCl_2)_3$ as precursors, a considerable effort had been put to maximize the yield of polydichlorophosphazene $(NPCl_2)_n$. Such trimer had been industrially produced and commercialized by various industries, typically Otsuka Chemical Co., Nippon Seika Co., and Nippon Chemical Ind. However, nowadays Otsuka Chemical Co. is the only industry to produce $(NPCl_2)_3$. The $(NPCl_2)_3$ precursors can be easily hydrolyzed by water or moisture. To prepare stable cyclic phosphazene compounds, the chlorine atoms of $(NPCl_2)_3$ have been substituted by nucleophilic reagents, viz., alkoxides, phenoxides, amines, and other derivatives. One notable application of cyclic organophosphazenes is the chemosterilant, e.g., aziridinophosphazene as a biologically active agent under the trade name, "Apholate". Recently, potential applicability of these clyclic organophospazenes in 'antitumor' has been demonstrated, e.g., the phosphazene compound supplied by Kajiwara to NIH (USA) has a possibility of treating cancer or AIDS patients. Another interesting development is that the alkoxy and phenoxy derivatized phosphazenes exhibit flame retardancy to cellulosic materials, synthetic fibers, epoxides, polyurethanes, polyester, and acrylic wool. These phosphazenes are potential materials that might replace the conventional flame-retardants containing halogen atoms such as Cl or Br that are no longer used in the market place because of the environmental concern. In Japan, cyclic organophosphazenes have been marketed under the commercial name PPZ (Kyoeishya Chemical Co.) for use as the hard coating materials. Likewise, Otsuka Chemical Co. industrially produced similar products under the trade names of SPR-100, SA-100, SP-100, SIS-100 and phosphoryl (NF-46, NF-68, NF-100).

Subsequently, we describe various cyclic organophosphazene derivatives prepared by substitution reaction. Especially we emphasize that some derivatives will serve as the functional materials for use as flame-retardants, cross-linking agents for organic resins, and highly thermally stable electric solvents for use in Li-secondary battery in near future.

2. PHOSPHAZENE POLYMERS

2.1 Electrical Conductivity

2.1.1 Chelating Polymers

Kajiwara (1)-(2) extensively studied chelating polymers based on cyclophosphazenes, viz., $(NP)_3(NHCSNHC_6H_5)_6$ (PNP), $(NP)_3(NHCSNHNHC_6H_5)_6$ (PNPS) and $(NP)_3(NHCSNH_2)_6$ (PTT). These chelated polymers were prepared according to the aforementioned method from (PNP), (PNPS) and (PTT) precursors in conjunction with copper, nickel or cobalt in methanol and pyridine. From chemical analysis and NMR spectra of the chelating polymers, the chemical structures have been proposed to have a coordination of number of six. However, these chelating materials can neither melt nor dissolve in most organic solvents including water, and thus it was not possible to determine their molecular weights. By varying temperature, the resistivity for the chelating polymers was measured and subsequently the energy gap of the polymer was calculated in the framework of the general relation deduced for organic semiconductors, viz., $\sigma = \sigma_o \exp(-\Delta E_G/RT)$. It was found in the (PNPS) system that the value of ΔE_G tends to decrease in the order of Cu $<$ Ni $<$Co. It is therefore hypothesized that the lowest energy observed for the (PNPS)-Cu chelating polymer arises from the spin interaction of copper atoms.

2.1.2 Ionomers

Poly(anilino sulphonic)phosphazene $[NP(NHC_6H_5)_{2-(x+y)}(NHC_6H_4SO_3H)_x(NHC_6H_4SO_3Li)_y]_n$ as prepared by Kajiwara(3)-(4) show proton and/or ionic conductivity. The resistivities of the ionic polymers thus determined were summarized in Table 1. The synthesized polymers of (i)-(iv) exhibit proton conductivity, whereas the polymers of (v) and (vi) reveal both proton and Li cationic conductivity. From Table 1, it is apparent that the conductivity increases with increasing x value, which may be attributed to the increase in the proton content in the polymers. Note that the starting (or mother) polymer of x = y = 0 represents the insulator. Concerning the polymers having proton and Li cation, the conductivity of the polymer(vi) was lower than that of the polymer(v). Furthermore, voltage-current (V-I) characteristics of the polymer(vi) were determined at various temperatures. It was found that V-I characteristics of the polymer(iv) showed a slight change with increasing temperature, but they followed Ohm's law in the range of 4×10^{-4} to 18×19^{-4} mA. Also, the relationship between voltage and time was determined at various currents and at two temperatures, 5 and 15 ^{o}C. However, the observed voltage is independent of time. Similarly, electrical conductivity of the polymer(iv) was determined for various currents during temperature rise and drop, in that the polymer(iv) showed a minor change in conductivity with temperature because of the low ionic conductivity. Moreover, the value of the energy gap, ΔE and logρ was estimated using the aforementioned equation. The activation energy estimated for the conduction in the polymer(iv) was about 1.1 eV, but it was lower than that of $[NP(NHC_6H_5)_2]_n$ reported by Hayashi(5).

Table 1. Resistivity of $[NP(NHC_6H_5)_{2-(x+y)}(NHC_6H_4SO_3H)_x(NHC_6H_4SO_3Li)_y]_n$

$[NP(NHC_6H_5)_{2-(x+y)}(NHC_6H_4SO_3H)_x(NHC_6H_4SO_3Li)_y]_n$			Resistivity
	x	y	(Ω cm)
(i)	0	0	1.0×10^{15}
(ii)	0.2	0	2.0×10^{10}
(iii)	0.4	0	2.5×10^{7}
(iv)	0.8	0	2.7×10^{4}
(v)	0.1	0.2	3.4×10^{7}
(vi)	0	0.8	1.6×10^{7}

2.1.3. Polybisaminophosphazenes

Eley and Wills(6) and Reucroft(7) described the electrical conductivity of single crystals and/or powdery forms of hexachlorocyclotriphosphazene $(NPCl_2)_3$ and octachlorocylotetraphosphazene $(NPCl_2)_4$. Subsequently, Allcock and Best(8) reported electrical conductivity of liquid $(NPCl_2)_3$ as well as its polymerization mechanism. Following the procedure of Allcock(9)-(10), various polybisaminophosphazenes involving $[PN(NHC_2H_5)_2]_n$(i), $[NP(NHC_3H_7)_2]_n$(ii), $[NP(NHC_4H_9)_2]_n$(iii), $[NP(NHC_6H_{13})_2]_n$(iv) $[NP(NHCH_2CH_2)_2]_n$(v), $[NP(NMe_2)_2]_n$(vi), $[NP(NH_2N(C_2H_5)_2)]_n$(vii), $[NP(NHC_6H_5)_2]_n$(viii) were prepared in the laboratory of Kajiwara(11). The electrical conductivity of polybisaminophosphazenes was measured using the apparatus described by Kajiwara(1). The glass transition temperatures of the polymer (i), (iv) and (v) obtained from the conductivity measurement were located at about 45 - 80 °C. Below the glass transition temperature at 25 °C, the resistivity decreases as the conductivity increases with increasing number of carbon atoms in the pendant group. The activation energies and the resistivity of secondary aminophosphazene polymers were lower than the values of primary amino phosphazene polymers, which may be ascribed to the inductive effect of secondary amine linked to a side group. Furthermore, the resistivity of the anilinophosphazene polymer at room temperature was unusually high that it was not possible to measure. However, the resistivity decreased on heating and the polymer showed a glass transition temperature of 91 °C. In the case of benzylaminophosphazene polymer, the resistivity was too high to be measured even at elevated temperatures.

2.1.4 Matrix Polymers

Kajiwara(12) reported electrical conductivity and dielectric constant of cyclophosphazene matrix polymer of $[NP(OCH_2CF_3)(\text{2-hyroxyethylacrylate})]_n$. It is seen that resistivity decreased with increasing temperature. Based on the relationship between the conductivity of the polymer and the reverse of the absolute temperature, the activation energy calculated to be 0.92 eV from the slope of the linear portion of the conductivity line. Specific dielectric constants (ε_r') and dielectric loss (ε_r") for the polymer are plotted as a function of logarithm of the frequency. It should be noted that the magnitudes of specific dielectric constant and dielectric loss increase with decreasing frequency; specially this trend becomes more pronounced at frequencies lower than 100 Hz. In the temperature scans, it is evident that both dielectric constant and dielectric loss are higher at lower frequencies over the whole temperature range studied. The dielectric constant increases during temperature increment, whereas the dielectric loss vs. temperature curves exhibit maxima at some frequencies: 120 Hz, 300 Hz and 1 kHz. Specific dielectric constant, dielectric loss, and conductivity of the monomer at 19.3 ^{o}C were measured at various frequencies of 120 Hz, 300 Hz, 1 kHz, 3 kHz and 10 kHz. These complex dielectric constants are higher at a lower frequency similar to that of the polymer. The conductivity is also higher at a lower frequency, and then it becomes almost constant with increasing frequency.

The volume resistivity of the polymer varies from 10^9 Ω cm to 10^5 Ω cm with increasing temperature. Over the temperature range studied, the electrical nature changes from an insulator to a semiconductor. Residual Na(23.5 ppm), presumably NaCl, is not enough to account for the lower resistivity at elevated temperatures. It should also be noted that the electrically conducting functional groups, such as conjugated bonds, are not included in the structure. The activation energy for the electrical conduction in the polymer is 0.92 eV. Such a high value suggests that the conduction in the polymer is probably ionic, because if it were of electronic origin, the activation energy would be much lower. There is a profound frequency dependence of the dielectric dispersion in the polymer. This could be explained in terms of three mechanisms; (i) interfacial polarization, (ii) ion transition from one site to another with the aid of the oscillating electric field, or (iii) the existence of a certain amount of dipoles which have a large response at a lower frequency oscillation. These results are the same regardless of whether silver painted or stainless steel plates are used as the electrodes. Thus, in view of the strong response at lower frequencies, interfacial polarization does not seem to be a major factor. Since the amount of residual Na ions is negligible, the existence of a certain amount of dipoles seems to be a plausible explanation. With increasing temperature, the viscosity of the polymer decreased, which favors dipole orientation. Therefore, an increase in obtained dielectric constant and dielectric loss could be expected when the temperature raises. However, the decline in volume resistance at high temperatures cannot be explained by the existence of dipoles, unless these dipoles are dissociated into positive and negative ions and drift to the electrodes. Moreover if the dipoles are composed of polar groups attached to the polymer chain, species separation is impossible and the mechanism is inadequate to explain the experimental results. An alternative explanation is the formation of a charge transfer complex. If a charge transfer complex exists in the polymer system, it is possible to rationalize both the higher dielectric constants at lower frequencies and the lower volume resistances at higher temperatures, i.e., the dielectric response probably could be slowed down because of the rigidity of the polymer backbone. The decrease of volume resistivity with increasing temperature can also be explained by the reduced chain rigidity. The structure of such complex has yet to be determined. Side groups of the linear polyphosphazenes have been reported to form a charge transfer complex with trinitrofluorenon. Recognizing the difficulty in identifying the structure of charge transfer complexes, one may speculate that an electron is transferred to –COO- from a lone pair on –O- or =N- to form the complex. If this type of charge transfer complex were formed in the polymer, the same mechanism is expected to occur in the monomer.

Dielectric constant, dielectric loss, and conductivity for the monomer, measured at 19.5 °C and at frequencies of 120 Hz, 300 Hz, 1 kHz, 3 kHz and 10 kHz, showed an increase with decreasing frequency. This trend is the same for the resulting polymer, except that the magnitude is appreciably larger than those of the polymer at comparable temperatures. The conductivity of the monomer is also found to be greater than that of the polymer, which is consistent with the aforementioned notion of the charge transfer complex. In an effort to characterize the complex, the UV photo absorption spectrum of the emerged polymer has been examined. The absorption was low at wavelengths longer than 500 nm. As the wavelength decreased, the absorption increased and showed a maximum in the range from 390 to 210 nm. Even though a charge transfer complex exhibits an absorption peak between 210 and 390 nm, it is not well resolved. Further studies in this area are currently under way to elucidate the behavior of frequency dependence of dielectric dispersion.

2.2 Gas Permeability and Selectivity

Gas permeability and separations of various organic polymer membranes have been investigated by many researchers, but there are only few papers available in the field of gas permeability and separations. The study of gas permeability and separations of poly(organophosphazene) derivatives membrane such as $(NPR_2)_n$ or $(NPR^1R^2)_n$ has received special attention in industrial and academic research.

Recently, there appear increasing numbers of reports and patents on gas permeability and separations in several poly(organophosphazene) derivatives. For instance, Bittrova(13) reported the gas permeability of O_2, N_2 and Ar in $[NP(OC_8H_{17})_2]_n$ transparent elastic films to be $12.84x10^{-7}$, $5x10^{-7}$ and $11.88x10^{-7}$ cm^3 cm/cm^{-2} s atm, respectively. McCaffrey (14) studied the gas transport behavior and membrane microstructure of poly(phenoxyphosphazene). Kajiwara(15) described oxygen and nitrogen gas permeability and oxygen to nitrogen selectivity for some poly(organophosphazene) membranes, that is, the oxygen and nitrogen gas permeability were determined using an instrument manufactured by the Yanagimoto Co.

It was found that a membrane having high oxygen permeability exhibited low selectivity, whereas those with high selectivity exhibited low permeability. For instance, silicon rubber, which has the highest oxygen permeability of any reported polymeric membrane, has oxygen to nitrogen selectivity of about 2. This implies that the maximum oxygen concentration that can be produced in a single pass through the membrane is only about 35 %. On the other hand, the $[NP(OC_6H_4Cl\text{-}p)_2]_n$ membrane has an oxygen to nitrogen selectivity of about 3, and thus the selectivity is higher relative to silicon rubber. In general, the relation between oxygen to nitrogen selectivity and oxygen permeability can be approximated with a linear but negative slope in the aforementioned polymers including silicon rubber, polystyrene, ethyl cellulose, polyvinyl chloride or polyethylmethacrylate. Instead, the gas permeability of $(NPR_2)_n$ follows the solution diffusion mechanism of the well-known polymers. If the gas permeability of poly(organophosphazene) membranes were to follow the solution diffusion mechanism, it is generally expected that an approximately linear relation having a positive slope should be valid between the glass transition temperature of the polymer membrane and diffusion constant. However, the linear relation was not observed between the glass transition and oxygen or nitrogen permeability.

In order to separate carbon dioxide or nitrogen in air, the relationships between nitrogen and oxygen, hydrogen, and carbon dioxide or between carbon dioxide and hydrogen, oxygen, and nitrogen of the gas permeability and selectivity were established. It was found experimentally that there appears a linear relationship between nitrogen and overall sum of the gas permeabilities of oxygen, hydrogen, and carbon dioxide, but no correlation was found between carbon dioxide gas permeability and the summation of the gas permeability of hydrogen, nitrogen, and oxygen. In the case of nitrogen gas separation, the membrane that showed the highest selectivity in air is poly(bis-p-ethylphenoxyphosphazene). Its selectivity is about 7 times that of ploy(dimethylsiloxane). On the other hand, poly(bis-trifluoroethoxyphosphazene) has the highest selectivity for carbon dioxide separation, with a selectivity about 35 times greater than that of poly(dimethylsiloxane).

To obtain basic data on the separation or recovery of hydrogen and helium in natural gas, the relation between hydrogen to the summation of gas permeability of O_2, N_2, CO_2, and He, or He to O_2, N_2, CO_2 and H_2 were calculated. However, it was found that the relation between hydrogen (or helium) and the sum of gas permeability of the constituents is nonlinear. As for gas separation applications, it is desirable for the membranes to have the highest selectivity and gas permeability rates.

In the case of hydrogen separation, it is difficult to find a membrane having the gas selectivity larger than unity. However, the poly(bis-p-ethylphenoxyphosphazene) membrane has about 5 times the selectivity of poly(dimethylsilioxane) membrane for hydrogen separation. Poly(bis-trifluoroethoxyphosphazene) exhibits the selectivity about 12 times greater than that of poly(dimethylsiloxane) membrane for helium. To understand the factors affecting the relation between gas permeability and the dielectric constant of poly(organophosphazene) membranes, the data reported by Murakami(16) were re-analyzed. It was found that the selectivity increased with decreasing gas permeability, exhibiting an approximately linear relation. Also, gas permeability is improved with increasing glass transition temperature. In addition, gas permeability can be correlated to the dielectric constant for each gas, that is, gas permeability increased with decreasing dielectric constant. The present relationship between gas permeability and dielectric constant (ε') of poly(organophosphazene) membranes was consistent with those reported by Kajiwara(17). On the other hand, helium having the lowest dielectric constant has the highest gas permeability, but it does not depend on the kind of membranes utilized.

The activation energies and the gas permeability of the membranes showed no relevance to their covalent radii or dielectric constants. At present, it is unclear as to why the activation energy of gas transport in $[NP(NHPr\text{-}n)(Et_2N)]_n$ or $[NP(OC_6H_4CH_3\text{-}p)_2]_n$, is negative and why the activation energy is independent of their covalent radii or dielectric constants.

2.3 Oxygen Gas Permeability in Water

Linear poly(organophosphazenes) are typical inorganic-organic hybrid polymers. A variety of organic side groups bound to each phosphorus atom can be easily introduced and modified. For example, poly(organophosphazenes) bearing suitable side-groups can be used as membrane separators in the field of gas permeation(18), organic liquid separation and evaporation(19)-(22), ion separation(23), ultra filtration(24) and micro filtration(25). As for oxygen gas permeability, many polymers have been determined under dry(26) or wet(27) conditions. Silicon polymer and silicon-methacrylate(28) co-polymers(60-70% silicon and 40-30 % methacrylate) have been tested for oxygen gas permeability under wet conditions, yielding $79x10^{-11}$ and $10x10^{-11}$ cm^2 s^{-1} $mmHg^{-1}$ for the polymer and co-polymer, respectively. At present, soft polymer membranes having higher oxygen gas permeability are needed for application for contact lenses. Kajiwara(29)-(34) measured the oxygen gas permeability in water for various poly(organophosphazene) membranes using a film oxygen permeater of Seikaken type at 20 °C. The dimensions of the films were 15 mm diameter x 0.2 mm thickness. Oxygen gas permeability (D_k) was subsequently evaluated in accordance with equations (1) and (2):

$$K = 14.2\ p/\ i\ l\ 760 \tag{1}$$

$$D_k = K\ i'\ l'\ 760/\ p' \tag{2}$$

where p (mm Hg) is the observed atmospheric pressure of the standard sample (Teflon), i (mA) the electric current, and l (mm) the thickness, whereas the prime denotes for those of the membrane samples such as poly(organophosphazenes).

It is found that the membrane having the highest D_k is $[NP(NEt_2)(NHBu\text{-}n)]_n$. In general, poly(aminophosphazenes) membrane displayed higher D_k values than poly(aryloxyphosphazenes). Also, membranes of varying thickness were prepared with $[NP(NEt_2)_2]_n$ and $[NP(NEt_2)(NHBu\text{-}n)]_n$ and the D_k values in water of these membranes were also determined. The results showed that the D_k is inversely proportional to the membrane thickness (1/L). Using the technique described by Magatani(35), Refojo(36), Huang(37) and Ng(38), the straight line from a plot of $1/D_k$ vs. 1/L can be extrapolated to zero, yielding the oxygen gas permeability value in the dry condition. The three straight lines were extrapolated to zero, and the results calculated from the point of intersection are summarized in Table 2. Also Rogers(39) found that D_k of polymers measured under gas/gas condition is independent of sample thickness. On the other hand, Hung(37) and Ng(38) reported that with silicon rubber poly(4-methylpentlene) membranes D_k in water was found to increase asymmetrically with sample thickness. Kajiwara(40) noticed that the difference between D_k in water and in dry conditions is apparent in these experimental results. It seems that the difference in behavior is due to the additional resistance to oxygen permeability at the boundary between water and the membrane as described by Refojo(36).

Next, poly(n-butylamino)$_{2-x}$(di-allylamino)$_x$phosphazene was prepared through the reaction of $(NPCl_2)_n$. Molecular weight, gas permeability, Young's modulus, glass transition temperature, and refraction index of the membrane were measured. M_w was found to be $5x10^4$ and of 10 x 10^6, respectively. The film prepared by the casting method is transparent, having a light transmission of 95 - 100 % in the visible region. The refractive index of the film is constant even if the n-butylamine content is increased, that is, the refractive index of the film is independent of the molecular weight and the side chain group content of the polymer. Also,

Table 2. D_k Values of Poly(Organophosphazenes) Under Air or Water

$(NPR_1R_2)_n$		D_k ($cm^2\ s^{-1}\ mmHg^{-1}$)	
R_1	R_2	Air	Water
OCH_2CF_3	OCH_2CF_3	$2.0x10^{-9}$	$2.0x10^{-10}$
$N(Et)_2$	HNPr-n	$2.8x10^{-7}$	$52.6x10^{-8}$
$N(Et)_2$	HNBu-n*	$83.3x10^{-8}$	

*Pressed membrane

T_g decreased with decreasing di-allylamine content in the polymer regardless of the molecular weight. T_g of the lower molecular weight polymer was strongly dependent on molecular symmetry as compared to that of the high molecular weight polymer. That is to say the lower molecular weight polymer is more flexible than the high molecular weight polymer. This means that the molecular symmetry is diminished with the side chain group content.

As for the films prepared from the high molecular weight polymer, the Young's modulus have the highest value for a film having, x = 0.9; however, elongation of this film is lower than those of the other films. In the case of the film prepared from the low molecular weight polymers, the Young's modulus and elongation have their maximum values for the film with x = 1.3, but the tensile strength, breaking strength, and elongation at break point are lower for this film than those for the other films. The mechanical properties of these polymer films prepared are inferior to those of PVA or PE films. It was found that D_k value of the high molecular weight film increases with increasing T_g. Whilst in the film having low molecular weights, the D_k value increased with decreasing di-allylamine content and T_g. This means that the gas permeability and glass transition temperature are influenced by the polymer molecular symmetry such as the symmetric or anti-symmetric structure of the polymer molecules. In order to prepare a film having a high D_k value, the polymer has to possess an antisymmetric topology in order to have a higher free volume.

Subsequently, the membranes prepared by the casting method were further irradiated by ultraviolet, electric beam, and ^{60}Co γ-rays. Ultraviolet light having 634 $mJcm^{-2}$ was irradiated on the polymer film, whereas the electron beam was irradiated on the films with an intensity of 150 keV. However, the $CH{=}CH_2/NH$ ratio for the irradiated films estimated from FT-IR data was not so dramatically different as compared with that of the original unirradiated polymer film. Then, the level of the electron beam irradiated on the film was raised to 250 keV. It was found from FT-IR that the $CH{=}CH_2/NH$ ratio decreased with increasing electron beam dose, indicating that the C=C bond is cleaved by the intensed electron beam. Also, ^{60}Co γ–ray were irradiated on the films and FT-IR measurements were taken. These irradiated films were insoluble in THF or other organic solvents.

The oxygen permeability value of electron beam irradiated film in water was determined. It was found that the D_k value of the original film is enhanced with decreasing x value, suggesting that the symmetric or antisymmetric structure of the polymer is closely related to gas permeability. On the other hand, the lowest D_k value was obtained for the high molecular weight film at irradiated 2.3 Mrad, whereas the highest D_k value was achieved for the film irradiated by ^{60}Co γ-ray for 2.55 Mrad and 3 h.

In addition, $[(NPNR_2)(NHR)_{1-x}(OCH_2CH_2COO{=}CH_2)_x]_n$ type polymers were prepared via the reaction of $(NPCl_2)_n$ with di-n-butylamine, n-butylamine and 2-hydroxyethylacrylate. After Irga-Cure 184 was added to the polymer dissolved in THF, the solution was kept at room temperature. The films obtained were irradiated at 160 Wcm^{-1} for 30 s. It is seen that D_k value of the film before or after ultraviolet radiation decreased with increasing 2-hydroxzyethylacrylate in comparison with that of the original film. On the other hand, breaking strength and elongation of the films before or after ultraviolet radiation decreased in comparison with those of the film. Furthermore, to enhance the mechanical properties of the polymer films such as $[NP(NHBu\text{-}n)(di\text{-}n\text{-}hexylamino)phosphazene]_n$, neopentyl ether (NPG) and trimethylol propanetriglycidyl ether (TMP) were added to the original polymer. The transparent films were obtained after the original cross-linking agents were heated at 60 and 120 °C. Also, the polymers cured with NPG and TMP cannot be dissolved in THF or other organic solvents. FT-IR spectra of the original as well as the cured films were acquired. It was found from FT-IR that –NH groups appeared near 3470 and 3370 cm^{-1} peak declines with increasing of –CH(OH) groups that appear near 1350-1260 cm^{-1} and 1100 cm^{-1}. It may be hypotheisized that the cross-linking reaction probably occur as given in Scheme (3), that is, the PNG or TMP epoxy ring is cleavaged by the HNBu-n proton containing in the polymer and the intermolecular cross-linking reaction proceeds by heating.

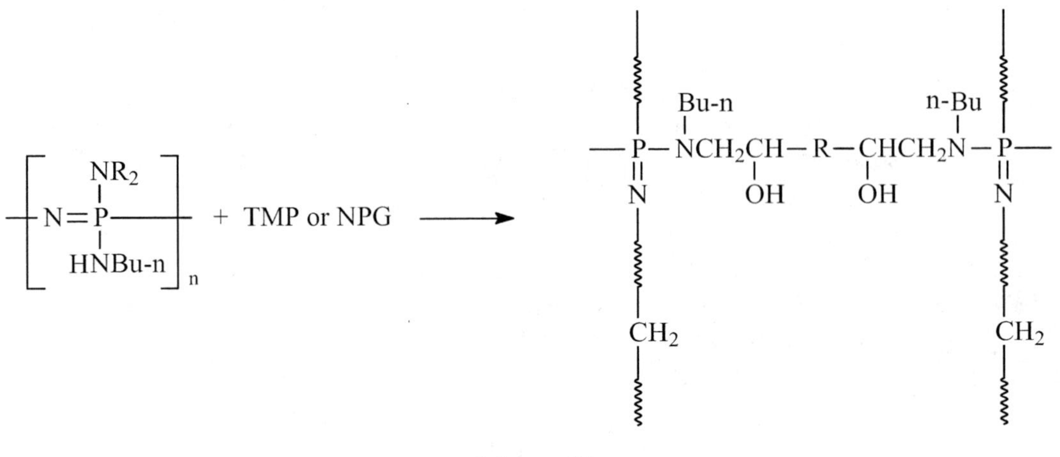

Scheme (3)

Mechanical properties of the film were determined by Toyo Boldwin Co. Ltd, UTMII-20(Tensilon). It was evident that the value of elongations at yield point and at break of the cured films are lower than those of the original film. On the other hand, the breaking strength of the film cured with 0.1 wt % NPG at 60 and 120 °C is greater than that of the original film. As for the films cured with TMP, the elongations at yield point and at break are lower, but breaking strength is higher relative to the original film. Also, the film cured with NPG and TMP showed higher Young's modulus or higher tensile strength.

As for D_k values, it was found that D_k values of the film cured by NPG and TMP are larger than those of the original film. Especially, the film cured with 0.05 wt% TMP is about three times higher compared with that of the original film. However, D_k values of the cured film are smaller than that of dimethylpolysiloxane.

2.4. Cultivation Bed

Cyclophosphazenes such as $N_3P_3Az_6$, $N_4P_4Az_8$ and $N_4P_4pyrro_8$ were reported by Labarre(41), Nabi(42), Cho(43) and Bovi(44) to have the function of anticancer agents. Allcock(45) added that some poly(organophosphazene) derivatives have special properties like biodegradable materials. These materials were produced mainly by means of macromolecular substitution reactions carried out on a reactive polymeric intermediate. Organic side groups such as amino acid esters, glucose, glycolate, lactate or imidazolyl sensitize the polymers to hydrolysis to benign materials such as phosphate, ammonia, and the free side unit.

Recently, polymeric anticancer drug conjugates have been studied in order to improve the side effects of the anticancer drugs. In comparison with low-molecular weight anticancer drugs, polymeric anticancer drug conjugates have been found to be accumulated more in tumor tissues than in normal tissues by virtue of the enhanced permeability and the retention effect. In addition, anticancer drug conjugates could afford to prolong the institutor activity by releasing the drug at a controlled rate. Among several polydoxorubicin conjugates hitherto studied, N-(2-hydroxypropyl)methacrylamide copolymer/doxorubicin conjugate has currently reached the clinical study, but there is no report on polyphosphazene/doxorubcin conjugate. Song(46) found that poly(organophosphazene)/doxorubicin conjugates exhibited high antitumor activity of three kinds of poly(organophosphazene)/doxorubicin conjugates having β-galactosyl, 2-(2-aminoethoxy)-ethanol and L-glutamonyl dibenzyl ester p-toluenesulfonate.

Goedemoed(47) demonstrated that some poly(organophosphazenes) were suitable for development of implantable antitumor devices. However, there are no available reports regarding the study of bioreactor materials using poly(organophsphazenes). Kajiwara(48) described the study of adhesive properties and the degree of colony formation of Chinese Hamster Ovary(V-79) cell and Bowes cell lines with poly(organophosphazenes).

Regarding a culture fluid and a bed of the cell cultivation, the culture fluid was prepared by adding minimum essential medium (MEM) to 10 wt % of fetal calf serum. The viscosity and density of this culture fluid were 0.01 poise and 1.01 g/cc, respectively. A bed for cell cultivation was prepared by dissolving 1 g of poly(organophosphazenes) in THF, and the film was formed by casting in Falcon Petri dishes, and then 5 ml of culture liquid having 4×10^4 of V-79 cell was added to the film dishes, where they were kept at 37 °C for 4 to 72 h. Bowes cells (Myeloma cell) and Chinese Hamster Ovary cells, which produced tissue plasminogen activator, were cultured using the cultivation on polystyrene, using Cytodex III (Pharmacia) as the standard sample. From a typical colony formation as observed under a microscope, the results of the percent colony formation and the adhesive properties based on V-79 cells for poly(organophosphazene) films are summarized in Table 3.

Table 3. Adhesive (A) and Percent Colony Formation (B) of Poly(Organophosphazenes) using V-79 Cells

$(NPR_1R_2)_n$		(A)	(B)	D_k
R_1	R_2	(%)	(%)	(10^{-11} cm^2 s^{-1} $mmHg^{-1}$)
OCH_2CF_3	OCH_2CF_3	11.7		
OC_6H_4F-p	OC_6H_4F-p			
$OC_6H_4CH_3$-p	$OC_6H_4CH_3$-p	58	50	5.5
OC_6H_4Cl-p	OC_6H_4Cl-p	4.5		
OC_6H_4Et-p	OC_6H_4Et-p	62	50	19.2
$OC_6H_4CH_3$-m	$OC_6H_4CH_3$-m	70	46	0.2
OC_6H_5	OC_6H_5	77	50	6.4
NHPr-n	NHPr-n	83	19	15.5
NHBu-n	NHBu-n	61	44.2	
NHC_5H_{11}	NHC_5H_{11}	27.7		
NHPr-n	$N(Et)_2$	55.2		
NHBu-n	$N(Et)_2$	33	55.2	
NHC_5H_{11}	$N(Et)_2$	2	16.0	
Falcon	88	65		

As can be seen in Table 3, poly(bis-phenoxyphosphazene) $[NP(OPh)_2]_n$, poly(bis-p-ethylphenoxyphosphazene) $[NP(OC_6H_4Et\text{-}p)_2]_n$, and poly(bis-butylaminophosphazene) $[NP(NHBu\text{-}n)_2]_n$ have good percent adhesiveness in comparison to other poly(organophosphazenes). On the other hand, percent colony formation on poly(bis-p-ethylphenoxyphosphazene) $[NP(OC_6H_4Et\text{-}p)_2]_n$ and poly(bis-phenoxyphosphazene) $[NP(OPh)_2]_n$ are about 50 % less. It was also observed that adhesiveness and percent colony formation of poly(organophosphazene) films in V-79 cells are lower than that of Falcon Petri dishes. It was also found that halogen-containing polymers including poly(bis-trifluoroethoxyphosphazene) $[NP(OCH_2CF_3)_2]_n$ and poly(bis-p-chlorophenoxyphosphazene) $[NP(OC_6H_4Cl\text{-}p)]_2]_n$ are poor in adhesiveness or colony formation since V-79 cells responded negative to the halogen. In the case of the polymers having $-NC_5H_{11}$ or $-NEt_2$ groups, adhesiveness and percent colony formation are lower than that of amino phosphazene polymers, viz., poly(bis-n-propylaminophosphazene) $[NP(NHPr\text{-}n)_2]_n$ or poly(bis-butylaminophosphazene) $[NP(NHBu\text{-}n)_2]_n$, owing to basicity. As colony formation of V-79 cells is related to oxygen permeability of poly(organophosphazene) films in water. Oxygen permeability (D_k) in water of these polymers was determined according to the method of Kajiwara(49). The results obtained are summarized in Table 3.

It seems that percent colony formations depend on oxygen permeability of the films, except for films having no halogen atoms or $-NC_5H_{11}$ and $-NEt_2$ groups. The films having D_k= 5–20 x10^{-11} $cm^2s^{-1}mmHg^{-1}$ give good colony formation. To prepare the films having the highest adhesiveness and percent colony formation, some modified poly(organophosphazenes) such as $[NP(OPh)_{2-x}(NHBu\text{-}n)_x]_n$ were synthesized.

Table 4. Adhesive (A) and Percent Colony Formation (B) of Poly(Organophosphazene) $[NP(OC_6H_5)_{2-x}(NHBu\text{-}n)_x]_n$

$[NP(OC_6H_5)_{2-x}(NHBu\text{-}n)_x]_n$ x	(A) (%)	(B) (%)
0	77	50
0.2	84	54
1.0	62	77
1.8	84	53
2.0	61	

In this study, it is evident that compositions of the best films having the highest percent colony formation are $[NP(OPh)_{1.8}(NHBu\text{-}n)_{0.2}]_n$, but the percent colony formation of these films is a little lower than that of the Falcon Petri dishes. There appears no relation between the x value of $[NP(OPh)_{2-x}(NHBu\text{-}n)_x]_n$ and percent adhesiveness or colony formation. One interesting observation is that the cell colonies thus grown on the film substrate can be removed, although its influence on the bioreactor materials such as poly(organophosphazenes) has yet to be resolved. Nevertheless, it may be suitable for applications as artificial skins because the films thus formed after applying poly(organophosphazene) on the damaged skin could hold for several days and can be removed easily from the regenerated skin. Regarding cultivation of Bowes and Chinese Hamster Ovary cell lines using polystyrene beads coated with poly(bis-phenoxyphosphazene) $[NP(OPh)_2]_n$, it was found from Table 4 that the percent adhesiveness and cultivation of V-79 cells for $[NP(OPh)_2]_n$ film was similar to the values of the Falcon Petri dishes. The cultivation of Bowes and Chinese Hamster Ovary cell lines were carried out on polystyrene beads coated with $[NP(OPh)_2]_n$. First, 4 x 10^4 cells were cultivated on polystyrene beads. The total cell number was estimated by counting them for the period of cultivation. In the case of Bowers cell line, after 4 h of cultivation, the cell mummer declined while closely approaching the steady state. When Cytodex III was used as the standard sample, cell numbers showed an increasing trend with increasing cultivation time. In the case of Chinese Hamster Ovary cell, after 4 h of the cultivation, the cell were decreased and degree of the percent cultivation reached the steady state after 2 day or compared with the results of the Cytodex III standard. However, the cells disappeared completely after 3 days.

2.5 Microlithography

The minimum feature size in ULSI (Ultra large-scale integrated circuit) fabrication continues to shrink and is expected to approach the 0.2 μm range by the end of this century. E-beam lithography is thought to have potential to afford to reach such a small-scale regime; however, it will be necessary to alleviate the resolution loss due to forward- and back-scattering effects. To overcome these difficulties, multilayer resist systems have been developed; a bilayer resist system, based on oxygen-reactive ion etching (O_2-RIE), has been regarded as the most promising method for realizing a practical multilayer resist system because of a small number of processing steps and its reliability. The system consists of a thin imaging layer and a thick planarizing layer. The pattern of the top imaging plate is transferred into the thick bottom layer via anisotropic pattern transfer such as O_2-RIE. In addition, the imaging layer acts as an important etching barrier to O_2-RIE; thus a radiation-sensitive polymer, durable to O_2-RIE, should be used as an imaging layer.

Various organosilicon polymers have been reported as candidates for the top-layer resist (50)-(54). Kajiwara (55)-(57) added organosilyl groups to the silica sheet in order to enhance solubility in appropriate solvents as well as sensitivity to E-beams. Poly(organophosphazenes) have been known not only to resist oxidative break down, but also have provided good uniform films for lithographic applications (58)-(59). Recently, with increased applications of multilayer lithographic processes, poly(organophosphazenes) should find applications as lithographic materials similar to poly(siloxanes) and poly(silanes), since they have high oxygen resistance. Kajiwara (60)-(63) further investigated the plasma stability of some ply(organophosphazene) films. The stability of poly(organophosphazene) films for plasma was evaluated for various film thicknesses. The pressure used was 1, 5 and 20 Pa for an etching time of 1 or 2 min under 0.08 Wcm^{-2} (power density). The ion-etching machine utilized was a parallel type reactor from DEM 451 (Aneruba Co., Ltd) and the ESCA was XSAM-800 (Kratos Co., Ltd). Refractive indices of poly(organophosphazene) films were measured using an ellipsometer. It was found that the refractive index of the film having aromatic side chains is higher than that of aliphatic side chains. The RIE of poly(organophosphazene) films spin coated onto silicon wafer were also measured under 5 Pa and 0.08 Wcm^{-2} using plasma (CF_4-O_2). The relation between the reduction in film thickness and exposure time was investigated. It was found that poly(organophosphazene) films having aliphatic side chains such as ethoxy or trifluoroethoxy show poor RIE in comparison with novolak resin (LMR), i.e., $[NP(OCH_2CH_3)_2]_n$ has a lower RIE value than that of $[NP(OCH_2CF_3)_2]_n$. Poly(organophosphazene) films with aromatic side chains such as phenoxy or aniline have a lower RIE relative to those with side chains such as p-chlorophenoxy and p-chloroaniline. Furthermore, the RIE values of $[NP(OPh)_2]_n$ and $[NP(HNC_6H_5)_2]_n$ are larger than that of LMR. It is apparent that the film of poly(organophosphazene) exhibits the lowest RIE as well as LMR. The films of poly(organophosphazenes) having aromatic amines reveal a high RIE in comparison with others. The RIE of the films was investigated by means of an ellipsometer under various pressures. It was found that the rate of disappearance of poly(bis-trifluoroethoxyphosphazene) film increased with increasing pressure, suggesting that poly(bis-trifluoroethoxyphosphazene) film has no resistance to plasma. On the other hand, the rate of the disappearance of films having aromatic side chains declined with increasing pressure except for poly(bis-anilinophosphazene) film which reached a steady state when the pressure was greater than 5 Pa. This means that poly(bis-anilinophosphazene) film has resistance to plasma under higher pressure. Poly(organophosphazene) films such as $[NP(OCH_2CF_3)_2]_n$, $[NP(OEt)_2]_n$, $[NP(NHC_6H_5)_2]_n$ and $[NP(NHC_6H_4Cl\text{-}p)_2]_n$ prepared without (A) and (B) plasma on a silicon wafer were investigated using ESCA. The maximum peaks of F_{1s}, O_{1s}, N_{1s}, C_{1s}, P_{2p} and Cl_{2p} core signals were determined. The relative atomic percent of F, O, N, C, P and Cl on films (A) and (B) were estimated from the peak area.

In the case of $[NP(OCH_2CF_3)_2]_n$ films, there appear films six peaks in films A and B. The maximum peak of the fluorine core signal (F_{1s}) appears at 690.6 and 690.8 eV for films A and B, respectively. The relative atomic per cent of F in film A is similar to that of film B, but the calculated atomic per cent of $[NP(OCH_2CF_3)_2]_n$ is not in accord between them. On the other hand, the maximum peak of oxygen core signal (O_{1s}) appears at 536.1 and 536.2 eV for films

A and B, respectively. The relative atomic percent of oxygen of film B increase about 20 % in comparison with that of film A though the spectrum in the film B is broad. This implies that the $–OCH_2CF_3$ side chain may be oxidized by plasma. In the case of nitrogen core signal (N_{1s}), the maximum peak appears at 400.6 and 401.2 eV for films A and B, respectively in which the latter peak is shifted to a higher energy side for ~ 0.6 eV. The relative atomic percent of nitrogen is almost the same for both films A and B. The maximum peaks of carbon core signal (C_{1s}) appear at 295.0 and 289.9 eV for film A or 295.3 and 290 eV for B. The relative atomic percent of carbon in the film B decreased by ~ 18 % in comparison with that of film A. It is postulated that the carbon atom must be removed by the plasma. In the phosphorus core signal (P_{2p}), the maximum peak is located at 136.6 and 136.8 eV for film A and B. It was found that the relative atomic percent of P in film B increases.

In $[NP(OCH_2CH_3)_2]_n$, the original film A and B of have five peaks from which the relative atomic percents of F, O, N, C and P were estimated. The maximum peak of fluorine core signal (F_{1s}) of the film can be discerned at 688.5 and 688 eV for films A and B. Although $[NP(OCH_2CH_3)_2]_n$ had no fluorine atoms, the original film A includes about 0.6 % F and the relative concentration of F of film B is about 5.3 at %. This may be attributed to the reaction between the film and plasma, that is, the film reacts with CF_3^+ or F^- ions originated by the plasma. The maximum peak arising from the oxygen core signal (O_{1s}) of the film A and B appears at 534.1 eV, and also the relative atomic percent of oxygen of the films A and B are almost the same, suggesting that C-O bond is not affected significantly by the plasma. The maximum peak of nitrogen core signal (N_{1s}) appears at 399.2 and 400.4 eV for films A and B, respectively. The latter peak is shifted to a higher energy side by about 1.2 eV. The relative atomic percent of oxygen of film A is also similar to that of film B. The maximum peak of carbon core signal of the film A appears at 286 eV, but the relative atomic percent of carbon of film B is lower than that of the film A by about 18 %, implying that carbon atoms must be eliminated by the plasma. On the other hand, the maximum peak of phosphorus core signal appears at 135 and 135.3 eV for film A and B, respectively. The relative atomic percent of phosphorus is similar in both films A and B which in turn suggests that the NP skeleton of the film is not affected by the plasma or oxygen. It seems that $–OCH_2CH_3$ side chain must be fluorinated with the CF_3^+ or F^- by the plasma. Even if the side chains can be oxidized with plasma, the rate of oxidation of $[NP(OCH_2CH_3)_2]_n$ must be lower than that of $[NP(OCH_2CF_3)_2]_n$, because the former is stable against the plasma. As for the fluorine core signal (F_{1s}) of $[NP(NHC_6H_5)_2]_n$, film A shows a broad peak from 690 to 687 eV, whereas film B reveals multiple peaks ranging from 698 to 680 eV. The relative atomic percent of fluorine is about 3.9 % in film B, suggesting that the $–NHC_6H_5$ side chain may react with the CF_3^+ or F^- originating from the plasma (CF_4-O_2). Although this film did not contain oxygen atoms, the maximum peak of the oxygen core signal appears at 533.4 and 534 eV for films A and B. In General, the peak appearing near 533 eV can be assigned to C-O, C=O or H_2O. It may be inferred that this peak belongs to the absorbed water in film A. On the other hand, the relative atomic percent of oxygen is higher than that of the film A by about 26 %, implying that film B must be oxidized with plasma. The maximum peak of the nitrogen core signal appears at 400.4 and 401.4 eV in film A and B in which the latter peak is shifted to the higher energy side by about 1.0 eV while the relative atomic percent of nitrogen becomes lower by about 43 % for film B. As for the carbon core signal, the maximum peak is located at 285.6 and 286.5

eV for films A and B due to C-H or C-N bonds. The maximum peak of phosphorus core signal also appears at 134.6 and 135.5 eV in films A and B.

In $[NP(NHC_6H_4Cl\text{-}p)_2]_n$, except for chlorine, the relative concentrations of nitrogen, carbon and phosphorus are in accord with their theoretical values. The fluorine core signal (F_{1s}) of the original film A showed multiple peaks from 698 to 683 eV. However, only one peak is evident at 687.8 eV in film B simply due to its reaction with the CF_3^+ or F^- ions originated from the plasma. The relative fluorine content was about 14.8 at %.

Although the film of them did not contain any oxygen atoms, the maximum peak of oxygen core signal (O_{1s}) can be discerned at 532.9 and 532.6 eV in films A and B. These peaks are due to absorbed water in the film. The relative concentration of oxygen was 13.7 at % in film B that is higher than that of film A. This means that the film B must be oxidized by plasma. Comparing the maximum peaks of nitrogen (N_{1s})) appearing at 399.5 and 399.4 eV for film A and B, the relative concentration of nitrogen in film B declines about 28 at % relative to that of film A. The maximum peak of carbon core signal (C_{1s}) appears at 284.9 and 284.7 eV for film B. There are no significant changes of the maximum peaks of nitrogen signal(N_{1s}) and phosphorus (P_{2p}) core signal in both films A and B, except that the relative concentrations in film B are lower than that of film A. It appears that the film is not stable for plasma since fluorination and oxidation reaction occurred with the plasma.

2.6 Liquid Crystalline Behavior of Phosphazene Polymers

2.6.1 Phase Transition of Poly(Bis(Octyloxybiphenoxy)Phosphazene) ($[PN(OC_6H_4C_6H_4OC_8H_{17})_2]_n$, (PBOP))

Poly(bis(4'-(4-octyloxy)biphenoxy)phosphazene) (PBOP) was prepared according to the substitution reaction of polydichlorophosphazene(PDCP) and 4-hydroxy(4'-octyloxy)biphenyl and NaH in the presence of tetra-n-butylammonium bromide in THF under reflux for 160 h. PDCP was polymerized via the ring opening polymerization of hexachlorocyclotriphosphazene using sulfur as a catalyst at 270 °C for 3 h under vacuum in a pyrex tube. The reaction was followed using P-NMR. The resulting polymers were extracted by means of Soxhlet apparatus by reprecipitating the polymer from THF solution in ethanol. The P-NMR showed a broad singlet around -14 ppm, but the elemental analysis confirmed that only 66 % of the side chains were attached to the polyphosphazene backbone. In the PBOP, which are not fully substituted, no mesomorphic phase was observed. The polymer backbone of the polyphosphazenes probably prevent the formation of the liquid crystalline phase. (64)-(65)

Moriya and Kajiwara(66) were the first to report that poly(bis(4-(N-(4-heptyloxyphenyl) iminomethylphenoxy)phosphazene)) ($[PN(OC_6H_4C{=}NC_6H_4OC_7H_{15})_2]_n$, PBHSP), having mesogenic Schiff side chains, shows the existence of mesomorphic phase between 463 and 486 K. Based on the polarizing microscope observations and the X-ray powder diffraction measurements, the smectic layer structure was identified in the mesophase.

3. Phosphazene Oligomers

3.1 Cross-Linking Agents

Epoxy resins have been used in paints, adhesive, and insulation materials. Recently, epoxy resins having higher thermal stability and lower expansion have been in great demand in the electrical and other fields. In particular, the flame-retardant epoxy resins containing non-halogen atoms need to be developed, because halogen atoms are susceptible to corrosion to lead wire and also a potential source of air pollutions. Also, it is believed that the chemical, electrical, and mechanical properties of epoxy resins depend strongly on the chemical structure of the curing agents. Kajiwara(67)-(70) investigated the mechanical, electrical and chemical properties of epoxy resins cured with hetero-ring amine such as diaminotetraorganophosphazenes $(NP)_3(NH_2)_2R_4$ (i) (R = OPh, OC_6H_4Cl-p, NHC_6H_5, CH_3O, etc)). (i) was prepared according to the method of Gerald and McBee, and pure (i) was obtained by recrystallization from diethylether. The epoxy resin used was Epikote-828 and 1045-H-70, which composed of diglycidol ether and bis-phenol A. After the addition of equivalent mole ratio of curing agents such as (i), and organic diamine such as ethylenediamine, the resin mixtures were heated and stirred to produce a homogeneneous solution. The solutions were then placed in the mould and precured at 80 –120 °C, and post cured at 120 –180 °C. The resins thus cured were insoluble in organic solvents. On the basis of Fourier transform infrared spectroscopy, it is proposed that the cured epoxy resins have the following structure in Scheme (4).

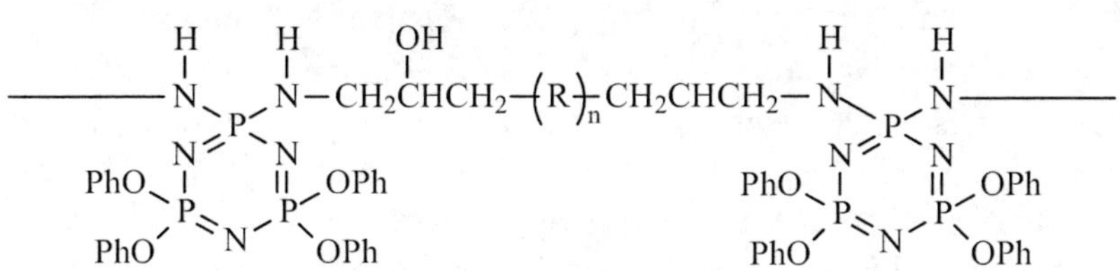

where R stands for

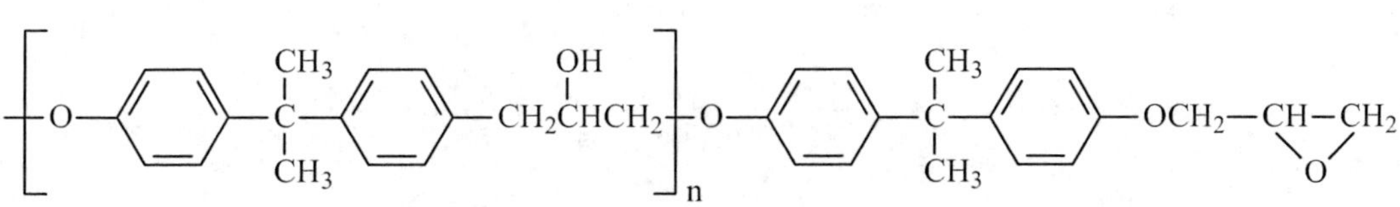

Scheme (4)

Next, the dynamic mechanical analysis and tensile measurements were undertaken on the cured epoxies. As for the glass transition temperature as obtained from stroge modulus E' and loss modulus E", the Tg of epoxy resins cured with (i) increased in comparison with those cured with aliphatic amines. Also, Young's modulus and tensile strength of epoxy resins cured with (i) reveal higher mechanical properties than epoxy resins cured with aliphatic amines.

Furthermore, the flammability, hardness and other properties of the cured epoxy resins were determined based on the standard methods. It was found that the pencil hardness of the cured epoxy resins is between 3H and 2H. The epoxy resins cured with (i), having aromatic substitutions groups, are self-extinguishable. Also, the epoxy resins cured with (i) having alkoxy or amino groups are non-flammable. Subsequently, the electrical properties of the cured resins were determined. The electrical resistance of the cured epoxy resins is $10^{15} – 10^{16}$ Ωcm suggesting that the insulation property of the epoxy resin is not disappeared when diaminotetraorganocyclotriphosphazenes were used as curing agents.

3.2 Flame Retardants

In general, the non-flammable resins or fibers were fabricated via addition of the halogen compounds including Cl, Br, P, Bi and Sb atoms (71). Godfrey(72), Rice(73) and Quinn(74) found that the non-flammable resins or fibers can be prepared through addition of cyclic organophosphazene or poly(organophosphazenes). Kajiwara(75) reported that cyclophosphazenes such as (i) and (ii) were found to be effective flame retardants. In the preparation of boric ester, $H_2N(CH_2)_2OB(OH)_2$ (EABP) was reacted with one mol ethanol amine and boric acid at 150 °C for 3 or 4 days. When the reaction was over, the reaction mixture was dissolved in water. To remove unreacted boric acid, 10 % of $Ba(OH)_2$ was added to the solution, and then the precipitated product was filtrated using a glass filter. After the filtrate was evaporated under reduced pressure, the residues were extracted with ethanol. Subsequently, ethanol, dehydrated with Na_2SO_4, was evaporated, and then an oily substance was formed. Also, $H_2N(CH_2)_3OB(O)OH$ (PABBP) was prepared with the same manner. The extrated oily substance was analyzed by means of H-NMR, IR and chemical analysis.

It was found from the data that the product was apparently EABP and PABBP. Also, both boric ester was reacted with $(NPCl_2)_3$ using THF and $(Et)_3N$ at 60 °C for 1 – 24 h. When the reaction was over, $(Et)_3N{:}HCl$ was filtered through a glass filter. Also, the filtrate was evaporated under reduced pressure, and the residues were purified with column chromatography using α-Al_2O_3. The purified product was determined using H-NMR, P-NMR and IR. It was confirmed that the structure of the product was (i) and (ii) in Scheme (5).

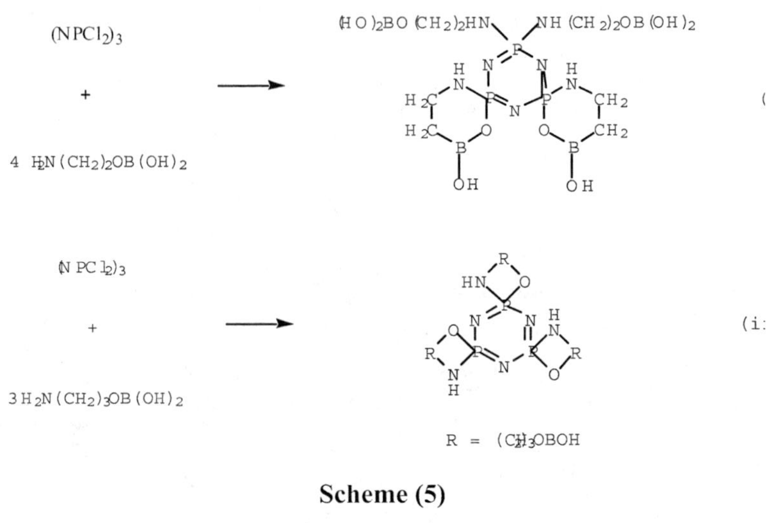

Scheme (5)

The product (i) and (ii) was added to phenol, formaldehyde, and HCl, and then the mixture was heated at 120 °C for 24 h. The thermal stability of phenolic resins including (i) and (ii) was determined using DTA and TG. No identifiable exothermic and endothermic reaction peaks appear in the range of about 500 °C in the case of both phenol resins. However, some difference was observed in the TG measurement, that is, the weight loss of phenol resin that contained 5 % of (i) was found to be ~ 20 % at 230 °C, whereas the weight loss of mother phenol resin occurred about 90 °C that continued to decline with increasing temperature. It is reasonable to conclude that the addition of 5 % of (i) into phenolic resin promotes thermal stablility. Also, LOI (Limited Oxygen Index) value of the phenolic resin was determined using Suga IN-1 and ON-1D instruments using the following equation.

$$LOI = [\ O_2\ (1/min)\ ]/[O_2\ (1/min) + N_2\ (1/min)]\ x\ 100$$

The value shown in Table 5 represents the average of three or five runs. The LOI value of the resins containing (i) and (ii) was higher than that of the pure resin. In both cases, the LOI value increased with increasing the amount of (i) and (ii). Of particular interest is that the LOI value of the resin including (ii) is higher relative to that of resin containing (i). It may be inferred that the additive (ii) is an ideal flame-retardant for phenolic resins because the phosphorus concentration was higher than that of (i). The smoke concentration generated during burning was determined. It was found that the total amount of smoke was less than that of the neat resin, while the combustion time was a shorter as compared with the resin without (i) and (ii). It may be concluded that additives (i) and (ii) presumably supress the formation of smoke and also shorten the combustion time.

Table 5. LOI Values of Phenol Resins Containing Additives (i) and (ii)

Additive	LOI	
(%)	(i)	(ii)
5	30	35
10	40	45
20	45	55
Blank	28	

3.3 Electrical Conductivity

The electrical conductivities of polycrystalline cyclophosphazene compounds such as $(NPCl_2)_3$, $(NP)_3(NH_2)_2R_4$ (R= OC_6H_5, OCH_2CF_3 NHC_6H_5, SC_6H_5), $[NPCl_{2-x}(NHC_6H_5)_x]_3$, $(NP)_3Cl_{6-x}(NHC_6H_4NH)_x$ and $[NPCl_{2-x}(NHC_3H_2)_x]_3$ were investigated by Kajiwara according to the procedure reported earlier(79)-(82). The electrical receptivity (ρ) was estimated from $\rho = RAV/L$, where R is the electrical resistance, A is the area of the plate (1.3773 cm^2), L is the sample thickness (cm), and V is the volume fraction. In general, the receptivity of molecular crystals like pthalocyanine is reduced with increasing in the molding pressure. However, the receptivity reached a constant value if the material is molded under a pressure above 80 kg/cm^2. The resistivity of polycrystalline $(NPCl_2)_3$, subjected to various pressures for 10 min and at different temperatures, is inversely propotional to the molding pressure, a phenomenon usually observed in molecular crystals. The activation energy ΔE was found to be the same although the receptivity was different at different pressures which has led to a speculation that the conductivity is produced by ions or by π electrons in the N_3P_3 rings.

To differentiate the aforementioned two mechanisms, temporal dependence of the conductivity was investigated under a constant pressure. The resistivity was found to increase with time and after 10 min reached a constant value, suggesting that the most suitable condition to be carried out is about 10 min after the sample was placed in the field. Such a behavior occurs due to ionic conductivity caused by polarization or Hall effect. If conductivity in the N_3P_3 ring were attributed to the π electrons, the excitation energy is expected to obey $\Delta E_{ph} = h_c/\lambda_{max}$; the theoretical value needed for this exciting state, according to the Mottogurney equation, would be ~5.04 eV at 250 nm. Ultraviolet absorption of $(NPCl_2)_3$ determined in heptane gave an excitation energy of 2.0 eV. The large difference in the excitation energy indicates that the conductivity is governed by ions rather than by π electrons. The electric current was determined using a dielectric voltage from 3.0 to 100 V. Ohm's law was found to hold up to 300 V/cm, but it failed above this value. Many reports describe similar results. It may be speculated that the cause is either an accumulation of a space (or a barrier) charge between the plate and specimen. It is important to investigate the reproducibility of receptivity. Although the plots were linear and obey the equation $\rho = \rho_o \exp^{(\Delta E/RT)}$, different lines are obtained. However, the activation energy ΔE to induce conductivity as calculated from these lines shows the same value.

In the case of hexaisothiocyanocyclotriphosphazene $[NP(NCS)_2]_3$, disks of $[NP(NCS)_2]_3$ were prepared under pressures of 400 kg/cm^2 and 100 kg/cm^2 for 0.5 h. The energy gaps (ΔE) of the trimer calculated from the general semiconductor equation are 1.94 eV at 400 kg/cm^2 and 1.03 eV at 100 kg/cm^2, respectively. The value of the resistivity of the trimer is three orders of magnitude smaller than that for $(NPCl_2)_3$, suggesting that $[NP(NCS)_2]_3$ is easily ionized and polymerized. Athough $(NPCl_2)_3$ shows strong temporal of dependence electrical conductivity, it is not pronounced for $[NP(NCS)_2]_3$ because the electrical conductivity of $[NP(NCS)_2]_n$ is appreciably lower than that of the trimer. The energy gap (ΔE) of the trimer calculated is about 0.6 eV. The value of the energy gap or the resistivity is about two to three order of magnitude smaller than the trimer. It appears that the mechanism of the electrical conductivity of the polymer may be different from that of the trimer, that is to say, the polymer shows predominantly the electron conductivity as opposed to the trimer where ionic conductivity is dominant.

It should be pointed out that the conductivity measured during the course of increasing temperature does not coincide with the values measured during decreasing temperature. This descrepany (or hysterisis) may be caused by deterioration of contact between the sample and the electrode on heating and also by the degrading rubbery polymer. In the case of diaminophosphazenes, ammonia is detected when $(NP)_3(NH_2)_2R_4$ is heated under N_2, but is not detected in the case of R = SC_6H_5. Anilline is also detected during heating of $[NP(NHC_6H_5)_2]_3$. The amounts of ammonia formed at 190, 220, 250 and 280 °C were determined by chemical analysis, and the reaction order was calculated. The initial reaction was found to be the first order with apparent activation energy of deammoniation to be about 134 kJ/mole. The products prepared when ammonia evaporation reaches a steady state are determined by chemical analysis shows that the energy of ammoniation of R = OC_6H_5 is the largest. All of these products are glasslike, soluble in most organic solvents, and the oligomers are stable to hydrolysis. Also, ultraviolet absorption and electrical conductivity of aminophosphazenes were measured. The resistance of the compound $(NP)_3(NH_2)_2(SC_6H_5)_4$ is lower than that of other compounds. This demonstrates that delocalization of π electrons in the N_3P_3 ring is increased by the influence of the SC_6H_5 radical.

In the case of $(NP)_3Cl_{6-x}(NHC_6H_5)_x$ and $(NP)_3Cl_{6-x}(NHC_6H_4NH)_x$, the resistivity was plotted against 1/ (f) for x = 2, 4, 6. The true resistivity is calculated from the intercept of the curve. The conductance of other phosphazene compounds was determined under various frequencies and the resistivity values were evaluated. It was found that $(NP)_3(NHC_6H_4NH)_3$ reveals the lowest resistivity. Other compounds, except for $(NP)_3(NHC_6H_4NH)_3$ and $(NPCl_2)_3$, may be used as semiconductors. The reasons as to why some phosphazene derivatives have low resistivity as compared to the trimer $(NPCl_2)_3$, appears due to electron delocalization in N_3P_3 ring. However, ultraviolet spectroscopy, which has been of great use in studying organic aromaticity, failed to provide convincing evidence of activity in the cyclophosphazenes. Furthermore, the relation between the resistivity and number of chlorine atoms in $(NP)_3Cl_{6-x}(NHC_6H_5)_x$ and $(NP)_3Cl_{6-x}(NHC_6H_4NH)_x$ is established. In the case of $(NP)_3Cl_{6-x}(NHC_6H_5)_x$, the resistivity decreased with increasing Cl content while the resistivity of $(NP)_3Cl_{6-x}(NHC_6H_4NH)_x$ increased with increasing Cl content. The frequency dependence of tan δ and dielectric constants for $(NPCl_2)_3$ and $(NP)_3Cl_{6-x}(NHC_6H_5)_x$ (x = 2, 4, 6) are determined. It is reported that the dissipation factor at high frequencies becomes,

$$\tan \delta = G/2fC_0 \times 10^{-12}, \; \varepsilon'_r = C_0 \times t$$

where tan δ and ε'_r are the intrinsic dielectric loss tangent and the storage dielectric constant of $(NPCl_2)_3$, respectively. Since ε'_r is approximately constant, tan δ decreases with frequency. In general, the dielectric constant does not depend on frequency for covalent compounds, but it shows strong dependence on frequency in compounds having weak ionic group. Therefore, it appears that $(NP)_3Cl_2(NHC_6H_5)_4$ is relatively ionic relative to other compounds.

3.4 Adhesive Properties

Kumar(83) investigated the synthesis and adhesive strength of organophosphazenes compounds having P-O-C linkages such as phtaloglucidol (i), benzoic acid (ii) and bisphenol A(iii) formed by the reaction of $(NPCl_2)_3$ with phtaloglucidol, benzoic acid and bisphenol A. It was reported that (i), (ii) and (iii) have good adhesive properties for steel and duralumin. According to Kobayashi(84), the substances formed from $(NPCl_2)_3$ or $HOCH_2CF_3$ and EMA showed good adhesive properties for tooth. According to Colclorgh(85), the material has good joining and adhesive properties for fuel and explosive. Also, Daimler-Benz AG (86) reported that the product of the general formula $[NP(A)a(B)b]_x$ had a good adhesion for use in automobiles. Kajiwara(87)-(89) studied adhesive properties for polyvinylchloride with various cyclophosphazene compounds such as $(NP)_3(NH_2)_2(OR)_4$ and $(NP)_3(NHR_1)_2(NHR_2)_4$. Cyclicphosphazene compounds, used as adhesive materials, were prepared by the reaction of primary or secondary amines and alkoxides or phenoxides. Also, the reaction of $(NP)_3(NH_2)_2(OR)_4$ with TDI was carried out in benzene by refluxing for 24 h. The molar ratio of the phosphazene compound to TDI was 1:1. When the reaction was over, the reaction mixture was filtered and the filtrate was evaporated under vacuum. The molecular weight of the purified product from the reaction of $(NP)_3(NHMe)_2(OPr\text{-}i)_4$ with TDI was determined by GPC. In addition, the products obtained from the reactions of various phosphazene compounds with TDI were dissolved in DMF solvent. The solutions are coated on polyvynilchloroide laminates, and then the bonding strengths were measured. The product prepared through the reaction of $(NP)_3(NH_2)_2(OMe)_4$ with TDI had a stronger bonding strength. The relationship between bonding strength and Mw/Mn was investigated but there appears no correlation between them.

In the field of phosphazene polymer chemistry, polycyclolinear or polycyclomatrix phosphazenes have attracted great interest, particularly, the reaction of diaminotetraalkoxy and phenoxycyclotriphosphazene with TDI and the adhesive properties of the products.

Next, the reaction of $(NP)_3(NH_2)_2(OCH_2CF_3)_4$ and TDI was carried out in various solvents by refluxing for 24 h. The molecular weight of the purified products was determined using benzene. The molecular weight of the product increased with reaction temperature and longer time. To determine the effects of pendant groups of the products, the other diaminoorganophosphazenes compounds were allowed to react with TDI by refluxing for 24 h using benzene as the solvent. It was found that the highest molecular weight was obtained in the product prepared through the reaction of TDI with compounds consisting of electron attracting groups such as trifluoroethoxy. Because of the reaction of $P(NH_2)_2$ with –NCO groups, the product have the bonds of HNP-NH and NH-C(O)NH. In order to identify these bonds, the IR absorption spectrum of the products obtained from the reaction of $(NP)_3(NH_2)_2(OCH_2CF_3)_4$ with TDI was recorded using KBr windows. According to Derkach(97), the N-H stretching of–PNHC(O)R appears to occur in a rather low range of 3,160 to 3,040 cm^{-1} adjacent to the C-O group stretching. The band at 1,600 cm^{-1} is assigned to –PNHCONH. The H-NMR of the product was measured in $CDCl_3$ using TMS. Also, $-NH_2$ in diaminotetrafluoroethoxytriphosphazene appears at 1.8 ppm. The peak appearing at 1.8 ppm may be attributed to –PNH and the one at 3.4 ppm to –CNH. The ratio of –PNH to –CNH is about 1:1. C_6H_4 and $-CH_3$ appear at 7.3 and 4.3 ppm, respectively, with the ratio of approximately 1:1. Consequently, the structure of the product is speculated to be as Scheme (6).

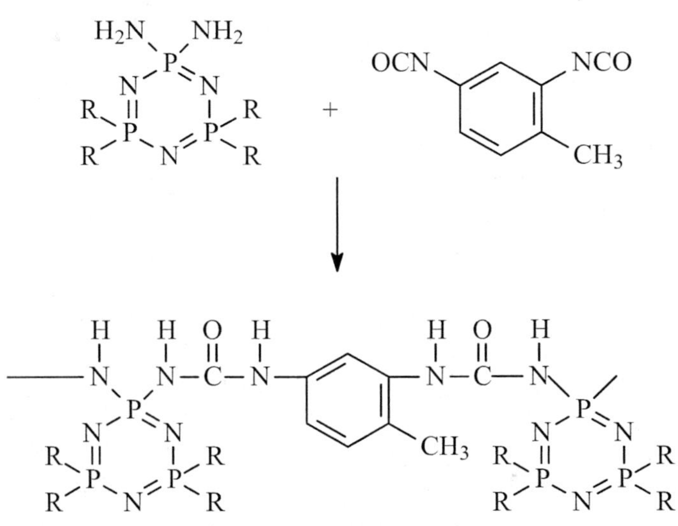

Scheme (6)

Bonding strength was measured after the products prepared from the reaction of various phosphazenes compounds with TDI were coated on polyvinylchloride laminates. The product prepared from the reaction of diaminotetraethoxycyclotriphosphazene with TDI yield the highest bonding strength compared to other products. Also, interfacial or cohesive failure was observed, but there is no correlation between the failure and bonding strength. Also, diaminotetraphenoxycyclotriphosphazene gem-$(NP)_3(NH_2)_2(OPh)_4$ (ATP) was reacted with KH_2PO_4, B_2O_3 or KH_2PO_4 and B_2O_3 at 150 – 250 °C. Evaporated gas from this reaction was determined by means of gas chromatography. It was found that main gas is NH_3 with a minor amount of water. It is found from Table 6 that the reaction rate is faster in the case of higher concentration of B_2O_3, however, the reaction rate is slow in the case of KH_2PO_4. They are soluble in THF or DMF solvents having high dipolmoment. In KH_2PO_4-B_2O_3 binary systems, ATP was reacted with changing of KH_2PO_4/B_2O_3 mol ratio at 250 °C. The reaction rate is determined with evaporated NH_3 gas, and yield of the product dissolved in THF is summarized in Table 6.

It is found that the reaction rate is faster in the case of high concentration of KH_2PO_4, and a total amount of evaporated NH_3 gas is increased with increasing of B_2O_3 content. Furthermore, molecular weight of the product obtained from a ternary system such as ATP-KH_2PO_4-B_2O_3 is significantly higher in comparison with other systems. Regarding the structural determination, H-NMR, IR, Raman and chemical analysis were employed. P-N(1,400-1,300 cm^{-1}), P-O-P(950 cm^{-1}), POH(2,000 cm^{-1}), P-NH-P(1,600 cm^{-1}) bonds observed from IR. The broad peaks appear at 890-980 cm^{-1} in the case of Raman spectra.

Table 6. Mole Ratio of NH_3/ATP and Yield of the Product under the Various Experimental Conditions

Reaction mixture(g)			Reaction conditions		NH_3/ATP	Product
ATP	KH_2PO_4	B_2O_3	Temp(°C)	Time (h)	Mole ratio	(g)
2.76			170	40	0.6	2.7
			200	20	0.8	2.7
			250	15	1.0	2.6
2.54	0.63		170	150	1.0	2.4
	0.65		200	110	1.2	2.5
	0.63		250	80	1.2	2.4
2.10		0.27	150	3.25	1.0	1.4
1.95		0.26	200	3.0	1.2	1.2
2.05		0.26	250	1.58	1.2	1.2
2.51	0.63	0.30	170	100	1.1	2.3
2.51	0.63	0.32	200	50	1.4	2.1
2.51	0.63	0.33	250	10	1.0	1.5

ATP;$(NP)_3(NH_2)_2(OPh)_4$

According to Ray(91), the peaks of BO_3 and BO_4 generally appeared near 890 cm^{-1} and 490 cm^{-1}, respectively. To determine the ratio of x:y:z as shown in Scheme (7), the product, reacted with trimethylchlorosilane $(Me)_3SiCl$, was examined using H-NMR. It was found that two peaks appear at 0.07 ppm and 7.1 ppm, in which the former peak belongs to –POSiMe and the latter is assigned to $-P(OPh)_2$. On the basis of these H-NMR, IR and Raman data, it may be postulated that the product to have the structure as depicted in Scheme (7).

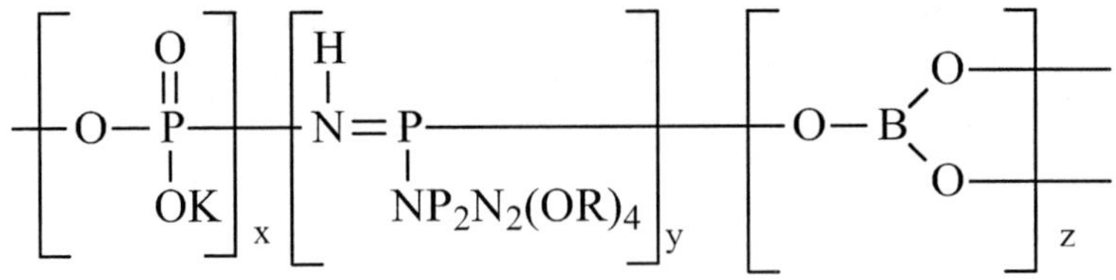

where $R = C_6H_5$

Scheme (7)

Moreover, the x:y:z ratio was estimated from the H-NMR data. The H-NMR data shows that KH_2PO_4 is reacted with ATP and B_2O_3. Bonding strength for laminated polyvinylchloride resin is determined. It turned out that the bonding strength of the product from ATP-KH_2PO_4 and ATP-KH_2PO_4-B_2O_3 is higher than that of other products. Especially, the product obtained from ATP-KH_2PO_4 at 250 °C has the highest bonding strength, which is comparable to that of the rubber adhesive. In the case of ATP-KH_2PO_4 binary system, bonding strength depends on the degree of molecular dispersion, that is, bonding strength increased with increasing the degree of molecular dispersion. On the contrary, bonding strength of the ATP-KH_2PO_4-B_2O_3 system decreased with increasing the molecular dispersion. According to the hydrolysis study, the product of ATP-KH_2PO_4-B_2O_3 system appears to be stable in water in comparison with other products.

3.5 Non-flamable Electrolyte for Lithium Battery

Lithium battery possesses high energy density compared to other secondary batteries. In general, lithium batteries were made up of cathode, anode, separator, and electrolyte such as lithium hexafluorophosphate $LiPF_6$ in ethylene carbonate (EC) and dimethyl carbonate (DMC). Small Li-ion batteries having a capacity of 1300-1900 mAh are currently commercially available to power portable electronic devices such as camcorders, computers, cameras, among others. In addition, Li-ion and Li-polymer batteries are being developed as power sources for electric vehicles (EV) to provide longer driving ranges, higher accelerations, and longer life times. However, safety concerns have limited the full utilization of Li batteries in EV applications. The primary challenge in designing the Li-ion and Li-polymer EV batteries is safety under abusive as well as normal operating conditions. Under abusive conditions, and occasionally even under normal conditions, Li-ion cells undergo thermal runway, producing exceedingly high temperatures, smoke, explosion, and fire due to the liquid organic electrolyte used in these batteries. Under certain conditions, the flash point of the electrolyte can be exceeded such that the Li-ion battery is overheated, thereby resulting in major safety problems. Manufacturers employ external safety devices in small consumer Li-ion batteries to overcome these problems.

Praskash(92) is the first to report a novel flame retardant additive such as hexamethoxycyclotriphosphazene $[NP(OMe)_2]_3$ for lithium batteries. The flame-retardant was synthesized by reaction of sodium methoxide NaOMe and hexachlorocyclotriphosphazene $(NPCl_2)_3$. The electrochemical and thermal properties were measured using cyclic voltammetry, electrochemical cycling, differential scanning calorimetry, and accelerating rate calorimetry. Also, the electrochemical stability of the electrolyte was examined using cyclic voltammetry. A mixture of 1 mol $LiPF_6$ and EC-DMC (50:50 wt %) was used as electrolyte. To study the electrochemical performance, $Li/LiNi_{0.8}Co_{0.2}O_2$ cells were fabricated with and without $[NP(OMe)_2]_3$. The main conclusions to be draw from these results was as follows; Cyclic voltammetry showed that the electrolyte containing the flame retardant additive was found to be electrochemically stable up to 5.0 V. The charge and discharge capacities of the cell containing 1.5 wt % flame retardant additive were noticeable, i.e., higher than those of the comparable cell without the flame retardant additive. Results of these thermal investigations strongly suggest that the addition of the flame retardant additive in the electrolyte significantly supresses the self-heat rate. The reduced self-heat rate, in turn, helps in improving the thermal stability and hence the non-flammability of the electrolyte.

Recently, Angell(93) reported about inorganic electrolyte solutions such as methane sulfonylchloride CH_3SO_2Cl (i), trichlorophosphazosulfonylchloride $Cl_3P{=}NSO_2Cl$ (ii) and trichlorophosphazophosphonylchloride $Cl_3P{=}NP(O)Cl_2$ (iii). These electrolytic solutions were prepared by dissolving calculated amounts of $LiAlCl_4$ in solvent (i), (ii) and (iii). The resultant solution is transparent and somewhat viscous. A class of inorganic electrolyte has been evaluated for use as electrolytic solvents in rechargeable lithium batteries. These showed much improved safety performance. The solvents (i), (ii) and (iii) are much volatile in compare with organic solvents, and among them, (i) is least sensitive to moisture. Although (ii) and (iii) are crystalline at room temperature, their binary solution with $LiAlCl_4$ has freezing points far below the room temperature. According to US.Pat(94) the solvents such as PC, EC and DEC have been used in Li batteries either in pure form or in solvent mixtures. These solvents provide high conductivities in the presence of suitable lithium salts. However, the chemical stability and safety of these solvents has recently been questioned. Improved liquid organic electrolytes with higher conductivities than these solvents are need to meet the power requirements for any consumer applications, especially at low temperatures. It was claimed that fire retardant electrolyte compound containing a lithium salt solution is phosphazene having the structure of $[NPR^1R^2]_n$ (iv), where n is an integer 3 to 6 and R^1 and R^2 are arbitrarily selected from the group containing of halogen, hydrogen. and $-OR^3$, where R^3 is C_1-C_6 alkyl terminally substituted with 0 to 3 halogen atoms and containing 0 to 3 ether linkages with the proviso that when one of R^1 or R^2 is $-OR^3$. Fire retardant electrolyte compositions having cyclophosphazene give concrete expression to (v) and (vi). According to Toyota central R and D Labs Inc.(95), it may be necessary to use non-flammable lithium secondary batteries in electric vehicle for safety. However, EC or DEC was not safe because they had a lower flashing point, especially the flashing point of DEC was lower. To make non-flammable electrolyte, Toyota reported that $[NP(OEt)_2]_3$ $[NP(OCH_2CF_3)_2]_3$, $[NP(OPr\text{-}n)_2]_3$ and $Cl_3P{=}NP(O)Cl_2$ may be added to EC or DEC. It was found from the study that the addition of phosphazene compounds to EC or DEC raised safety margin for lithium secondary batteries without any power loss.□Japan Chemical Ind(96) reported that lithium salts such as $LiPF_6$ can be dissolved in $(R^1O)_3P{=}SO_3R^2$(vii) and $(R^1O)_2P{=}NSO_2(OR^2)_3$(viii) type solvents. The solution prepared by arbitrarily mixing(vii) and (viii) with DEC or EC and PC shows non-flammability, suggesting that (vii), (viii), and their electrolyte mixtures are safe electrolyte solvents. Furthermore, the electrolyte solvent containing (vii) or (viii) supressed the dissolution of Mn from the $Li_xMn_{2-y}Mn_yO_{4-z}$ cathode. Solvents innovated by Kajiwara was patented in Japan and US through Bridgestone Inc.(97) which presumably solved the aforementioned problems. The typical compounds with improved performance is summarized in Table 7.

Table 7. The Typical Phosphazene Compounds of the Liquid Electrolyte Solvent

Cyclic phosphazene $(NPR_1R_2)_3$ R1	R2	Wt.loss 50 % (°C)	Linear phosphazene $(R_1O)_3P{=}N{-}P(O)(OR_2)_2$ R1	R2	Wt.loss 50 % (°C)
(A) OPh	OCH_2CF_3	350	(E) CH_2CF_3	$(CH_2)_2OCH_3$	225
(B) $OCH_2CF_2CF_3$	R_1	300			
(C) $O(CH_2)_3CH_3$	R_1	300			
(D) $O(CH_2)_2CH_3$	R_1	250			

All of these liquid electrolyte solvents may dissolved in lithium salts by mixing with approximately portions of non-proton organic solvents. In the study of thermal stability, PC-EC (1:1) shows a 50 % weight loss at 260 °C. However, the liquid electrolyte solvents except (E) appear very stable upon heating. Also, the temperatures at which the 50 % weight loss occurred were summarized in Table 7.

Next, the cyclic behavior and safety data of batteries prepared by various phosphazene compounds is given in Table 8. It was found from Table 8 that phosphazene electrolyte solvents is very safe relative to that of ordinary organic electrolyte solvents such as PC or EC. In the case of phosphazene solvents, even if short circuit occurs, there is no anomalous behavior. Batteries using the liquid electrolyte solvent (E) have a lower internal resistance, and excellent characteristics. However, the liquid electrolytes having viscosity of 300 cP can cause the short circuit resulting in appreciably decreased discharge capacity under higher current density. Also, the cyclic behavior tends to drop.

Table 8. The Cyclic Behavior of Batteries Prepared by Various Phosphazene Compounds

Comp.	Viscosity	Initial property		Cyclic property			High discharge			Estim-ation
	(cP)	(1)	(2)	(3)	(4)	(5)	(6)	(7)	(8)	
(I)	60	3.32	210	745	708	85	745	660	89	S
(II)	35	3.28	220	50	705	94	750	660	88	S
(III)	100	3.28	400	750	585	85	690	470	68	S
(IV)	400	3.33	190	755	725	96	755	685	91	S
PC+EC (1:1)	20	3.33	180	760	730	96	760	690	91	D

(1) Voltage. (2) Internal resistance (mΩ). (3) Initial (mAH) (4) 50 (mAH). (5) Retention (%). (6) 50mA (mAH). (7) 500 mA (mAH). (8). Retention (%). S: Safety. D: Danger.
(I). $[NP(OCH_2CH_2OCH_3)_2]_n$ (II). $[NP(OCH_2CF_2CF_3)_2]_n$. (III).$[NP(OCH_2(CF_2)_3)_2]_n$ n = 3
(IV). $(CF_3CH_2O)_3P{=}N{-}P(O)(OEt)_2$. PC: $(C_3H_7)_2C{=}O$. EC: $(CH_2O)_2C{=}O$.

Phosphazene electrolytes such as $[NP(OCH_2CH_2OCH_3)_2]_3$ and $(CF_3CH_2O)_3P{=}N{-}P(O)(OEt)_2$ were mixed with DME. DME-PC mixture is commonly used as a standard solvent. Five kinds of batteries were prepared by using various electrolyte solvent mixtures. The cyclic behavior and safety of batteries as determined is summarized in Table 9. It can be seen from Table 9 that DME-PC mixture solvent results in short circuit owing to breakage and leakage through of the solvents. However, such problems were not encountered in the phosphazene-DME mixture solvent. This finding enables to make high quality safety batteries from the aforementioned phosphazene-DME mixture solvents without any power loss. The estimated results strongly suggest that the performance of phosphazene-based batteries is superior to that of ordinary batteries.

Table 9. The Cyclic Properties of Batteries Prepared with a Mixture Electrolyte Solvents of Phosphazene and 1,2-Dimethoxyethane (DME)

Solvent	Viscosity	Initial property		Cyclic property			High discharge			Esti-
	(cP)	(1)	(2)	(3)	(4)	(5)	(6)	(7)	(8)	mation
(I)/DME										
(60/40)	18	3.28	150	765	730	95	765	705	92	S
(75/25)	24	3.32	170	760	720	95	760	685	90	S
(85/15)	32	3.24	185	760	705	93	760	675	89	S
(IV)/DME										
(85/15)	13	3.30	140	760	730	96	760	705	93	S
PC/DME										
(50/50)	10	3.30	140	770	740	96	770	725	94	D

(1). Voltage (2). Internal resistance (mΩ) (3). Initial(mAH) (4). 50(mAH) (5). Retention (%).
(6). 50mA(mAH) (7).500mA(mAH) (8). Retention (%).
S: Safety. D:Danger. DME: $CH_3OCH_2CH_2OCH_3$
(I).$[NP(OCH_2CH_2OCH_3)_2]_n$ n = 3 (IV). $(CF_3CH_2O)_3P{=}N{-}P(O)(OEt)_2$

3.6 Liquid Crystalline State of Phosphazene Oligomers

3.6.1 Cyclotriphosphazene Derivatives

3.6.1.1 Phase transition in hexakis(4-(4'-biphenoxy)cyclotriphosphazene) ($[PN(OC_6H_4C_6H_4OC_nH_{2n+1})_2]_3$, n = 6-12, HBCTP)

Phosphazene-based inorganic liquid crystalline monomers and oligomers were extensively studied by Moriya and Kajiwara(64) (98)-(100), who found that HBCT P with n=6 exhibits monotropic nematic (N) and smectic C phases during the course of cooling in DSC. However, enantiotropic N and SmC phases were found in the case of n=7 member, whereas only the enantiotropic SmC phase was detected in the HBCTP with n=8 - 12. In the series of HBCTP, the temperature range of the mesomorphic phase increased with an increase in n. The odd-even effect was not pronounced, which may be a consequence of the peculiar shape of the hexakis(4-(4'-alkyloxy)biphenoxy)cyclotriphosphazene with three side chains pointing three upwards and three downwards in both sides of the cyclotriphosphazene ring. In the crystalline to smectic C phase transition of HBCTP with n=8, the infrared bands of P=N and P-O-(C) stretching vibrations were found to shift to the lower frequency, indicating the weakening of the cyclotriphosphazene ring in the liquid crystalline phase. According to the P-NMR measurement, the local orientational order parameter around the P atom is estimated ca. 0.45 in the smectic A phase of hexakis(4-(4'-dodecyl)biphenoxy)cyclotriphosphazene(101). This value is smaller than that found for the usual smectic A liquid crystals (102), suggesting a relatively high disordering of the cyclotriphosphazene ring in the smectic A phase. The similar situation is considered to occur in the SmC phase of the n=8 in the series of HBCTP. The interlayer spacing in the SmC phase as revealed by X-ray diffraction is 36 A, which is smaller than the value expected for the fully extended molecular model, indicating that the tilt angle of the molecules is about 35° from the normal to the layer plane. The SmC diffraction pattern shows the molecular arrays perpendicular to the director which reflects the peculiar shape of the cyclotriphosphazene(103).

3.6.1.2 Phase transition in hexakis(4-(N-(4-alkoxyphenyl)iminomethylphenoxy)-cyclotriphosphazene) ($[PN(OC_6H_4C{=}NC_6H_4OC_nH_{2n+1})_2]_3$, n = 3-12, HSCTP)

A series of cyclotriphosphazenes bearing Schiff bases with different carbon number alkoxy end groups(HSCTP) in the side chains were synthesized, and their phase transitions and liquid crystallinity were studied using DSC and polarizing microscopy. Moriya and Kajiwara(104) reported that a monotropic Sm1 phase exists in HSCTP with n = 3. The Sm1 phase represents a low-order uniaxial smectic phase, whereas Sm2 phase is a low-order biaxial smectic phase. With increasing carbon number, the authors found an enantiotropic Sm1 phase for n = 4 - 5, enantiotropic Sm1, SmC, and SmA phases for n= 6 - 7, enantiotropic SmC and SmA phases for n = 8 - 10, and enantiotropic Sm2, SmC and SmA phases for n=11 and 12 in the HSCTP series. The melting point decreased with an increase in the end alkoxy carbon number n of the cyclotriphosphazenes, but the clearing point increased with an increase in n. As a result, the mesomophic region expands with increasing n. The melting and clearing points did not show distinct odd-even effects, which may caused by the peculiar shape of the cyclotriphosphazenes.

3.6.1.3 Phase Transition in Cyclotriphosphazenes with Different Mesogenic Moieties in the Side Groups

Moriya and Kajiwara(64) (105)-(106) found that cyclotriphosphazenes having similar end groups, but with different central groups have different mesomorphic phases. As shown in Table 10, chemical shifts of P-NMR peaks are different for cyclotriphosphazenes having different mesogenic side chains, suggesting that these materials have different electronic and structural disparity. In comparing the compounds having alkyl and alkyloxy groups in the end of the side chains, alkyloxy compound 2 has more than 20 K higher melting and clearing points than alkyl compound 3. The order of the thermal stability in cyclotriphosphazene is phenylazobenzene > benzylideneaniline > biphenyl. This order of thermal stability of liquid crystalline phase is similar to those found usually in liquid crystals having mono-mesogenic groups in the core.

Table 10. Liquid Crystalline Phase Transition and P-NMR Chemical Shifts for the Compounds $[(C_{12}H_{25}X)_2PN]_3$

Compound	X	Phase transition temperature / ° K					^{31}P NMR chemical shift / ppm
		Cr	S1	SmC	SmA	I	
1	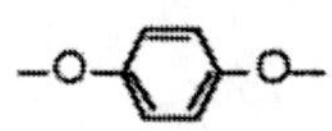	• 346				•	11.1
2	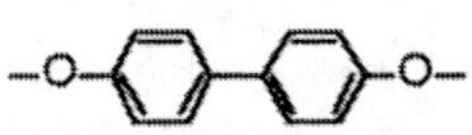	• 425		• 453		•	10.6
3	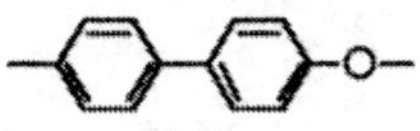	• 406		• 424	• 426	•	10.6
4	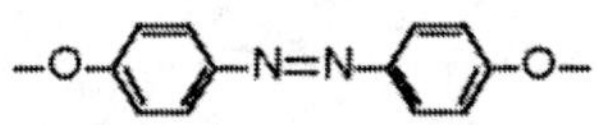	• 416		• 443		•	9.7
5	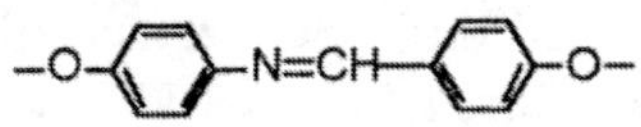	• 435	• 452	• 486	• 500	•	9.7

3.6.2 Cyclotetraphosphazene Derivatives

3.6.2.1 Phase transition in octakis(4-(4'-biphenoxy)cyclotetraphosphazene) ($[PN(OC_6H_4C_6H_4OC_nH_{2n+1})_2]_4$, n = 7-12, OBCTP)

Based on the DSC measurements and polarizing microscope observations, Moriya and Kajiwara(107) found in the series of OBCTP that only a monotropic SmA phase was observable for the samples with n = 11 and 12, but no mesophase was discerned between n = 7 and 10. It is seen that the OBCTP series has relatibly low mesogenicity, which is probably caused from the molecular structure and electronic environment in the OBCTP.

3.6.2.2 Phase transition in octakis(4-(N-(4-alkoxyphenyl)iminomethylphenoxy)-cyclotetraphosphazene) ($[PN(OC_6H_4C{=}NC_6H_4OC_nH_{2n+1})_2]_4$, n = 6-12, OSCTP)

A series of cyclotetraphosphazenes with meogenic Schiff side groups (OSCTP) was prepared, and their phase transitions and mesogenicity were studied using polarizing microscopy, X-ray diffraction, and differential scanning calorimetry (DSC). When n = 6, no mesomorphic phase was observable. However, an enantiotropic smectic A phase was found for OSCTP liquid crystals having n = 7 - 12. The thermal stability of the SmA - I phase transition temperature increased with increasing alkoxy carbon number n of the side groups. The mesomorphic temperature range expands with the increasing carbon number. In the X-ray measurements, new broad peaks were found which may be attributed to reduction in the LC orderness presumably caused by the peculiar shape of the cyclotetraphosphazene(107)-(108).

3.6.3 The Comparison of Phase Transition in Cyclotriphosphazene and Cyclotetraphosphazene

The alkoxy chain length dependence of the phase transition in the series of HBCTP and OBCTP was studied, where n is the alkoxy carbon number of the side chain. In the series of the trimer with an increase in the end alkoxy carbon number, the clearing points increased and the melting points decreased with an increase in the carbon number. As a result, the mesomorphic temperature region in HBCTP increased with the increasing carbon number. In the OBCTP series, only monotropic SmA phase was observed at n = 11 and 12 in the cooling cycle. Therefore for biphenoxy moieties cyclotriphosphazene reveals higher mesomorphic stability relative to that in cyclotetraphosphazene. This is probably driven by the difference in the molecular structure of the cyclotriphosphazenes versus cyclotetraphosphazenes(107).

Similarly, the alkoxy chain length dependence of the phase transition in the series of HSCTP and OSCTP was studied. In the series of the tetramer with an increase in the end alkoxy carbon number, the clearing points increased and the melting points declined with increasing cabon number. Consequently, the mesomorphic temperature region in the OSCTP expands with increasing carbon number. In the cyclotriphosphazenes with similar mesomorphic side chains, the same tendency of the expansion of the mesomorphic temperature region was discerned. However, there are some difference in thermal stability between cyclotriphosphazene and cyclotetraphosphazene. That is to say, the mesomorphic phase in cyclotriphosphazene is thermally more stable, because both the melting and clearing points in the cyclotriphosphazene are higher than those in the cyclotetraphosphazene.

References

1. M.Kajiwara, M.Hashimoto and H.Saito, *Polymer*, 14 (1973) 488.
2. M.Kajiwara, M.Hashimoto and H.Saito, *Polymer*, 16 (1975) 861.
3. Y. Kurachi, K. Shiomoto and M. Kajiwara, *J. Mater. Sci.*, 25 (1990) 2036.
4. Y. Kurachi and M. Kajiwara, *J. Mater. Sci.*, 26 (1991) 1799.
5. T. Hayashi and H. Saito, *Bull. Chem. Soc. Jpn.*, 50 (1977) 1023.
6. D.D. Eley and M.R. Wills, *J. Chem. Soc.*, (1963) 1534.
7. P.J. Reucroft, P.L. Kronick, M. Scott and M. Labes, *Nature*, (1964) 4919.
8. H.R. Allcock and R. Best, *J. Can. Chem.*, 42 (1964) 447.
9. Allcock, H. R.; Kugel, R. L.; Valan, K. J. *Inorg. Chem.* **1966**, *5*, 1709.
10. Allcock, H. R.; Kugel, R. L. *Inorg. Chem.* **1966**, *5*, 1716.
11. M. Kajiwara and H.Saito, *Polymer*, 17 (1976) 1813.
12. T. Kimura and M. Kajiwara, *J. Inorg. Organomet. Polym.*, 2(4) (1992) 431.
13. Bittirova, F. A.; Kireev, V. V.; Mikitaev, A. K. *Vysokomol. Soedin, Ser. B.* **1981**, *23*, 30, Chem. Abstr. 94, 122013 (1981)F.A. Bittrova,
14. R.R. McCaffrey, and D.G. Cummings, *Sep.Sci.Technol.*, 23(12-13) (1988) 1627.
15. M. Kajiwara, *J. Mater. Sci.*, 23 (1988),1360; *J. Mater. Sci. Letters.*, 7 (1988) 102.
16. I. Murakami, H. Ochiai, S. Tamura, K.Takeda, Y. Tsurusaki, Y. Fujii and M. Kajiwara., *Rep. Prog. Polym. Phys. Jpn.*, 30(493) (1987) 473.
17. M. Kajiwara, *Sep. Sci. Technol.*, 26(6) (1991) 841; *Synthetic Polymeric Membranes*, Waltwer de Gruyter and Co., Berlin: New York., p 347(1987).

18. E. Drioli, S.M. Zhang, A. Basile, G. Golemme, S.E. Gaeta and H.C. Zhang, *Gas. Separation Purification*, 5 (1991) 252.
19. R.R. McCaffrey, R.E. McAtee, A.E. Grey, C.A. Allen, D.G. Cummings and A.D. Appelhans., *J. Membrane. Sci.*, 28 (1986) 47.
20. R.R. McCaffrey, R.E. McAtee, A.E. Grey, C.A. Allen, D.G. Cummings, A.D. Appelhans, R.B. Wright, and J.G. Jolley, *Sep. Sci.Technol.*, 22 (1987) 873.
21. F. Suzuki, K. Onozato, H. Taegashi and T. Masko, *J. Appl. Polym. Sci.*, 34 (1987) 2187.
22. Kraus, A. K. US Patent, 4,783,202 (1988), Chem. Abstr. 103, 179417m (1985), assigned to Monsanto Co.
23. C.A. Allen, D.G.Cummings, A.E. Grey, R.E. McAtee and R.R. McCaffrey, *J. Membrane. Sci.*, 33 (1987) 181.
24. S.N. Gaeta, H.C. Zhang and E. Drioli, in Proceeding of *the First International Conference on Inorganic Membranes (ICM 89)*, Montpellier, 3-6 July, (1989) p.65.
25. S.M. Zhang, A. Basile, S.N. Gaeta and E.Drioli, in *Third European Polymer Fed Symposium on Polymer Materials*, Sorrento 1-5 October, (1990)
26. J. Brandrup. and E.H. Immergut, *Polymer Handbook*, 2nd End., Wiley-Interscience, New York, (1975) p.229; M. Kajiwara., *J. Mater.*, in press; T. Kojima, and N. Kajiwara, *Kokai Tokkyo Koho JP*, 62 06,217 [87 06,217] (1987), *Chem. Abstr.* 107 (1987) 64922n, assigned to Tome Sangyo K.K.
27. N.J. Balloy, *Contact Lens Forum*, (1984) p.31.
28. Schmidt, H.; Philipp, G., *NATO ASI Ser. (E)*, (1985), 92 (Glass) 580, Chem. Abstr. 103, 14700 (1985).
29. M. Kajiwara, *Polymer*, 30 (1989) 1536.
30. M. Kajiwara, *J. Mater. Sci.*, 29 (1994) 4339.
31. M. Kajiwara, *J. Mater. Sci.*, 29 (1994) 6268.
32. M. Kajiwara and T. Kimura, *J. Mater. Sci.*, 31 (1996) 5457.
33. M.Kajiwara and T. Kimura *J. Mater. Sci.*, 31 (1996) 6659.
34. M. Kajiwara, and T. Kimura, *Phosphorus, Sulfur and Silicon.*, 147 (1999) 975.
35. H. Magatani, A. Hirano and T. Matsumura, *Nippon Contact Lens Gakkai-shi.*, 26 (1984) 10, Chem. Abstr. 101, 157632 (1984)
36. M.F. Refojo and F.L. Leong, *J.Membr. Sci.*,4 (1979) 415.
37. S.T. Hwang, T.E. Tang and K. Kammermeyer., *J.Macromol Sci. (B),* 5 (1971) 1.
38. O. Ng, D.G. Pedley and B.J. Tighe, *Br. Polym.J.*, 124 (1976) 24.
39. C.F. Rogers, *Permeability and Chemical Resistance in Engineering Design for Plastic* (Ed. E. Baer), Van Nostrand Reinhold., London, (1964).
40. N. Kajiwara, T. Kusuda, and S. Hakozaki, *Jpn. Kokai Tokkyo Koho JP*, 06 136,272[94 136,272] (1994), *Chem. Abstr.* 121 (1994) 302264k, assigned to Nippon Gousei Kagaku Co., Ltd.; N. Kajiwara, T. Kusuda, S. Hakozaki, and Y. Koishi, Jpn. Kokai Tokkyo Koho JP, 06 145,362[94 145,362] (1994), *Chem. Abstr.* 121 (1994) 302587t. assigned to Nippon Gousei Kagaku Co., Ltd.
41. J.F. Labarre, J.P. Faucher, G. Levy, F. Sournies, S. Cros and G. Francois, *Eur. J. Cancer*, 15 (1979) 637; J.F. Labarre, G. Levy, and F. Sournies, *J. Mol. Struct.*, 63 (1980) 127.
42. S.N. Nabi, M. Biddlestone and R.A. Show, *J. Chem. Soc. Dalton Trans.*, (1975) 2634.
43. F.T. Cho, A.H. Khan and J.S. Driscoll, *J. Med. Chem.*, 19 (1976) 1302.
44. J.O. Bovin, J. Galy, J.F. Labarre, and F. Sournies, *J. Mol. Struct.*, 49 (1978) 421.

45. H.R. Allcock, *Macromol. Symp.*, 144 (1999) 33.
46. S.C. Song, C.O. Lee and Y.S. Sohn, *Bull. Korean. Chem. Soc.*, 20(2) (1999) 250; Y.S. Sohn, H. Baek, Y.H. Cho, Y.A. Lee, O.S. Jung, C.O. Lee, and Y.S. Kim, *Int. J. Pharm.*, 153 (1997) 79; S.C. Song and S.Y. Soo, *J. Control. Rel.*, 55 (1998) 161.
47. J.H. Goedemoed, and K. DeGroot, *Makromol. Chem. Macromol. Symp.*, 19 (1988) 341.
48. M. Kajiwara, *Phosphorus. Sulfur and Silicon*, 76 (1993) 423; Kajiwara, M., *App.Biochem. Biotech.*, 33 (1992) 43.
49. N. Kajiwara, and S. Aotani, *Jpn. Kokai Tokkyo Koho JP*, 63 84,489 [88 84,489] (1988), *Chem. Abstr.* 110 (1989) 73856h, assigned to Japan Synthetic Rubber, Co., Ltd.; Y. Furukawa, and N. Kajiwara, *Jpn. Kokai Tokkyo Koho* JP, 63 194,339 [88 194,339] (1988), *Chem. Abstr.* 110 (1989) 86889w, assigned to Nippon Telegraph and Telephone Public Corp.
50. M. Hatzakis, J. Paraszczak and J. Shaw, in *Proc. Conf. Microlithorg.*, (1981) p. 386.
51. M. Suzuki, K. Saigo, H. Gokan and Y. Ohnishi, *J. Electrochem .Soc.*, 130 (1983) 1962.
52. M. Morita, S. Imamura, A. Tanaka and T. Tamamura, *J. Electrochem. Soc.*, 131(1984) 2402.
53. A.S. Gozdz, C.C. Carnazza and M.J. Bowden, *Proc. SPIE-Int. Soc. Opt. Eng.,* 631 (1986) 2.
54. S. Sugito, S. Ishida and Y. Iida, in *Eztend. Abstr, 20th Conf. SSDM.*, (1988) 561.
55. Y. Yamashita and M. Kajiwara, *J. Inorg. Organomet. Polym.*, 2(1) (1992) 129.
56. Y. Yamashita and M. Kajiwara, *Polymer for Microelectronics-Sci and Tech.*, 31 (1990) 317.
57. Y. Yamashita and M. Kajiwara, *J. Electrochem. Soc.*, 137(10) (1990) 3253.
58. H. Hiraoka, *J. Vac. Sci. Technol.*, B5 (1987) 386.
59. H. Hiraoka, W.Y. Lee, L.W. Welsh and R.W. Allen, *Macromolecules*, 12 (1979) 753.
60. M. Kajiwara and Y. Yamashita, *J. Mater. Sci.*, 26 (1991) 2797.
61. M. Kajiwara, *J. Inorg. Organomet. Polym.*, 1 (1991) 277.
62. Y. Yamashita, T. Ito, and N. Kajiwara, *Jpn. Kokai Tokkyo Koho JP*, 01 44,935 [95 44,935] (1989), *Chem. Abstr.* 111 (1989) 48165v, assigned to Oki Electric Industry Co., Ltd.
63. Y. Yamashita, T. Ito, and N. Kajiwara, *Jpn. Kokai Tokkyo Koho JP*, 01 44,927 [89 44,927] (1989), *Chem. Abstr.* 111 (1989) 48164u, assigned to Oki Electric Industry Co., Ltd.Kokai Tokkyo Koho Jap.
64. T. Kimura, M. Kajiwara, *J. Inorg. Organomet. Polym.*, 2 (1992) 431.
65. K. Moriya, T. Suzuki, Y.Kawanishi, T. Masuda, H. Mizusaki, S. Nakagawa, H. Ikematsu, K. Mizuno, T. Suzuki, S. Yano and M. Kajiwara, *Appl. Organomet. Chem.*, 12 (1998) 771,
66. K. Moriya, Y. Kawanishi, T. Suzuki, S. Yano and M. Kajiwara, *Kobunshi Ronbunsyu*, 56 (1999) 390, *Chem. Abstr.* 131 (1999) 185317.
67. K. Takahashi, T. Yamamoto and M. Kajiwara, *J. Soc. Mater. Soc. Jpn.*, 37(421) (1987) 1223; K. Takahashi, T. Yamamoto, M. Kajiwara, *Zairyo* **1988**, *37*, 1223, *Chem. Abstr.* 110 (1989) 174285b.
68. K. Takahashi, T. Yamamoto, S. Itoh, K. Harakawa and M. Kajiwara, *J. Soc. Mater. Soc. Jpn.*, 37(415) (1988) 454; K. Takahashi, T. Yamamoto, S. Itoh, K. Harakawa, and M. Kajiwara, *Zairyo* 37, (1988) 454, Chem. Abstr. 109 (1988) 111367.
69. M. Kajiwara, *J. Mater. Sci.Lett.*, 13 (1994) 842.

70. Kojima, T.; Kajiwara, N. *Jpn. Kokai Tokkyo Koho JP*, 62 06,217 [87 06,217] (1987), *Chem. Abstr.* 107 (1987) 64922n, assigned to Tome Sangyo.
71. Comp. de St-Gobain: *Belg. Pat.*, 715,792 (1968); J. Simon, B. Androsits, T. Kozma, Therm. Anal., 2 (1980) 469.
72. L.E. Godfrey and J.W. Schapped, *Ind. Eng. Chem. Prod. Develop.*, 9 (1970) 426.
73. R.G. Rice, R.M. Murch and D.C. DeVore, *US. Pat.*, 3,545,942 (1970), *Chem. Abstr.* 74 (1971) 54398, assigned to R.W. Grace & Co.; Harrington, R. C.; Bond, J. H. *US Patent*, 3,530,204 (1970), *Chem. Abstr.* 73 (1970) 99963h, assigned to Eastman Kodak Co.
74. E.J. Quinn and R.L. Dieck., *J. Fire & Flammability*, 7 (1976) 5.
75. M. Kajiwara and Y. Kurachi., *Br. Polym. J.*, 15 (1983) 104; Kajiwara, M.; Saito, H. *Shikizai Kyokai shi* 57 (1984) 7, *Chem. Abstr.* 100 (1984) 175823f.
76. D.D. Eley, and M.R. Wills, J. *Chem. Soc.*, (1963) 1534.
77. P.J. Reucroft, P.L. Kronick, H.Scoot and M.Labes, *Nature*, 201(1964) 609.
78. H.R. Allcock, and R.L. Kugel, *J. Am. Chem.Soc.,* 87 (1965) 4216.
79. M.Kajiwara and H. Saito, *Polymer*, 18 (1977) 351.
80. M. Kajiwara and H. Saito, *J. Macromol. Sci., Chem.*, A11(5) (1977) 1084.
81. M. Kajiwara, K. Shimono, and H. Saito, *J. Macromol. Sci., Chem.*, A12 (1978) 945.
82. M. Kajiwara and H. Saito, *J. Inorg. Nucl. Chem.*, 42 (1980) 1378.
83. D. Kumar, *Int. J. Adhesion and Adhesive*, 18 (1998) 109.
84. H. Kobayashi, *Shika Zairyo, Kikai*, 17 (1998) 145.
85. M.E. Colclorgh, Private letter.
86. T. Raith and W. Nuding, Ger. DE 19 645,350, *Chem. Abstr,* 128 (2000) 1,805,248, assigned to Daimler Benz A.-G., Germany
87. M. Kajiwara, *J. Adhesion. Soc. Jpn.*, 35 (1999) 587.
88. M. Kajiwara, *J. Macromol. Sci., Chem.*, A16(2) (1981) 587; M. Kajiwara and H. Saito, *J. Macromol. Sci., Chem.*, A16(4) (1981) 873.
89. M. Kajiwara, K. Hori and J. Saito, *Kobunshi Ronbunshi*, 37(10) (1980) 691, *Chem. Abstr.,* 94 (1981) 31394b.
90. G.I. Derkach, E.S. Gubnitskaya, V.A. Shokol and A.A. Kisilenko, *Zh. Obsch. Kim.*, 34 (1964) 82, *Chem. Abstr.*, 60 (1964) 11502d.
91. N.H. Ray, *Inorganic Polymer*, Academic Press, London, New York, San Francisco, (1978) p. 77.
92. C.W. Lee, R. Venkatachalapathy, and J. Prakash, *Electrochem. Solid-State Lett.*, 3(2), (2000) 63.
93. K. Xu, N.D. Day and C.A. Angell, *J. Electrochem. Soc.*, 143(9) (1996) L209.
94. S. Narang, S. Ventura, M. Zhao, S. Smedley, G Koolpe, and B.J. Dougherty, *PCT Int. Appl. WO*, 97 44,842 (1997), *Chem. Abstr.* 128 (1998) 50771j, assigned to SRI International, USA.
95. A. Shiga, Y. Aoki, and K. Takeichi, *Jpn. Kokai Tokkyo Koho JP*, 2000 30,740 (2000), *Chem. Abstr.* 132 (2000) 95813y, assigned to Toyota Central Research and Development Laboratories, Inc.
96. H. Tsutiya, H. Kawakabe, A. Wakui, T. Kamata, and H. Sam, PCT Int. Appl. WO 0033,410, *Chem. Abstr.* 133 (2000) 20130d, assigned to Nippon Chemical Industrial Co., Ltd. and to Sony Co.); S. Fui, Y. Narita, T. Saito, N. Ohara, A. Wakai, and T. Kamata, Jpn. Kokai Tokkyo Koho JP, 338, 682 (2001), *Chem. Abstr.* 136 (2002) 21959m, assigned to Sony Co. and to Nippon Chemical Industrial Co., Ltd.; M. Kikuchi,

F. Yonekawa, A. Wakai, and T. Kamata, Jpn. Kokai Tokkyo Koho JP, 338, 683 (2001), *Chem. Abstr.* 136 (2002) 21960e, assigned to Nippon Chemical Industrial Co., Ltd.

97. N. Kajiwara, T. Ogino, T. Miyazaki, and T. Kawagoe, *Jpn. Kokai Tokkyo Koho JP*, 06 13,108[94 13,108] (1994), Chem. Abstr. 120 (1994) 303349h, assigned to Brigiston Co. Ltd.
98. K. Moriya, H. Mizusaki, M. Kato, S. Yanoand M. Kajiwara, *Liq. Cryst.,* 18 (1995) 795.
99. K. Moriya, H. Mizusaki, M .Kato, T. Suzuki, S.Yano, M.Kajiwara and K.Tashiro, *Chem. Mater.*, 9 (1997) 255.
100. K. Moriya, T. Suzuki, H. Mizusaki, S. Yano and M. Kajiwara, *Chem. Lett.*, 1997, 1001.
101. K. Moriya, T. Suzuki, S.Yano and S. Miyajima, *J. Phys. Chem.*, 105 (7920) 7920.
102. J.W. Doane, R.S. Parker, B. Cvikl, D.L. Johnson, D.L.Fishel, *Phys. Rev. Lett.*, 28 (1972) 1694.
103. A.M. Levelut and K. Moriya, *Liq. Cryst.,* 20 (1996) 119.
104. K. Moriya, T. Masuda, T. Suzuki, S.Yano and M. Kajiwara, *Mol. Cryst. Liq. Cryst.*, 318 (1998) 267.
105. K. Moriya, T. Suzuki, S.Yano and M. Kajiwara, *Liq. Cryst.*, 19 (1995) 711.
106. K. Moriya, T. Suzuki, S. Yano and M. Kajiwara, *Trans. Mater. Res. Soc. Jpn.*, 24 (1999) 481.
107. K. Moriya, T. Yamane, T. Suzuki, T. Masuda, H. Mizusaki, S. Yano and M. Kajiwara, *Phosphorus, Sulfur and Silicon*, 177 (2002) 1427.
108. K. Moriya, Y. Kawanishi, S. Yano and M. Kajiwara, *J. Chem. Soc., Chem. Comm.*, (2000) 1111.

ABOUT THE EDITORS

MARIO GLERIA

Dr. Mario Gleria is “Researcher” at the National Research Council (C.N.R.) of Italy, and Adjunct Professor of Materials Science at the Padova University, from which he received his degree in Chemistry in 1969.

He carried out most of his scientific activity at the C.N.R. Institute of Photochemistry and High Energy Radiation Chemistry (now Institute for the Organic Synthesis and the Photoreactivity, ISOF), and he is presently working at the CN.R. Institute of Sciences and Molecular Technologies (I.S.T.M.).

He is in the Editorial Board of the Journal of Inorganic and Organometallic Polymers, and member of the Italian Association of Macromolecules (A.I.M.).

He is author or co-author of 125 scientific articles and of 22 patents on phosphazene compounds.

His research interests are in phosphazene materials, both trimers and polymers, with particular concern in the synthesis, characterization, functionalization, photochemistry, blend formation, and hybrid material preparation of these substrates.

Roger De Jaeger

Roger De Jaeger is Professor of Chemistry at the University of Sciences and Technologies of Lille (France), from which he received his degree in Chemistry in 1973.

He is the author or co-author of 75 scientific articles and of 13 patents in the phosphazene domain.

He is member of the French Chemical Society (SFC), and was President of the "Nord-Pas de Calais-Picardie" Section of this Society from 1999 to 2003. He also belongs to the French Group for Polymer Studies and Applications (GFP).

His research activity concerns synthesis, characterization and functionalization of phosphazene materials (monomers, oligomers and polymers), and his work in this field, in cooperation with Atochem (Elf Aquitaine Group), led in 1988 to the pilot-scale production of flame retardant and elastomeric polyphosphazenes.

List of Contributors

Allen C.W.
University of Vermont, USA
cwallen@zoo.uvm.edu

Bortolus P.
Istituto per la Sintesi Organica e la Fotoreattività, Legnaro, Padova, Italy
Pietro.Bortolus@lnl.infn.it

Brusatin G.
Padova University, Italy
Giovanna.Brusatin@unipd.it

Canova L.
Polimeri Europa, Novara, Italy
Luciano.Canova@polimerieuropa.com

Carenza M.
Istituto per la Sintesi Organica e la Fotoreattività, Legnaro, Padova, Italy
Carenza@lnl.infn.it

Diré S.
Trento University, Italy
dire@ing.unitn.it

Drioli E.
Calabria University, Italy
E.Drioli@unical.it

Facchin G.
Istituto di Scienze e Tecnologie Molecolari, CNR, Padova, Italy
gfacchin@virgilio.it

Fiocca L.
Polimeri Europa, Novara, Italy
luisa.fiocca@polimerieuropa.com

Gettlemann L.
Louisville University, Kentucky, USA
gettleman@louisville.edu

Giannotta G.
Polimeri Europa, Novara, Italy
giorgio.giannotta@polimerieuropa.com

Gleria M.
Istituto di Scienze e Tecnologie Molecolari CNR., Padova, Italy
gleria@chin.unipd.it

Golemme G.
Calabria University, Italy
ggolemme@unical.it

Guglielmi M.
Padova University, Italy
massimo.guglielmi@unipd.it

Harrup M.K.
Idaho National Engineering and Environmental Laboratory, Idaho, USA
Harrmk@inel.gov

Hernandez-Rubio D.
University of Vermont, USA
dhernand@zoo.uvm.edu

Heyde M.
Ghent University, Belgium
Mieke.Heyde@rug.ac.be

Inoue K.
Ehime University, Japan
inoue@en3.ehime-u.ac.jp

Kajiwara M.
Nagoya University, Japan
kajiwara@se.starcat.ne.jp

Lora S.
Istituto per la Sintesi Organica e la Fotoreattività, Legnaro, Padova, Italy
lora@lnl.infn.it

Luther T.A.
Idah National Engineering and Environmental Laboratory, Idaho, USA
Luthta@inel.com

Minto F.
Istituto per la Sintesi Organica e la Fotoreattività, Legnaro, Padova, Italy

Moriya K.
Gifu University, Japan
moriya@apchem.gifu-u.ac.jp

Orme C.J.
Idaho National Engineering and Environmental Laboratory, Idaho, USA
Cjo2@inel.gov

Pertici P.
Centro di Studio Macromolecole Stereoregolate ed Otticamente Attive, CNR, Pisa, Italy
pervit@dcci.unipi.it

Pintauro P.N.
Department of Chemical Engineering
Case Western Reserve University Cleveland, USA
pnp3@po.cwru.edu

Pitzalis E.
Istituto Processi Chimico-Fisici, CNR, Pisa, Italy
pitzalis@icas.pi.cnr.it

Po' R.
Polimeri Europa, Novara, Italy
riccardo.po@polimerieuropa.com

Salvadori P.
Centro di Studio Macromolecole Stereoregolate ed Otticamente Attive, CNR, Pisa, Italy
Psalva@dcci.unipi.it

Schacht E.
Ghent University, Belgium
etienne.schacht@rug.ac.be

Stewart F.F.
Idaho National Engineering and Environmental Laboratory, Idaho, USA
FSF@inel.gov

Vitulli G.
Centro di Studio Macromolecole Stereoregolate ed Otticamente Attive, CNR, Pisa, Italy
pervit@dcci.unipi.it

Wycisk R.
Department of Chemical Engineering
Case Western Reserve University Cleveland, USA
rjw19@cwru.edu

INDEX

#

A

B

C

R

S

T

V